Glencoe Earth Science
Features and Benefits

AF697064

		Pages
Dynamic Instructional Strategies	**. . . present clear and comprehensive coverage of middle school science.**	
	■ Each engaging chapter opener includes a *Launch Lab* and *Foldables*™.	7, 35
	■ *National Geographic Visualizing* features illustrate important concepts in middle school science.	10, 38
A Strong Reading Strand	**. . . encourages active reading and learning for students of all reading levels. In the *Student Edition:***	
	■ as you read gives students a preview of learning objectives and vocabulary for each section;	8, 36
	■ *Reading Checks* help students check their reading comprehension; and	9, 15, 39
	■ *Caption questions* ask students to interpret visuals.	8, 17, 47
	■ *Reading Essentials, An Interactive Student Textbook* is designed to help struggling readers comprehend core content. It is written at a reading level of about two to three grades below the *Student Edition.*	
Meeting the Needs of All Students	**. . . facilitates understanding of science concepts for students of all learning levels. In the *Teacher Wraparound Edition:***	
	■ *Differentiated Instruction* strategies help meet the needs of students with learning disabilities and physical challenges, or create opportunities to enrich and extend students' learning;	17, 41
	■ *Daily Intervention* provides intervention strategies for struggling students; and	12, 18, 39
	■ *Identifying Misconceptions* helps uncover and address common science misconceptions.	64F
	■ The *English/Spanish Glossary*, also in the *Student Edition,* helps English-language learners comprehend science terms.	227–233
Extensive Standardized Test Practice	**. . . gives students the opportunity to practice for state and national exams.**	
	■ Each chapter ends with a variety of standardized test practice questions, including *Multiple Choice, Short Response/Grid In*, and *Open Ended.*	152–153
A Variety of Labs	**. . . gets students excited about being involved in science. The *Student Edition* provides:**	
	■ *MiniLABs,* traditional labs, and *Design Your Own, Model and Invent,* and *Use the Internet* labs; and	9, 37
	■ *Extra Try at Home Labs* provide opportunities for students to practice their science skills at home with adult supervision using materials from the kitchen, junk drawer, or backyard.	198–200
	■ *Virtual Labs* CD-ROM contains an interactive virtual lab for each chapter.	17, 42
	■ *Video Labs* (VHS) reinforce lab techniques and safety skills, offer troubleshooting tips, and give expected outcomes.	18, 50
	■ The *Science Lab Manual, Probeware Lab Manual,* and *Science Inquiry Lab Manual* provide additional opportunities to practice laboratory techniques.	
Multi-Level Review	**. . . presents multiple opportunities for all students to review and master content.**	
	■ Each section ends with a review that contains a *Summary* of the section's major concepts and a *Self Check* that has questions to assess student learning and practice math or science skills.	12, 18, 43
	■ The *Study Guide* at the end of each chapter can preview, review, summarize, and visualize the chapter's main ideas.	29, 59
	■ *Study Guide* and *Reinforcement* help students grasp core content.	
Teacher Resources	**. . . provide innovative strategies to help new and experienced teachers.**	
	■ *Chapter Resources Fast File* ™ contains important reproducible masters.	64B
	■ Section Focus, Assessment, and Teaching transparencies accompany each chapter.	124C
	■ *Performance Assessment in the Science Classroom* has assessment guidelines, strategies, sample rubrics, and more.	
Online Resources	**. . . enrich the learning experience with the click of a mouse.**	
	■ For prescreened Web links, standardized test practice, self-check quizzes, chapter tests, *Vocabulary PuzzleMaker,* extra math practice, science career information, current science news, and *WebQuest* interactive projects, visit **bookf.msscience.com**.	
	■ The complete interactive *Student Edition* is available at The McGraw-Hill Learning Network Web site, **mhln.com**.	
Technology	**. . . provides timesaving products to help teachers creatively engage their students.**	
	■ *MindJogger Videoquizzes* (VHS & DVD) provide a game-show style interactive quiz for each chapter.	
	■ Easy to edit *Interactive Chalkboard* Microsoft® PowerPoint® presentations include step-by-step lessons, an image bank, chapter and section review questions, standardized test practice, and transparencies.	
	■ *ExamView® Pro Testmaker* CD-ROM in English or Spanish allows you to customize assessments.	
	■ *TeacherWorks* CD-ROM is your all-in-one resource center that helps you plan and organize lessons.	
	■ *StudentWorks* CD-ROM solves the heavy backpack problem.	

SAFETY SYMBOLS

SAFETY SYMBOLS	HAZARD	EXAMPLES	PRECAUTION	REMEDY
DISPOSAL	Special disposal procedures need to be followed.	certain chemicals, living organisms	Do not dispose of these materials in the sink or trash can.	Dispose of wastes as directed by your teacher.
BIOLOGICAL	Organisms or other biological materials that might be harmful to humans	bacteria, fungi, blood, unpreserved tissues, plant materials	Avoid skin contact with these materials. Wear mask or gloves.	Notify your teacher if you suspect contact with material. Wash hands thoroughly.
EXTREME TEMPERATURE	Objects that can burn skin by being too cold or too hot	boiling liquids, hot plates, dry ice, liquid nitrogen	Use proper protection when handling.	Go to your teacher for first aid.
SHARP OBJECT	Use of tools or glassware that can easily puncture or slice skin	razor blades, pins, scalpels, pointed tools, dissecting probes, broken glass	Practice common-sense behavior and follow guidelines for use of the tool.	Go to your teacher for first aid.
FUME	Possible danger to respiratory tract from fumes	ammonia, acetone, nail polish remover, heated sulfur, moth balls	Make sure there is good ventilation. Never smell fumes directly. Wear a mask.	Leave foul area and notify your teacher immediately.
ELECTRICAL	Possible danger from electrical shock or burn	improper grounding, liquid spills, short circuits, exposed wires	Double-check setup with teacher. Check condition of wires and apparatus.	Do not attempt to fix electrical problems. Notify your teacher immediately.
IRRITANT	Substances that can irritate the skin or mucous membranes of the respiratory tract	pollen, moth balls, steel wool, fiberglass, potassium permanganate	Wear dust mask and gloves. Practice extra care when handling these materials.	Go to your teacher for first aid.
CHEMICAL	Chemicals can react with and destroy tissue and other materials	bleaches such as hydrogen peroxide; acids such as sulfuric acid, hydrochloric acid; bases such as ammonia, sodium hydroxide	Wear goggles, gloves, and an apron.	Immediately flush the affected area with water and notify your teacher.
TOXIC	Substance may be poisonous if touched, inhaled, or swallowed.	mercury, many metal compounds, iodine, poinsettia plant parts	Follow your teacher's instructions.	Always wash hands thoroughly after use. Go to your teacher for first aid.
FLAMMABLE	Flammable chemicals may be ignited by open flame, spark, or exposed heat.	alcohol, kerosene, potassium permanganate	Avoid open flames and heat when using flammable chemicals.	Notify your teacher immediately. Use fire safety equipment if applicable.
OPEN FLAME	Open flame in use, may cause fire.	hair, clothing, paper, synthetic materials	Tie back hair and loose clothing. Follow teacher's instruction on lighting and extinguishing flames.	Notify your teacher immediately. Use fire safety equipment if applicable.

Eye Safety
Proper eye protection should be worn at all times by anyone performing or observing science activities.

Clothing Protection
This symbol appears when substances could stain or burn clothing.

Animal Safety
This symbol appears when safety of animals and students must be ensured.

Handwashing
After the lab, wash hands with soap and water before removing goggles.

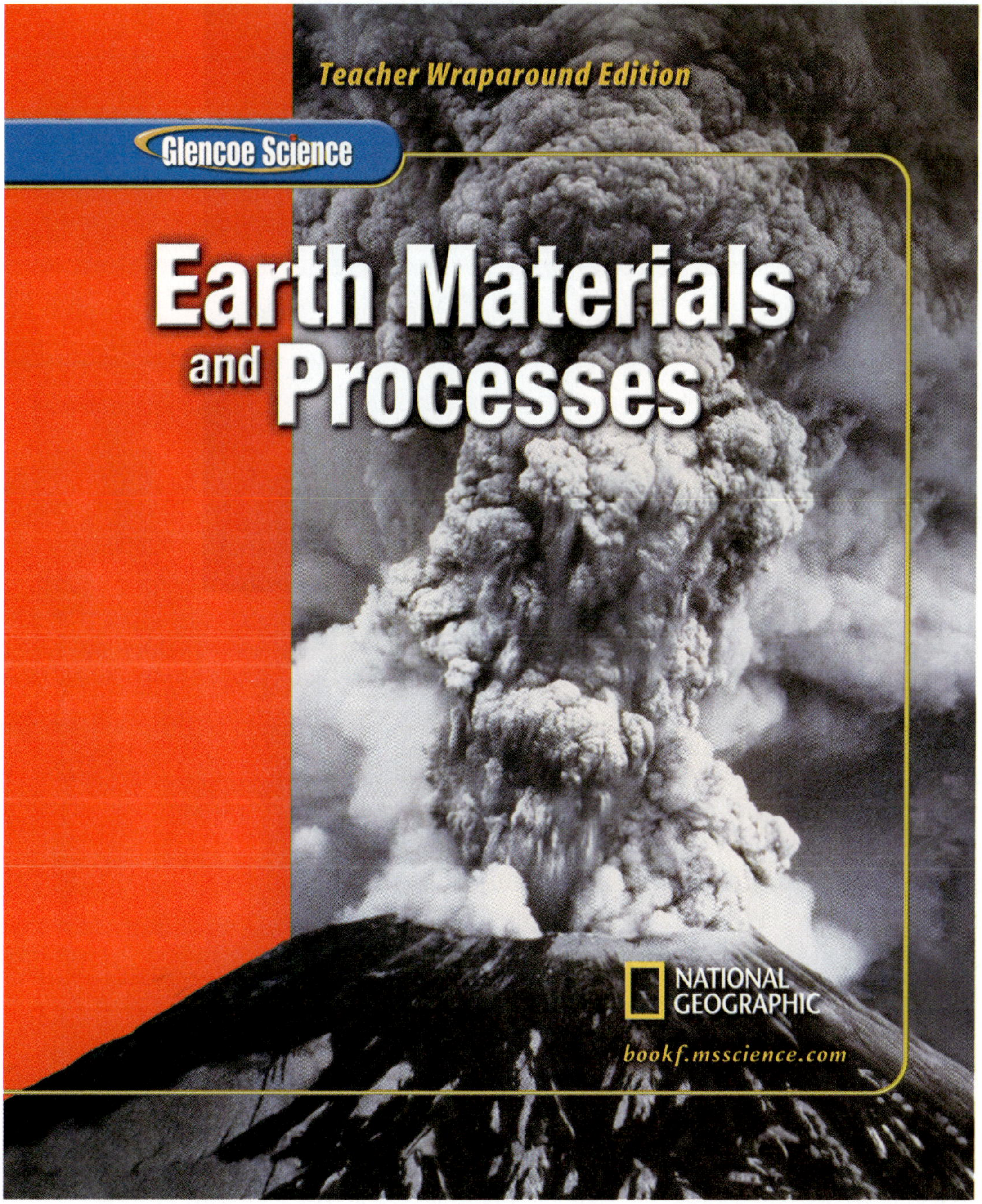

McGraw Hill Glencoe

New York, New York Columbus, Ohio Chicago, Illinois Peoria, Illinois Woodland Hills, California

Glencoe Science

Earth Materials and Processes

The eruption column above Mount St. Helens, Washington, as it exploded on May 18, 1980, rose thousands of feet skyward and drifted downwind, dumping dark, gray ash over eastern Washington and beyond. The eruption lasted nine hours, but the landscape was changed within moments.

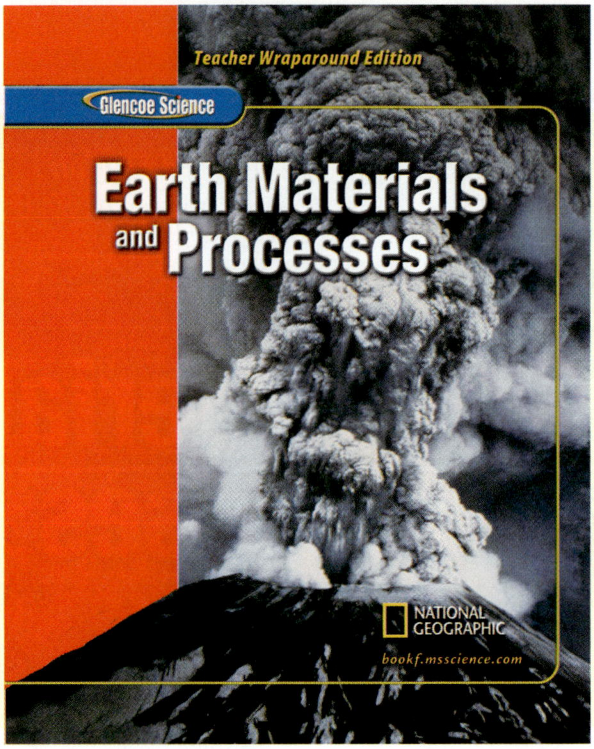

Glencoe

The McGraw·Hill Companies

Send all inquiries to:
Glencoe/McGraw-Hill
8787 Orion Place
Columbus, OH 43240-4027

ISBN 0-07-861749-9 (Student Edition)
ISBN 0-07-861750-2 (Teacher Wraparound Edition)

Printed in the United States of America.

1 2 3 4 5 6 7 8 9 10 027/043 09 08 07 06 05 04

Authors

NATIONAL GEOGRAPHIC
Education Division
Washington, D.C.

Ralph M. Feather Jr., PhD
Assistant Professor
Geoscience Department
Indiana University of Pennsylvania
Indiana, PA

Dinah Zike
Educational Consultant
Dinah-Might Activities, Inc.
San Antonio, TX

Series Consultants

CONTENT

William C. Keel, PhD
Department of Physics and Astronomy
University of Alabama
Tuscaloosa, AL

Robert Nierste
Science Department Head
Hendrick Middle School, Plano ISD
Plano, TX

MATH

Michael Hopper, DEng
Manager of Aircraft Certification
L-3 Communications
Greenville, TX

Teri Willard, EdD
Mathematics Curriculum Writer
Belgrade, MT

READING

Carol A. Senf, PhD
School of Literature, Communication, and Culture
Georgia Institute of Technology
Atlanta, GA

SAFETY

Aileen Duc, PhD
Science 8 Teacher
Hendrick Middle School, Plano ISD
Plano, TX

Sandra West, PhD
Department of Biology
Texas State University-San Marcos
San Marcos, TX

ACTIVITY TESTERS

Nerma Coats Henderson
Pickerington Lakeview Jr. High School
Pickerington, OH

Mary Helen Mariscal-Cholka
William D. Slider Middle School
El Paso, TX

Science Kit and Boreal Laboratories
Tonawanda, NY

Series Reviewers

Lois Burdette
Green Bank Elementary-Middle School
Green Bank, WV

Marcia Chackan
Pine Crest School
Boca Raton, FL

Mary Ferneau
Westview Middle School
Goose Creek, SC

Annette D'Urso Garcia
Kearney Middle School
Commerce City, CO

Nerma Coats Henderson
Pickerington Lakeview Jr. High School
Pickerington, OH

Sharon Mitchell
William D. Slider Middle School
El Paso, TX

Joanne Stickney
Monticello Middle School
Monticello, NY

Glencoe Science 15-Book Series

Teach science your way!

With the 15 Life, Earth, and Physical Science titles in our modular series, you can select the science topics you want to cover and customize your science curriculum in any way you choose.

F Earth Materials and Processes

1 Minerals
2 Rocks
3 Earth's Energy and Mineral Resources
4 Plate Tectonics
5 Earthquakes
6 Volcanoes

G The Changing Surface of Earth

1 Views of Earth
2 Weathering and Soil
3 Erosional Forces
4 Water Erosion and Deposition
5 Clues to Earth's Past
6 Geologic Time

H The Water Planet

1 Water
2 Freshwater at Earth's Surface
3 Groundwater Resources
4 Ocean Motion
5 Oceanography

I The Air Around You

1 Atmosphere
2 Weather
3 Climate
4 Air Pollution

J Astronomy

1 Exploring Space
2 The Sun-Earth-Moon System
3 The Solar System
4 Stars and Galaxies

Multi-Level Review and Assessment

Each chapter provides five pages of review and testing to help you evaluate students' knowledge and ability to apply science concepts.

Section Review

- Summary pinpoints important concepts in the section.
- Skill-based questions promote critical thinking skills.

Study Guide

- Main idea summary of each section
- Concept mapping activity to help students visualize the main ideas

Chapter Review

- Using Vocabulary
- Checking Concepts
- Thinking Critically
- Performance Activities
- Applying Math

Standardized Test Practice

- Multiple Choice
- Short Response/ Grid In
- Open-Ended Questions

Dynamic Instruction

The consistent instructional strategies in each chapter strengthen students' learning—from the beginning of each chapter where students see "Chapter Preview," to the end where they have a chance to test the knowledge they have acquired and prepare for the next lesson.

Chapter Opener

- *Chapter Preview* introduces the main concepts.
- *Science Journal* promotes critical-thinking skills
- *Dinah Zike's Foldables*™ let students create interactive study guides.
- *Launch Labs* give students an opportunity to explore new ideas.

Section Opener

- *What You'll Learn* introduces main concepts.
- *Why It's Important* provides an answer to "Why do we have to learn this?"
- *Review Vocabulary* reviews a term that helps students better understand section content.
- *New Vocabulary* highlights new terms students will learn in the section.

Labs

- *Labs* allow students to design their own experiments or follow well-tested procedures, enabling them to learn and practice science processes.
- *MiniLABS* offer students quick and easy-to-do ways to clarify concepts and reinforce skills.

Assessment

Glencoe Science offers the Glencoe Assessment Advantage, a system designed to give you all the tools you need to prepare your students for success in any testing situation.

In the *Student Edition*

Section Review and **Applying Math** questions appear in every chapter.

Chapter Review questions help you evaluate students' knowledge and ability to apply science concepts.

Standardized Test Practice questions at the end of each chapter provide students with additional opportunities to practice their test-taking skills.

In the *Teacher Wraparound Edition*

Assessments located throughout the *Teacher Wraparound Edition* provide methods for assessing students' comprehension with Performance, Process, and Content exercises.

Teacher Classroom Resources

Performance Assessment in the Science Classroom

- Guidelines for assessing the performance of a task
- Reproducible activities for evaluating students
- Sample rubrics and checklists

***Fast File* Chapter Resources** provides six pages of assessement for every chapter including *Testing Concepts, Applying Concepts,* and *Writing Skills.*

Technology Support

MindJogger Videoquizzes are interactive video quizzes set in game show format. Each is designed for the full range of student learning styles.

Exam*View*® Pro Testmaker CD-ROM for Windows® and Macintosh® provides an easy way to create, edit, and customize your tests. Select your own test items by objective from two different levels of difficulty, or write and edit your own. Translate tests from English to Spanish and vice versa.

Rubrics

The following rubrics are sample scoring devices for short response and open-ended questions.

Short Response

Points	Description
2	The student demonstrates a thorough understanding of the science of the task. The response may contain minor flaws that do not detract from the demonstration of a thorough understanding.
1	The student has provided a response that is only partially correct.
0	The student has provided a completely incorrect solution or no response at all.

Open Ended

Points	Description
4	The student demonstrates a thorough understanding of the science of the task. The response may contain minor flaws that do not detract from the demonstration of a thorough understanding.
3	The student demonstrates an understanding of the science of the task. The response is essentially correct and demonstrates an essential but less than thorough understanding of the science.
2	The student demonstrates only a partial understanding of the science of the task. Although the student may have used the correct approach to a solution or may have provided a correct solution, the work lacks an essential understanding of the underlying science concepts.
1	The student demonstrates a very limited understanding of the science of the task. The response is incomplete and exhibits many flaws.
0	The student provides a completely incorrect solution or no response at all.

Time-Saving Teacher Resources

Glencoe Science provides an extensive array of support materials and resources designed to help you create and customize your science course quickly and easily.

FAST FILE Chapter Resources

For each chapter, Chapter Resources contain key reproducible masters along with additional teaching strategies, teacher support, and answer keys.

Teacher Wraparound Edition

The *Teacher Wraparound Edition* is your key to the teaching resources available. In addition to teaching strategies and suggestions, the *Teacher Wraparound Edition* provides a guide for all print and software materials available for each lesson.

Transparencies

Color Transparencies provides three types of transparencies for use while teaching each chapter. The *Section Focus Transparencies* are designed to generate interest and focus students' attention on the topic being presented in the section. The *Teaching Transparency* for each chapter addresses a major concept that will benefit from an extra visual learning aid. The *Assessment Transparency* for each chapter is set up to resemble standardized tests.

Exam*View*® Pro Testmaker CD-ROM

This CD-ROM will help you create, edit, and customize tests. In addition, it will help you create multiple versions of tests, translate tests from English to Spanish and vice versa, and build tests aligned with state standards.

Video Labs

These VHS cassettes contain step-by-step procedures for selected *Student Edition* labs. They also contain lab safety skills, teacher support, and troubleshooting advice.

TeacherWorks™

This CD-ROM is your all-in-one teacher resource center. Personalize a lesson plan, access resources from the *Teacher Wraparound Edition*, connect to the Internet, or make a to-do list. These are only a few of the many features that can assist you in the planning and organizing of your lessons.

StudentWorks™

This CD-ROM is a valuable resource for students to access content online and use online resources to continue learning chapter concepts.

Interactive Chalkboard with Image Bank — PowerPoint® Presentations

This CD-ROM brings Microsoft® PowerPoint® presentations right to your door. With the large number of graphics provided, students can use a visual approach to learning chapter content.

Virtual Labs CD-ROM Program

The Virtual Labs CD-ROM contains a collection of labs that allow students to complete labs that are too expensive, take too long to complete, or might be too dangerous in a classroom laboratory.

Science Online

This website is a portal to hundreds of prescreened Internet sites (Web links) that correlate to content in the text. Visit this link to find interactive activities that review chapter concepts and to access the *Student Edition* online.

McGraw Hill Learning Network mhln.com

mhln.com is an online teaching and learning space for teachers, students, and parents.

Differentiated Instruction

Teaching Strategies

Following each suggested assessment and activity, ability levels are supplied to accommodate all students. For a key to the Teaching Strategies designations, see the C page before each chapter.

Identifying Misconceptions

These short, diagnostic, and perscriptive lessons target common science misconceptions.

Multiple Learning Styles

Look for these italicized designations under various activities to help you target your lessons to each student's preferred learning style.

- ***Kinesthetic*** learners learn through touch, movement, and manipulating objects.
- ***Visual-Spatial*** learners think in terms of images, illustrations, and models.
- ***Interpersonal*** learners understand and work well with other people.
- ***Intrapersonal*** learners can analyze their own strengths and weaknesses and may prefer to work on their own.
- ***Linguistic*** learners write clearly and easily understand the written word.
- ***Logical-Mathematical*** learners understand numbers easily and have highly-developed reasoning skills.

Daily Interventions

Found at the end of each chapter section, this feature is designed to intercept students who are struggling and prescribe a system to help them get back on track. *Reteach* provides reinforcement of the section's concepts through visual activities.

Differentiated Instruction

These activities present various teaching strategies designed to help you meet the special needs of students with learning disabilities, physical challenges, visual impairment, and hearing impairment. *Challenge* activities provide opportunities for students who excel to engage in activities and research projects that extend the chapter's concepts. English-language learners in the classroom will also find exercises that bridge the gap between language barriers and the chapter content.

Cultural Diversity

These readings provide insights into the unique ways in which people of different ethnicities and cultural heritage have approached science. The intent of these features is to build awareness and appreciation for the global community in which we live.

Inquiry-Based Science

The call for more inquiry-based science by the *National Science Education Standards* has been met by Glencoe Science.

Glencoe Science recognizes the importance of conducting inquiry-based science activities in the classroom. The process of doing inquiry models actual science practice, encouraging problem-solving strategies and developing critical thinking skills. Inquiry gets students actively involved in the learning process by allowing them to determine materials, procedures, or the topics and questions they want to investigate.

Inquiry can range from a very structured activity for those students who need more guidance to a more open-ended approach where students lead the investigations. Glencoe Science recognizes that the inquiry activities suggested will not look the same in every classroom. We encourage teachers to modify the suggested activities in a manner that best supports your students.

Glencoe also provides teachers with *Alternative Inquiry Labs,* teaching strategies or suggestions for making existing labs more inquiry-based.

Research-Based Learning Strategies

Glencoe Science incorporates the most current and applicable educational research on science learning and follows recommendations from the American Association for the Advancement of Science and the National Science Teachers Association. The following research-based strategies can be found throughout the text.

Learning Strategies

The following research-based strategies can be found throughout the text:

- **Using Prior Knowledge** Glencoe Science encourages students to use their prior knowledge to learn information because this adds relevance to the material. Students are referred back to other parts of the text or to their own real-life experiences.
- **Practicing Important Tasks** By offering students an opportunity to practice important tasks using a variety of labs and activities in the *Student Edition, Teacher Wraparound Edition,* ancillaries and technology, Glencoe Science makes learning fun and relevant for students.
- **Using Visuals to Communicate, Organize, and Reinforce Learning** High-quality art and photos throughout the text communicate concepts more efficiently and reinforce learning, while allowing students to organize information.
- **Motivating Students to Achieve** Active strategies and real-world experiences motivate students to achieve. Throughout Glencoe's programs, students are encouraged to apply their knowledge in ways that will motivate them to learn.
- **Developing Decoding and Reading Comprehension Strategies** Throughout the text, students are supplied with caption questions, reading checks, and other strategies to aid in comprehension.
- **Using Study Strategies** Through the use of highlighting, outlining, note-taking, summarizing, and other such strategies, students can monitor their own progress and organize information more effectively, thereby increasing their scientific literacy. These strategies are found throughout the text and ancillaries.

The use of these strategies within Glencoe Science will help teachers to achieve the goals set forth by the *National Science Education Standards.*

White Paper

The Glencoe Science White Paper outlines the educational strategies on which this program was based. This document provides specific examples from the *Student Edition, Teacher Wraparound Edition,* ancillary program, and technology resources, highlighting extensive use of educationally sound strategies that help students learn science.

Field Research and Testing

Feedback from students, teachers, curriculum supervisors, department chairpersons, parents, learning specialists, and science content experts was invaluable in the development of this program. The following pre-publication and post-publication research was conducted.

Prior to Publication

- Detailed classroom teacher and curriculum supervisor **surveys** were conducted by independently contracted researchers.
- A **nationwide panel** of science teachers, curriculum supervisors, and department chairpersons provided countless hours of feedback and assistance throughout program development.
- A wide range of **educator and content reviewers** provided in-depth reviews of and suggestions for manuscripts and pre-publication versions of the program.
- **Face-to-face interviews** with science teachers provided insight into teachers' day-to-day challenges.

Field-Test Results

- Field-test research indicates that test scores increased among students using Glencoe Science programs.
- Nine out of ten students earned higher scores after using Glencoe programs.
- Scores improved among both male and female students.
- Scores improved among both minority and non-minority students.
- Overall, the gap between the average pre-test score and a perfect score closed by 33 percent. Stated differently, on average, **scores increased 77 percent after students used the Glencoe program.**

After Publication

- Field tests were conducted in which students and teachers used a pre-publication manuscript in the classroom.
- Follow-up interviews, observations, and surveys of Glencoe Science users provide ongoing opportunities for program development and verification of program success.

KS Studios

National Education Standards

Correlation of *Glencoe Science* to the National Science Education Standards.

Content Standard	Chapter and Section
(UCP) Unifying Concepts and Processes	
1. Systems, order, and organization	F1-1, F1-2, F1-3, F2-1, F2-2, F2-3, F2-4, F3-1, F3-2, F3-3, F4-1, F4-2, F4-3, F5-1, F5-2, F5-3, F6-1, F6-2, F6-3, G1-1, G1-2, G1-3, G2-1, G2-2, G2-3, G3-1, G3-2, G3-3, G4-1, G4-2, G4-3, G5-1, G5-2, G5-3, G6-1, G6-2, G6-3, H2-1, H2-2, H2-3, H2-4, H3-1, H3-2, H3-3, H4-1, H4-2, H4-3, H5-1, H5-2, H5-3, I1-1, I1-2, I1-3, I2-1, I2-2, I2-3, I3-1, I3-2, I3-3, I4-1, I4-2, I4-3, J1-1, J1-2, J1-3, J2-1, J2-2, J2-3, J3-1, J3-2, J3-3, J3-4, J4-1, J4-2, J4-3
2. Evidence, models, and explanation	F1-1, F1-2, F1-3, F2-1, F2-2, F2-3, F2-4, F3-1, F3-2, F3-3, F4-1, F4-2, F4-3, F5-1, F5-2, F5-3, F6-1, F6-2, F6-3, G1-1, G1-2, G1-3, G2-1, G2-2, G2-3, G3-1, G3-2, G3-3, G4-1, G4-2, G4-3, G5-1, G5-2, G5-3, G6-1, G6-2, G6-3, H1-1, H1-2, H1-3, H2-1, H2-2, H2-3, H2-4, H3-1, H3-2, H3-3, H4-1, H4-2, H4-3, H5-1, H5-2, H5-3, I1-1, I1-2, I1-3, I2-1, I2-2, I2-3, I3-1, I3-2, I3-3, I4-1, I4-2, I4-3, J1-1, J1-2, J1-3, J2-1, J2-2, J2-3, J3-1, J3-2, J3-3, J3-4, J4-1, J4-2, J4-3
3. Change, constancy, and measurement	F1-1, F1-2, F1-3, F2-1, F2-2, F2-3, F2-4, F3-1, F3-2, F3-3, F4-1, F4-2, F4-3, F5-1, F5-2, F5-3, F6-1, F6-2, F6-3, G1-1, G1-2, G1-3, G2-1, G2-2, G2-3, G3-1, G3-2, G3-3, G4-1, G4-2, G4-3, G5-1, G5-2, G5-3, G6-1, G6-2, G6-3, H1-1, H1-2, H1-3, H2-1, H2-2, H2-3, H2-4, H3-1, H3-2, H3-3, H4-1, H4-2, H4-3, H5-1, H5-2, H5-3, I1-1, I1-2, I1-3, I2-1, I2-2, I2-3, I3-1, I3-2, I3-3, I4-1, I4-2, I4-3, J1-1, J1-2, J1-3, J2-1, J2-2, J2-3, J3-1, J3-2, J3-3, J3-4, J4-1, J4-2, J4-3
4. Evolution and equilibrium	G6-1, G6-2, J4-3
5. Form and function	F1-1, F1-2, F1-3, F2-1, F2-2, F2-3, F2-4, F3-1, F3-2, F3-3, F4-1, F4-2, F4-3, F5-1, F5-2, F5-3, F6-1, F6-2, F6-3, G1-1, G1-2, G1-3, G2-1, G2-2, G2-3, G3-1, G3-2, G3-3, G4-1, G4-2, G4-3, G5-1, G5-2, G5-3, G6-1, G6-2, G6-3, H1-1, H1-2, H1-3, H2-1, H2-2, H2-3, H2-4, H3-1, H3-2, H3-3, H4-1, H4-2, H4-3, H5-1, H5-2, H5-3, I1-1, I1-2, I1-3, I2-1, I2-2, I2-3, I3-1, I3-2, I3-3, I4-1, I4-2, I4-3, J1-1, J1-2, J1-3, J2-1, J2-2, J2-3, J3-1, J3-2, J3-3, J3-4, J4-1, J4-2, J4-3
(A) Science as Inquiry	
1. Abilities necessary to do scientific inquiry	F1-1, F1-2, F1-3, F2-1, F2-2, F2-3, F2-4, F3-1, F3-2, F3-3, F4-1, F4-2, F4-3, F5-1, F5-2, F5-3, F6-1, F6-2, F6-3, G1-1, G1-2, G1-3, G2-1, G2-2, G2-3, G3-1, G3-2, G3-3, G4-1, G4-2, G4-3, G5-1, G5-2, G5-3, G6-1, G6-2, G6-3, H1-1, H1-2, H1-3, H2-1, H2-2, H2-3, H2-4, H3-1, H3-2, H3-3, H4-1, H4-2, H4-3, H5-1, H5-2, H5-3, I1-1, I1-2, I1-3, I2-1, I2-2, I2-3, I3-1, I3-2, I3-3, I4-1, I4-2, I4-3, J1-1, J1-2, J1-3, J2-1, J2-2, J2-3, J3-1, J3-2, J3-3, J3-4, J4-1, J4-2, J4-3
2. Understandings about scientific theory	F1-1, F1-2, F1-3, F2-1, F2-2, F2-3, F2-4, F3-1, F3-2, F3-3, F4-1, F4-2, F4-3, F5-1, F5-2, F5-3, F6-1, F6-2, F6-3, G1-1, G1-2, G1-3, G2-1, G2-2, G2-3, G3-1, G3-2, G3-3, G4-1, G4-2, G4-3, G5-1, G5-2, G5-3, G6-1, G6-2, G6-3, H1-1, H1-2, H1-3, H2-1, H2-2, H2-3, H2-4, H3-1, H3-2, H3-3, H4-1, H4-2, H4-3, H5-1, H5-2, H5-3, I1-1, I1-2, I1-3, I2-1, I2-2, I2-3, I3-1, I3-2, I3-3, I4-1, I4-2, I4-3, J1-1, J1-2, J1-3, J2-1, J2-2, J2-3, J3-1, J3-2, J3-3, J3-4, J4-1, J4-2, J4-3
(B) Physical Science	
1. Properties and changes of properties in matter	G2-1, G5-3, H1-1, H4-1, H5-2, I4-1, J4-3
2. Motions and forces	F1-2, F4-1, F4-2, F4-3, F5-1, F5-2, F6-1, F6-2, G2-2, G2-3, G3-3, I1-2, I1-3, I2-2, I3-3, J2-1, J2-2, J3-1, J4-3
3. Transfer of energy	H5-2, I1-2, I3-3, J4-3
(C) Life Science	
1. Structure and function in living systems	H2-1, H2-2, H2-3, H5-2
2. Reproduction and heredity	H5-2
3. Regulation and behavior	H5-1, H5-2
4. Populations and ecosystems	H2-1, H2-2, H2-3, H5-1, H5-2, I4-2
5. Diversity and adaptations of organisms	H2-1, H2-2, H2-3, H5-2
(D) Earth and Space Science	
1. Structure of the Earth system	F1-1, F2-1, F2-2, F2-3, F2-4, F3-1, F3-2, F3-3, F4-1, F4-2, F4-3, F5-1, F5-2, F5-3, F6-1, F6-2, G1-1, G1-2, G1-3, G2-1, G2-2, G4-1, G4-2, G4-3, G6-1, G6-2, G6-3, H2-1, H2-2, H3-1, H3-2, H3-3, H4-1, H4-2, H4-3, H5-1, H5-2, I1-1, I1-2, I1-3, I2-1, I2-2, I2-3, I3-1, I3-2, I3-3, J2-1, J2-3
2. Earth's history	F1-1, F1-2, F2-1, F2-2, F2-3, F2-4, F3-1, F4-1, F5-1, F6-1, G1-1, G1-3, G2-1, G2-3, H3-3, I2-3, I3-3
3. Earth and the solar system	F1-3, F2-1, F2-2, F2-3, F2-4, F3-1, F3-3, H4-3, J1-1, J1-2, J1-3, J2-3, J3-2, J4-2

Content Standard	Chapter and Section
(E) Science and Technology	
1. Abilities of technological design	H4-3
2. Understandings about science and technology	H4-3
(F) Science in Personal and Social Perspectives	
1. Personal health	F3-2, G2-2, I4-2
2. Populations, resources, and environments	F3-1, F3-2, F3-3, H1-3, H2-4, H3-2, H5-3, I4-1, I4-2, I4-3
3. Natural hazards	F4-1, F4-2, F4-3, F5-1, F5-2, F5-3, F6-1, F6-2, G3-3, G4-1, G4-3, I2-3, I4-1, I3-3
4. Risks and benefits	G2-1, G3-3, H3-2
5. Science and technology in society	F3-1, G5-1, G5-3, I4-3, J1-3
(G) History and Nature of Science	
1. Science as a human endeavor	F1-3
2. Nature of science	F1-2, F1-3, F4-3, F6-3, I4-3
3. History of science	F1-2, F1-3, F5-2, G5-3, H5-3, I4-3, J2-3, J3-1, J3-4, J4-3

How Glencoe Science Aligns with the National Science Education Standards

The correlations at the left and above show the close alignment between Glencoe Science and the grade-appropriate standards. Glencoe Science allows students to discover concepts within each of the content standards and gives students opportunities to make connections among the science disciplines. Hands-on activities and inquiry-based lessons reinforce the science processes emphasized in the standards.

How Glencoe Science Aligns with the NCTM Standards for Grades 6–8

Throughout Glencoe Science, each Applying Math activity provides students with the opportunity to practice and apply some of the mathematical concepts and applications described in the NCTM Standards. These activities serve to reinforce mathematical skills in real-life situations, thus preparing students to meet their needs in an ever-changing world.

Correlation of *Glencoe Science* to NCTM Standards

Math Standard	Page
1. Number and Operations	F-31, F-54, F-61, F-121, F-143, F-151, F-172, F-181, G-31, G-46, G-59, G-87, G-119, G-149, G-173, H-47, H-63, H-95, H-123, H-130, I-31, I-39, I-61, I-114, I-121, J-35, J-65, J-80, J-99, J-130
2. Algebra	F-54, F-61, F-143, F-151, F-172, F-181, G-31, G-119, G-149, H-47, H-95, H-123, H-130, I-39, I-61, I-114, I-121, J-65
3. Geometry	G-31, H-95, J-16, J-65
4. Measurement	G-31, H-95, I-91, I-121, J-35, J-65
5. Data Analysis and Probability	F-31, F-93, F-151, F-181, G-46, G-149, G-181, H-31, H-63, H-123, H-153, I-31, I-39, I-61, I-121, J-35, J-99, J-130
6. Problem Solving	F-31, G-46, G-87, G-149
7. Reasoning and Proof	F-54, F-143, F-172, G-46, G-105, G-173, H-108, H-130, I-39, J-16, J-80
8. Communication	F-117, F-147, G-83, I-87, J-31
9. Connections	F-54, F-61, F-121, F-143, F-151, F-172, F-181, G-31, G-59, G-119, H-47, H-63, H-95, H-123, H-130, H-153, I-31, I-39, I-61, I-91, I-114, I-121, J-35, J-65, J-80, J-99, J-130
10. Representation	G-31, G-149

Foldables™

Foldables™ are easy-to-make, three-dimensional, interactive graphic organizers that students create out of simple sheets of paper. These unique hands-on tools for studying and reviewing were created exclusively for Glencoe by education specialist Dinah Zike.

Research Behind Foldables™

According to research (Bransford, 1979; Corno, 1994), study strategies help students understand, organize, remember and apply new information presented in science textbooks. Some study strategies include concept mapping, highlighting, outlining, note taking, summarizing, and underlining (Peverly, Brobst, Graham & Shaw, 2003). Glencoe Science offers Dinah Zike's Foldables™ Study Organizers as an organizational tool and study guide for students.

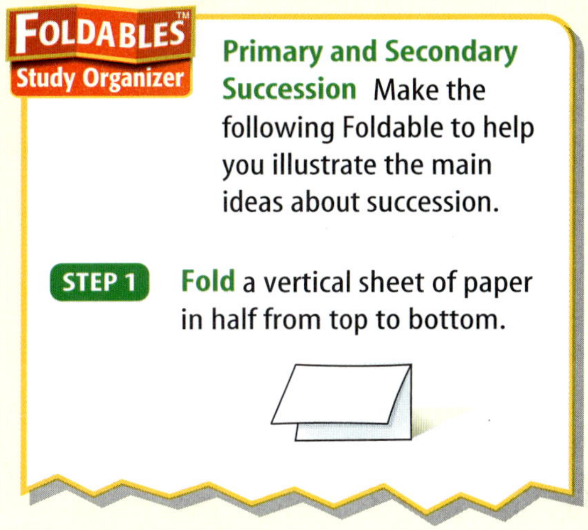

Foldables™

- Build prereading skills
- Encourage active reading and writing
- Summarize content for review

For more ideas on how to incorporate Foldables™ into your lessons consult **Dinah Zike's *Teaching Science with Foldables*™**

Educational Partnerships

NATIONAL GEOGRAPHIC

Some topics in the chapter either require or benefit from a larger, more detailed visual explanation. The National Geographic Society has created *Visualizing* features that call out an important concept from the chapter and illustrate it in a way that will inform, excite, and motivate your students.

TIME

TIME magazine brings science topics and history together to further explain the chapter's main ideas and show how science relates to real life.

Safety in the Laboratory

All activities are designed to minimize dangers in the laboratory. Careful laboratory planning and management by both the instructor and the student are essential to a safe laboratory experience. **Local, state, and federal laboratory safety laws and regulations must be strictly followed.** The information provided here is one of the many resources to which you can refer for information about laboratory safety.

Classroom and Laboratory Preparation

1. Store equipment properly and securely and other thing.
   a. Clean and dry all equipment before storing.
   b. Protect electronic equipment and microscopes from dust, humidity, and extreme temperatures.
   c. Number, catalog, and organize equipment.
2. Ensure adequate work space for each student.
3. Ensure adequate classroom and storeroom ventilation.
4. Explain and post safety and evacuation guidelines along with expectations of conduct.
5. Ensure that all safety equipment is functioning properly and is clearly visible.
6. Provide hot plates as a heat source whenever possible. If gas burners are used, know where the central gas supply shutoff valve is located.
7. Ensure that each workstation has a GFCI-protected electrical source.
8. Provide safety goggles consistent with ANSI Standard Z87.1 for each student, including students who wear corrective lenses.

Before Each MiniLAB or Lab

1. Arrange the lab in such a way that equipment and supplies are clearly labeled and easily accessible.
2. Have available only equipment and supplies needed to complete the assigned investigation.
3. Review the procedure with students, emphasizing any caution statements or safety symbols that appear.
4. Be sure all students know the proper procedures to follow if an accident should occur.

After the MiniLAB or Lab

1. Be certain that students have returned all equipment and disposed of broken glassware and chemicals properly.
2. Be sure that all hot plates and electrical connections are off.

Storage of Chemicals

Be sure to store all chemicals properly. The following are guidelines commonly used. Your school, city, county, or state may have additional requirements for handling chemicals. It is the responsibility of each teacher to become informed of the rules or guidelines in effect in his or her area.

1. Separate chemicals by reaction type. Strong acids should be stored together. Likewise, strong bases should be stored together and should be separated from acids. Oxidants should be stored away from easily oxidized materials, and so on.
2. Be sure all chemicals are stored in labeled containers indicating contents, concentration, source, date purchased (or prepared), any precautions for handling and storage, and expiration date.
3. Hazardous chemicals require special storage containers and conditions. Be sure to know which chemicals those are and the accepted practices for your area. Some substances must be stored outside the building.

Disposal of Chemicals

Local, state, and federal laws regulate the proper disposal of chemicals. These laws should be consulted before chemical disposal is attempted. Although many substances encountered in the science classroom can be flushed down the drain with plenty of water, it is not safe to assume that this is always true.

DISCLAIMER

Glencoe Publishing Company makes no claims to the completeness of this discussion of laboratory safety and chemical storage. The material presented is not all-inclusive, nor does it address all of the hazards associated with handling, storing, and disposing of chemicals, or with laboratory management.

Preparation of Solutions

It is important to use safe laboratory techniques when handling all chemicals. Always check the MSDS (Material Safety Data Sheet) for each chemical before using it in the classroom. Many substances might appear harmless, but might be toxic, corrosive, or very reactive. Chemicals should never be ingested. Use proper techniques to smell any chemical, wear safety goggles and an apron in the laboratory, and observe the following precautions.

1. **Dilution of Acids and Bases** When diluting acids with water, always add the acids to the water. Never add water to acids. When sulfuric acid and sodium hydroxide are added to water, a large amount of thermal energy is released. Use extra care when handling these substances.

2. **Poisonous and Corrosive Liquids or Vapors** Use a fume hood if possible. Examples include hydrochloric acid, acetic acid, nitric acid, and ammonium hydroxide.

3. **Poisonous and Corrosive to Eyes, Lungs, and Skin** Examples include acids, bases, silver nitrate, iodine, and potassium permanganate.

Bromthymol blue: Add 0.5 g bromthymol blue powder to 500 mL distilled water to make a BTB stock solution. Dilute 40 mL BTB stock solution to 2 L with distilled water. Solution should be bright blue. If not, add one drop of NaOH at a time, swirling to mix. Check color.

Hydrochloric acid (HCL) solution: To make a 5% solution, add 13.6 mL concentrated HCl to 73 mL water while stirring. To make a 0.1*M* solution, add 1 mL concentrated hydrochloric acid to 100 mL water while stirring.

Iodine solution/Iodine stain: Dilute 1 part Lugol's solution with 15 parts water.

Lugol's solution: Dissolve 10 g potassium iodide in 100 mL distilled water. Then add and dissolve 5 g iodine. Store in dark bottle. Keeps indefinitely.

Phenolphthalein indicator: From a drug store, buy a package of any laxative that contains phenolphthalein. To make 1% solution, mash 4 tablets and pour the powder into 10 mL of rubbing alcohol. Let mixture soak for 15 minutes. Pour liquid into and store in a dropper bottle.

Potassium permanganate: For a 0.01*M* solution of potassium permanganate, dissolve 0.15 g $KMnO_4$ in 100 mL water.

Red cabbage concentrate: Put 5 leaves of red cabbage in a pot. Add 1 L of water, bring to a boil, and simmer until water turns a deep purple. Pour liquid through a strainer or piece of cheesecloth into a storage bottle. Keep refrigerated.

Salt solution: For a 3.5% salt (NaCl) solution that simulates the concentration of ocean water, dissolve 35 g of salt (NaCl) in 965 mL of water. For a 1% solution (weak), dissolve 1 g of salt (NaCl) in 99 mL of water. For a 6% solution, dissolve 6 g of salt (NaCl) in 94 mL of water.

Silver nitrate solution: To make a 10% solution, put 5 g of silver nitrate in 50 mL of distilled water.

Sugar solution: Add 1 tablespoon of sugar to 1 cup of warm water in a deep jar or flask. Stir to dissolve.

Sodium hydroxide (dilute): To make a 1% solution, dissolve 1 g NaOH in 99 mL of water.

Equipment and Materials List

Refer to the Chapter Organizer in front of each chapter for a list of equipment and materials used for each laboratory activity in the chapter.

Material	Launch Lab (Chapter)	MiniLAB (Chapter-Section)	Lab (Chapter-Section)
Consumables			
ball, foam	6		
balloon, small			6-3
cardboard			1-1, 3-3
clay or modeling dough		4-1	
flour			6-3
food coloring		4-3	
hydrochloric acid solution, 5%			1-3, 2-4
magazine, old	4		
newspaper			6-3
olive oil		6-1	
plaster of paris		6-2	
plastic tubing			6-3
salt	1	1-1	
sand		2-1, 6-2	
sandpaper	5		
sediment samples		2-4	
soil, black, brown, and white sandy			3-2
sugar, granulated			1-1
wooden stick, flat			1-1
Nonconsumables			
balance, pan			1-3
basalt, gabbro, obsidian, pumice, rhyolite, vesicular basalt			2-2
building blocks		5-3	
calcite	1	1-2	
casserole dish, clear, colorless		4-3	
clamp for tubing			6-3
cooking oil	3		
dishes—glass, plastic			1-3, 3-2
dissecting probe		2-4	
file, steel			1-3
globe			5-2
granite	1		2-2
gypsum		1-2	
halite		1-2	
hot plate		4-3	1-1
lamp, gooseneck, 200-watt			3-2
meter, electric		3-1	
mica	1		
mica, muscovite		1-2	
mineral samples			1-3
nail, small iron			1-3
pan, shallow	3		1-1
protractor		6-2	
quartz crystals	1		
rock samples	2		2-4
sandstone	1, 3		
schist	1		
shale	3		
streak plate			1-3
thermal cup holder		3-3	
thermos bottle		3-3	

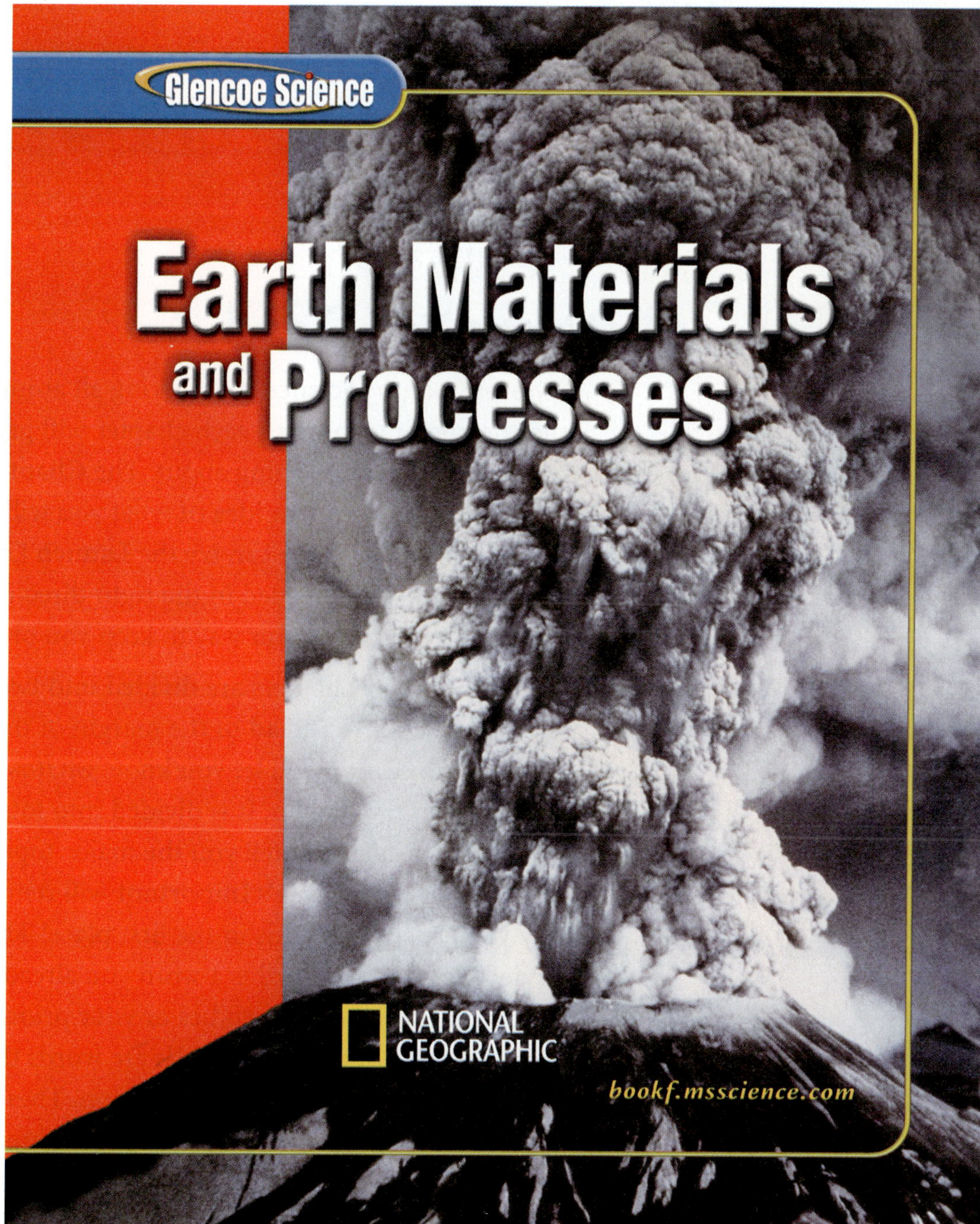

McGraw Hill Glencoe

New York, New York Columbus, Ohio Chicago, Illinois Peoria, Illinois Woodland Hills, California

HOW TO...

Use Your Science Book

Why do I need my science book?

Have you ever been in class and not understood all of what was presented? Or, you understood everything in class, but at home, got stuck on how to answer a question? Maybe you just wondered when you were ever going to use this stuff?

These next few pages are designed to help you understand everything your science book can be used for . . . besides a paperweight!

Before You Read

- **Chapter Opener** Science is occurring all around you, and the opening photo of each chapter will preview the science you will be learning about. The **Chapter Preview** will give you an idea of what you will be learning about, and you can try the **Launch Lab** to help get your brain headed in the right direction. The **Foldables** exercise is a fun way to keep you organized.
- **Section Opener** Chapters are divided into two to four sections. The **As You Read** in the margin of the first page of each section will let you know what is most important in the section. It is divided into four parts. **What You'll Learn** will tell you the major topics you will be covering. **Why It's Important** will remind you why you are studying this in the first place! The **Review Vocabulary** word is a word you already know, either from your science studies or your prior knowledge. The **New Vocabulary** words are words that you need to learn to understand this section. These words will be in **boldfaced** print and highlighted in the section. Make a note to yourself to recognize these words as you are reading the section.

As You Read

- **Headings** Each section has a title in large red letters, and is further divided into blue titles and small red titles at the beginnings of some paragraphs. To help you study, make an outline of the headings and subheadings.
- **Margins** In the margins of your text, you will find many helpful resources. The **Science Online** exercises and **Integrate** activities help you explore the topics you are studying. **MiniLabs** reinforce the science concepts you have learned.
- **Building Skills** You also will find an **Applying Math** or **Applying Science** activity in each chapter. This gives you extra practice using your new knowledge, and helps prepare you for standardized tests.
- **Student Resources** At the end of the book you will find **Student Resources** to help you throughout your studies. These include **Science, Technology,** and **Math Skill Handbooks,** an **English/Spanish Glossary,** and an **Index.** Also, use your **Foldables** as a resource. It will help you organize information, and review before a test.
- **In Class** Remember, you can always ask your teacher to explain anything you don't understand.

FOLDABLES™ Study Organizer

Science Vocabulary Make the following Foldable to help you understand the vocabulary terms in this chapter.

STEP 1 **Fold** a vertical sheet of notebook paper from side to side.

STEP 2 **Cut** along every third line of only the top layer to form tabs.

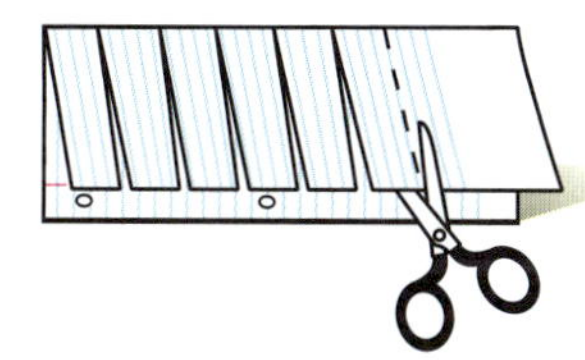

STEP 3 **Label** each tab with a vocabulary word from the chapter.

Build Vocabulary As you read the chapter, list the vocabulary words on the tabs. As you learn the definitions, write them under the tab for each vocabulary word.

In Lab

Working in the laboratory is one of the best ways to understand the concepts you are studying. Your book will be your guide through your laboratory experiences, and help you begin to think like a scientist. In it, you not only will find the steps necessary to follow the investigations, but you also will find helpful tips to make the most of your time.

- Each lab provides you with a **Real-World Question** to remind you that science is something you use every day, not just in class. This may lead to many more questions about how things happen in your world.
- Remember, experiments do not always produce the result you expect. Scientists have made many discoveries based on investigations with unexpected results. You can try the experiment again to make sure your results were accurate, or perhaps form a new hypothesis to test.
- Keeping a **Science Journal** is how scientists keep accurate records of observations and data. In your journal, you also can write any questions that may arise during your investigation. This is a great method of reminding yourself to find the answers later.

a or b? ? T or F?

Before a Test

Admit it! You don't like to take tests! However, there *are* ways to review that make them less painful. Your book will help you be more successful taking tests if you use the resources provided to you.

- Review all of the **New Vocabulary** words and be sure you understand their definitions.
- Review the notes you've taken on your **Foldables,** in class, and in lab. Write down any question that you still need answered.
- Review the **Summaries** and **Self Check questions** at the end of each section.
- Study the concepts presented in the chapter by reading the **Study Guide** and answering the questions in the **Chapter Review.**

Look For...

- **Reading Checks** and **caption questions** throughout the text.
- the **Summaries** and **Self Check questions** at the end of each section.
- the **Study Guide** and **Review** at the end of each chapter.
- the **Standardized Test Practice** after each chapter.

Scavenger HUNT

Let's Get Started

To help you find the information you need quickly, use the Scavenger Hunt below to learn where things are located in Chapter 1.

1. What is the title of this chapter?
2. What will you learn in Section 1?
3. Sometimes you may ask, "Why am I learning this?" State a reason why the concepts from Section 2 are important.
4. What is the main topic presented in Section 2?
5. How many reading checks are in Section 1?
6. What is the Web address where you can find extra information?
7. What is the main heading above the sixth paragraph in Section 2?
8. There is an integration with another subject mentioned in one of the margins of the chapter. What subject is it?
9. List the new vocabulary words presented in Section 2.
10. List the safety symbols presented in the first Lab.
11. Where would you find a Self Check to be sure you understand the section?
12. Suppose you're doing the Self Check and you have a question about concept mapping. Where could you find help?
13. On what pages are the Chapter Study Guide and Chapter Review?
14. Look in the Table of Contents to find out on which page Section 2 of the chapter begins.
15. You complete the Chapter Review to study for your chapter test. Where could you find another quiz for more practice?

Teacher Advisory Board

The Teacher Advisory Board gave the authors, editorial staff, and design team feedback on the content and design of the Student Edition. They provided valuable input in the development of the 2005 edition of ***Glencoe Science.***

John Gonzales
Challenger Middle School
Tucson, AZ

Rachel Shively
Aptakisic Jr. High School
Buffalo Grove, IL

Roger Pratt
Manistique High School
Manistique, MI

Kirtina Hile
Northmor Jr. High/High School
Galion, OH

Marie Renner
Diley Middle School
Pickerington, OH

Nelson Farrier
Hamlin Middle School
Springfield, OR

Jeff Remington
Palmyra Middle School
Palmyra, PA

Erin Peters
Williamsburg Middle School
Arlington, VA

Rubidel Peoples
Meacham Middle School
Fort Worth, TX

Kristi Ramsey
Navasota Jr. High School
Navasota, TX

Student Advisory Board

The Student Advisory Board gave the authors, editorial staff, and design team feedback on the design of the Student Edition. We thank these students for their hard work and creative suggestions in making the 2005 edition of ***Glencoe Science*** student friendly.

Jack Andrews
Reynoldsburg Jr. High School
Reynoldsburg, OH

Peter Arnold
Hastings Middle School
Upper Arlington, OH

Emily Barbe
Perry Middle School
Worthington, OH

Kirsty Bateman
Hilliard Heritage Middle School
Hilliard, OH

Andre Brown
Spanish Emersion Academy
Columbus, OH

Chris Dundon
Heritage Middle School
Westerville, OH

Ryan Manafee
Monroe Middle School
Columbus, OH

Addison Owen
Davis Middle School
Dublin, OH

Teriana Patrick
Eastmoor Middle School
Columbus, OH

Ashley Ruz
Karrer Middle School
Dublin, OH

The Glencoe middle school science Student Advisory Board taking a timeout at COSI, a science museum in Columbus, Ohio.

Contents

Contents

In each chapter, look for these opportunities for review and assessment:
- Reading Checks
- Caption Questions
- Section Review
- Chapter Study Guide
- Chapter Review
- Standardized Test Practice
- Online practice at bookf.msscience.com

Student Resources

Cross-Curricular Readings/Labs

available as a video lab

NATIONAL GEOGRAPHIC VISUALIZING

TIME SCIENCE AND Society

TIME SCIENCE AND HISTORY

Oops! Accidents in SCIENCE

Science and Language Arts

SCIENCE Stats

Launch LAB

Mini LAB

Mini LAB Try at Home

Labs/Activities

One-Page Labs

Two-Page Labs

Design Your Own Labs

Model and Invent Labs

Use the Internet Labs

Applying Math

Applying Science

INTEGRATE

Science Online

Standardized Test Practice

Content Details

Monitoring Volcanoes

Introduction

This features explores the reasons why people live near active volcanoes and offers examples of past eruptions to point out their destructive power. Through this information students are led to the importance of using science to try to predict volcanic eruptions. Volcanoes display signs when an eruption is imminent. The feature describes the instrumentation used by volcanologists to detect these signals. Students can also gain a sense of an ordinary workday for a volcanologist.

1 Motivate

Write answers on the board as students brainstorm the actions that would be necessary to evacuate their community in the event of a nearby volcanic eruption. Possible responses: closing schools and businesses, implementing some effective means of notifying citizens of the pending eruption, an evacuation plan in place, public transportation set up to quickly move people and equipment out of harm's way.

The NATURE OF SCIENCE

Science Today

Monitoring Volcanoes

Figure 1 The May 18, 1980, eruption of Mount St. Helens blew tons of ash, rock, and steam into the air when it erupted.

Volcanic eruptions can cause incredible destruction, yet many people continue to live near active volcanoes. One approach to protect lives and property is to look for signs that a volcano is about to erupt. This was done on Mount St. Helens in the state of Washington prior to its eruption on May 18, 1980.

Mount St. Helens exploded after 123 years of inactivity. Over 600 km^2 of surrounding land was devastated. More than 300 m of the volcano's north face blew away, creating a huge crater and sending a cloud of hot steam and ash roaring down the flanks of its north slope.

On the island of Hawaii, the Mauna Loa and Kilauea volcanoes erupt more quietly than Mount St. Helens, but they still have the potential to cause great damage. In 1990, lava flows from Kilauea destroyed property in Kalapana Gardens. In 1984, an eruption of Mauna Loa sent lava to within 6.5 km of Hilo, the largest city on the island of Hawaii.

Figure 2 The eruption of Mount St. Helens killed 57 people and caused hundreds of millions of dollars in damage. The force of the blast knocked down millions of trees.

Fun Fact

The eruption of the Indonesian volcano Tambora in 1815 released so much ash and dust into the atmosphere that it cooled climates worldwide. In New England there was snow in June, and crops around the nation were beset by unseasonable frosts. The chilly time after the eruption was called "the year without a summer."

Living Near a Volcano

Volcanoes are natural environmental hazards because of their potentially destructive power and their proximity to populated areas. Many people are reluctant or unwilling to move from their homes near active volcanoes even though there is no way to prevent volcanic eruptions. Such regions often enjoy rich soils of volcanic origin. Consequently, scientists have been working for many years to find the best ways to monitor various volcanoes around the world. They suggest that the data they gather will enable them to better forecast when a quiet volcano might erupt again, allowing people to evacuate a region before an eruption.

Figure 3 Kilauea has erupted continuously for more than 15 years. This lava flow encroached on property in Kalapana Gardens in 1990.

Science

Some advances in the study of volcanoes came about as scientists first attempted to solve the problem of how to forecast eruptions. Solving problems to help make people's lives safer and better is a benefit of science. When you solve a problem by finding a better way to do something, you are doing science.

Volcanology is part of Earth science, the scientific study of the solid part of Earth, the oceans, the atmosphere, and bodies in space. In this book, you will learn about the materials of which Earth is made. You also will learn about processes, such as volcanic eruptions, that shape and change Earth's surface.

Figure 4 Hilo, Hawaii, sits in the path of volcanic lava flows.

2 Teach

Content Background

The lava, hot ash, and gases emitted by erupting volcanoes pose dangers to people living nearby. But volcanoes at high altitudes are often mantled in ice and snow, posing the additional threat of volcanic mudflows called lahars. Eruptions of high-elevation volcanoes such as Mount Rainier in Washington can suddenly melt the glaciers and snowfields. This melting can generate floods and mudflows that sweep down the mountain without warning at speeds of 30 to 60 kilometers per hour. Virtually everything in their paths is destroyed.

Discussion

Why would people be unwilling to move from an eruption-prone area? Possible answers: Rich volcanic soils allow farmers to prosper; families that have lived for many generations in one place feel connected to the land; people might live in denial of the dangers.

How could people avoid getting caught in a volcanic mudflow? Possible answers: They can heed warnings of imminent eruptions and evacuate; they can look at geologic evidence of past mudflows and avoid building homes in their paths.

Curriculum Connection

Mathematics Have students calculate how long it would take for a volcanic mudflow traveling at 40 km/h to reach a town situated 5 km from the volcano. The mudflow would reach the town in 0.125 hours, or 7.5 minutes. This timing leaves little opportunity for escape.

Activity

Have each student pair research the location and details of one past major volcanic eruption. Place a large world map on the wall and provide pushpins and string. Have each pair place a pushpin into the map to mark the location of their volcano, and then run a string from the pushpin to a sheet of paper on which they have summarized important information specific to their eruption. Details could include the date, casualties, type of eruption, and whether there was advance warning. Have pairs present the information about their volcanoes to the class. When presentations are complete, discuss any trends students notice in terms of volcano locations. Students may notice that many volcanoes are located around the Pacific Ocean in the region known as the Ring of Fire. COOP LEARN LS **Interpersonal**

Extension

Challenge students to find several examples of volcanoes around the world that are currently being monitored for signs of eruption. Have them research why these volcanoes were chosen for monitoring projects, the danger signs displayed by each volcano, and the specific equipment volcanologists are using at each site. Have them write summaries of their findings and present them to the class. LS **Linguistic**

The NATURE OF SCIENCE

Science Today

For most of human history, volcanic eruptions have caught people off-guard. Eruptions have poured out lava, hot ash, and gas, often trapping people before they could escape. Today, although eruptions still cause great destruction, fewer people die because volcanologists—scientists who study volcanoes—can forecast many eruptions. For instance, workers knew that Mount St. Helens would explode thanks to advances in volcano monitoring techniques. They were able to warn people in the area and save many lives.

Where Volcanologists Work

Some scientists who monitor volcanoes work at the United States Geological Survey (USGS) volcano observatories, such as:

1. **Alaska Volcano Observatory:** Monitors Alaska's volcanoes and sends out warnings about eruptions in eastern Russia.
2. **Hawaii Volcano Observatory:** Monitors the active volcanoes on the island of Hawaii.
3. **Cascades Volcano Observatory:** Monitors and assesses hazards from volcanoes of the Cascade Range.
4. **Long Valley Observatory:** Monitors activity from the large and potentially hazardous calderas system near Mammoth Lakes, California.

Looking for Signs

Monitoring is reading the signs of activity generated by a volcano before an eruption. For example, prior to a volcanic eruption, magma moves toward Earth's surface. This movement causes earthquakes, changes in a volcano's shape, and the release of certain gases. Volcanologists use specialized instruments to measure changes in the ground surface, the amounts and types of gases emitted, and seismic waves released by earthquakes.

One sign that a volcano might erupt is an increase in the number of earthquakes in the region. Magma and gases force their way up through cracks deep in a volcano, causing the earthquakes. For example, two months before the eruption of Mount St. Helens, about 10,000 quakes occurred in the mountain. Seismographs placed on or near volcanoes can record such earthquakes.

Volcanologists also know that changes in the shape of a volcano can mean an eruption might soon occur. As magma moves upward, parts of a volcano might rise or sink. Mount St. Helens formed a huge bulge in the weeks prior to its eruption.

Science Journal

Signs of Volcanic Activity Tell students that volcanoes are usually considered dormant if they have not erupted within historic time. However, volcanoes can suddenly spring to life hundreds of years after they last erupted. Have students describe in their Science Journals why they think this might occur. What signs might offer clues that this is occurring?

Using Technology

Figure 5 This USGS solar-powered seismograph records small earthquakes on the flank of the Augustine volcano in Alaska.

Besides seismographs, volcanologists use tiltmeters, electronic distance meters (EDMs), spectrometers, and strainmeters. A tiltmeter measures changes in the slope of the ground caused by moving magma. Like a carpenter's level, it consists of a bubble inside a fluid-filled container. If the slope changes, the bubble moves and the difference is measured electronically. An electronic distance meter uses a laser beam to measure the distance between two points on a volcano. If magma moves rocks or widens cracks, the targets will move and the EDM will record a change in distance.

Spectrometers measure gases released from magma. The rate at which volcanoes release carbon dioxide and sulfur dioxide, for example, might change before an eruption.

The strainmeter (or dilatometer) is being used in Hawaii to monitor Mauna Loa and Kilauea. It consists of a small canister filled with liquid silicon that is placed deep in a hole drilled into a volcano. Any movement in the volcano that changes the shape of the ground squeezes the strainmeter and the measurements are recorded on instruments at the surface.

Working on a Volcano

Although some volcanoes are monitored using radio-controlled instruments, volcanologists also must work in dangerous conditions on active volcanoes. They install instruments, take readings, or collect gas escaping from volcanic vents.

Volcanologist Cynthia Gardner enjoys her work in Washington, Oregon, and Alaska because she's helping to save lives. When she's not in the field, she collects data, writes reports, and sets up emergency procedures in communities near volcanoes.

Figure 6 Volcanologist Cynthia Gardner uses advanced equipment to monitor volcanoes.

You Do It

Airplanes and satellites are tools that help volcanologists forecast the eruption of volcanoes. Research in your local library or by visiting **bookf.msscience.com** to find out how volcanologists employ these tools. How would their work be more difficult without the aid of airplanes and satellites?

Visual Learning

Figure 5 Ask students to list reasons why volcanologists are studying earthquakes on this volcano. Possible answers: This volcano has shown other signs of activity that prompted installation of the seismograph; swarms of noticeable earthquakes spurred scientists to determine whether imperceptible earthquakes were also occurring. **How might scientists label the x- and y-axes when graphically displaying data from the seismograph?** Possible answer: The *x*-axis would show time; the *y*-axis would show earthquake magnitude.

3 Assess

Use the question in You Do It to assess understanding. Answers should relate to volcanologists and should be as specific as possible.

You Do It

Lead students through this exercise as a class to clarify what they are being asked to do. Discuss the key words they might use in searching the Internet for information. Ask how they might narrow a library search for this information. Encourage students to enlist help from the librarian, and have them discuss ways to effectively present their findings.

chapter 1 Organizer

Section/Objectives	Standards		Labs/Features
	National	**State/Local**	
Chapter Opener	See pp. 9T–10T for a Key to Standards.		**Launch Lab:** What's the difference between a rock and a mineral?, p. 7 **Foldables,** p. 7
Section 1 Minerals 2 sessions 1 block 1. **Describe** characteristics that all minerals share. 2. **Explain** how minerals form.	National Content Standards: UCP.1, UCP.2, UCP.3, UCP.5, A.1, A.2, D.1, D.2		**MiniLAB:** Inferring Salt's Crystal System, p. 9 **Visualizing Crystal Systems,** p. 10 **Integrate Physics,** p. 11 **Lab:** Crystal Formation, p. 13
Section 2 Mineral Identification 2 sessions 1 block 3. **Describe** physical properties used to identify minerals. 4. **Identify** minerals using physical properties such as hardness and streak.	National Content Standards: UCP.1, UCP.2, UCP.3, UCP.5, A.1, A.2, B.2, D.1, G.2, G.3		**Applying Science:** How can you identify minerals?, p. 16 **MiniLab:** Observing Mineral Properties, p. 18
Section 3 Uses of Minerals 2 sessions 1 block 5. **Describe** characteristics of gems that make them more valuable than other minerals. 6. **Identify** useful elements that are contained in minerals.	National Content Standards: UCP.1, UCP.2, UCP.5, A.1, A.2, D.2, G.1, G.2, G.3		**Science Online,** p. 22 **Integrate Social Studies,** p. 23 **Lab:** Mineral Identification, p. 26 **Science and History:** Dr. Dorothy Crowfoot Hodgkin, p. 28

Glencoe Exclusive! **TeacherWorks™** All-In-One Planner and Resource Center

Lab Materials	Reproducible Resources	Section Assessment	Technology
Launch Lab: magnifying lens, quartz crystal, table salt, samples of sandstone, granite, gypsum, mica, schist	**Chapter *FAST FILE* Resources** Foldables Worksheet, p. 13 Directed Reading Overview, p. 15 Note-taking Worksheets, p. 29, 30	GLENCOE'S ASSESSMENT ADVANTAGE	**TeacherWorks** includes: • Interactive Teacher Edition • Lesson Planner with calendar • Access to all program blacklines • Correlations to standards • Web links
MiniLAB: magnifying lens, table salt, construction paper **Lab:** 250-mL beakers (2), cardboard, large paper clip, table salt, flat wooden stick, granulated sugar, cotton string, hot plate, magnifying lens, thermal mitt, shallow pan, spoon	**Chapter *FAST FILE* Resources** Transparency Activity, p. 40 MiniLAB, p. 3 Enrichment, p. 26 Reinforcement, p. 23 Directed Reading, p. 16 Lab Worksheets, pp. 5–6 Lab Activity, pp. 9–10 **Cultural Diversity,** p. 37 **Mathematics Skill Activities,** p. 47	**Portfolio** Activity, p. 11 **Performance** MiniLAB, p. 9 Applying Skills, p. 12 **Content** Section Review, p. 12	Section Focus Transparency Virtual Labs CD-ROM Guided Reading Audio Program Interactive Chalkboard CD-ROM
MiniLAB: samples of clear minerals such as gypsum, muscovite mica, halite, and calcite *Need materials?* Contact Science Kit at 1-800-828-7777 or www.sciencekit.com on the Internet.	**Chapter *FAST FILE* Resources** Transparency Activity, p. 41 MiniLAB, p. 4 Enrichment, p. 27 Reinforcement, p. 24 Directed Reading, p. 17 Transparency Activity, pp. 43–44 **Science Inquiry Lab,** pp. 31–32	**Portfolio** Activity, p. 15 **Performance** Applying Science, p. 16 MiniLAB, p. 18 Applying Skills, p. 18 **Content** Section Review, p. 18	Section Focus Transparency Teaching Transparency Virtual Labs CD-ROM Guided Reading Audio Program Interactive Chalkboard CD-ROM Video Lab
Lab: mineral samples, magnifying lens, pan balance, graduated cylinder, water, penny, glass plate, small iron nail, steel file, streak plate, 5% HCl with dropper, Mohs scale of hardness, Minerals Appendix, safety goggles	**Chapter *FAST FILE* Resources** Transparency Activity, p. 42 Enrichment, p. 28 Reinforcement, p. 25 Directed Reading, pp. 17, 18 Lab Worksheets, pp. 7–8 Lab Activity, pp. 11–12 **Lab Management and Safety,** pp. 54–57 **Home and Community Involvement,** p. 34	**Portfolio** Use Science Words, p. 20 Differentiated Instruction, p. 22 **Performance** Applying Math, p. 25 **Content** Section Review, p. 25	Section Focus Transparency Virtual Labs CD-ROM Guided Reading Audio Program Interactive Chalkboard CD-ROM Video Lab

End of Chapter Assessment

GLENCOE'S ASSESSMENT ADVANTAGE

Blackline Masters	Technology	Professional Series
Chapter *FAST FILE* Resources Chapter Review, pp. 33–34 Chapter Tests, pp. 35–38 **Standardized Test Practice,** pp. 7–10	MindJogger Videoquiz Virtual Labs CD-ROM Exam*View*® Pro Testmaker TeacherWorks CD-ROM Interactive Chalkboard CD-ROM	**Performance Assessment in the Science Classroom (PASC)**

chapter 1 Minerals

Transparencies

Section Focus

Section 1 Section Focus Transparency — A Study in Patterns

In a repeating pattern, the same structure occurs over and over again. Nature is full of repeating patterns similar to the pattern in the kaleidoscope image below. Crystals, for example, are patterns of repeating elements.

1. What basic structure is repeated and how is it grouped to form the total design?
2. How might a kaleidoscope image relate to the structure of a mineral?

L2

Section 2 Section Focus Transparency — Sorting It Out

About 4,000 different kinds of minerals can be found on Earth. Some minerals exist in a variety of forms. You'll need skill and knowledge to identify that interesting rock you found at the beach. What can you tell about the three mineral samples shown here?

1. What are some similarities and differences in appearance among the three samples?
2. Do you think that the photos show the same mineral or different ones? Explain.
3. What other traits might help you identify different minerals?

L2

Section 3 Section Focus Transparency — Golden Days

Gold is a valuable metal that occurs in small amounts in all igneous rocks. It's unusual, however, to find rock that is very rich in gold. In that case, the rock is called gold ore. The gold ore must be hauled to the surface and refined before it looks like the metal you see in jewelry.

1. Why do you think some minerals and metals are considered valuable, while others are not?
2. Describe the gold vein pictured above. Does the gold look the same as it does in jewelry?
3. What other metals can you name that are mined like gold?

L2

This is a representation of key blackline masters available in the Teacher Classroom Resources. See Resource Manager boxes within the chapter for additional information.

Assessment

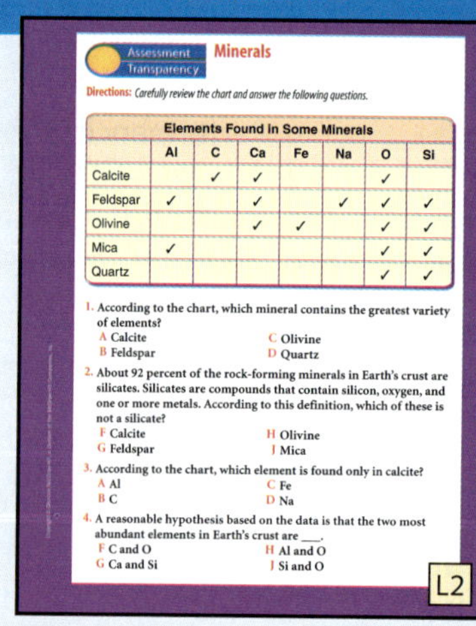
Assessment Transparency — Minerals

Directions: Carefully review the chart and answer the following questions.

Elements Found in Some Minerals

	Al	C	Ca	Fe	Na	O	Si
Calcite		✓	✓			✓	
Feldspar	✓		✓		✓	✓	✓
Olivine			✓	✓		✓	✓
Mica	✓					✓	✓
Quartz						✓	✓

1. According to the chart, which mineral contains the greatest variety of elements?
A Calcite C Olivine
B Feldspar D Quartz
2. About 92 percent of the rock-forming minerals in Earth's crust are silicates. Silicates are compounds that contain silicon, oxygen, and one or more metals. According to this definition, which of these is not a silicate?
F Calcite H Olivine
G Feldspar J Mica
3. According to the chart, which element is found only in calcite?
A Al C Fe
B C D Na
4. A reasonable hypothesis based on the data is that the two most abundant elements in Earth's crust are ___.
F C and O H Al and O
G Ca and Si J Si and O

L2

Teaching

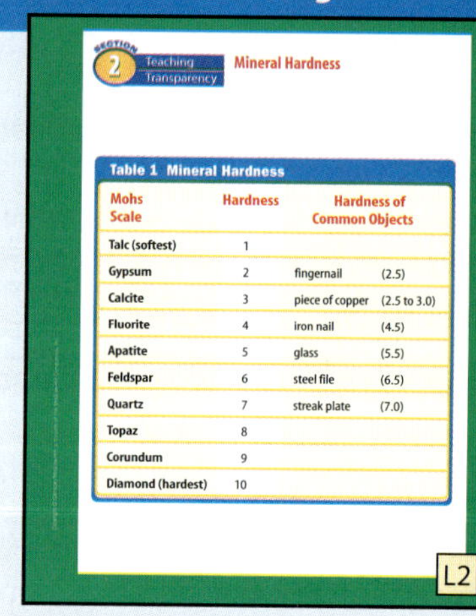
Section 2 Teaching Transparency — Mineral Hardness

Table 1 Mineral Hardness

Mohs Scale	Hardness	Hardness of Common Objects	
Talc (softest)	1		
Gypsum	2	fingernail	(2.5)
Calcite	3	piece of copper	(2.5 to 3.0)
Fluorite	4	iron nail	(4.5)
Apatite	5	glass	(5.5)
Feldspar	6	steel file	(6.5)
Quartz	7	streak plate	(7.0)
Topaz	8		
Corundum	9		
Diamond (hardest)	10		

L2

Key to Teaching Strategies

The following designations will help you decide which activities are appropriate for your students.

- L1 Level 1 activities should be appropriate for students with learning difficulties.
- L2 Level 2 activities should be within the ability range of all students.
- L3 Level 3 activities are designed for above-average students.
- ELL ELL activities should be within the ability range of English Language Learners.
- COOP LEARN Cooperative Learning activities are designed for small group work.
- LS Multiple Learning Styles logos, as described on page 6T, are used throughout to indicate strategies that address different learning styles.
- P These strategies represent student products that can be placed into a best-work portfolio.
- PBL Problem-Based Learning activities apply real-world situations to learning.

Hands-on Activities

Student Text Lab Worksheet

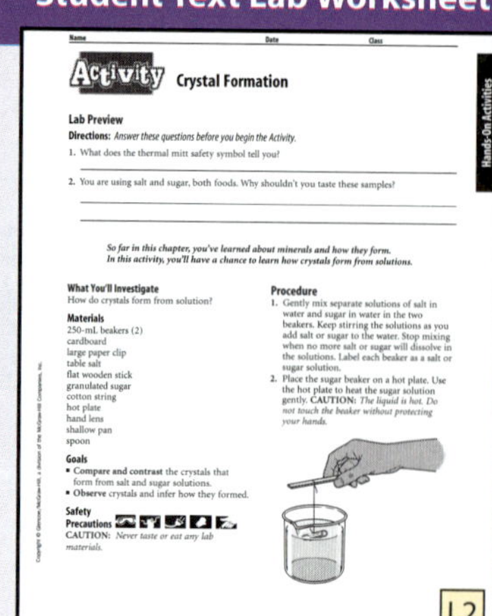
Activity — Crystal Formation

Lab Preview

Directions: Answer these questions before you begin the Activity.

1. What does the thermal mitt safety symbol tell you?
2. You are using salt and sugar, both foods. Why shouldn't you taste these samples?

L2

Laboratory Activities

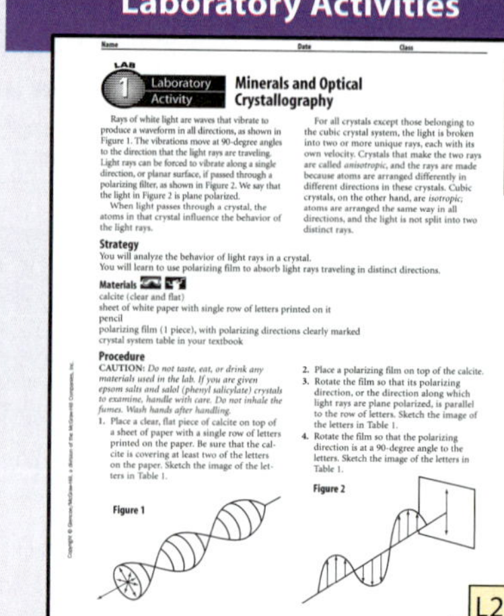
Laboratory Activity 1 — Minerals and Optical Crystallography

L2

Meeting Different Ability Levels

Content Outline

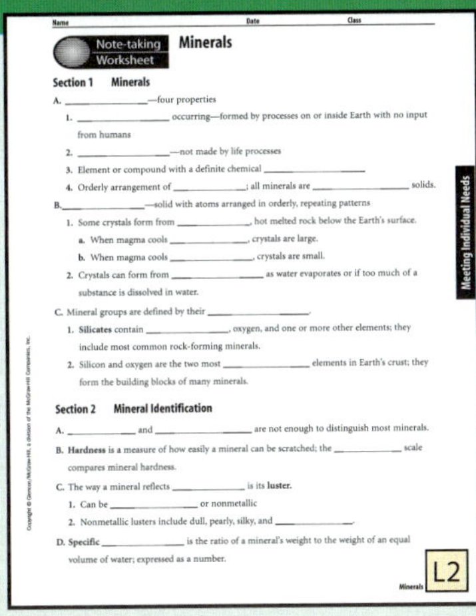
Note-taking Worksheet — Minerals

Section 1 Minerals

A. ____________—four properties
1. ____________ occurring—formed by processes on or inside Earth with no input from humans
2. ____________—not made by life processes
3. Element or compound with a definite chemical ____________
4. Orderly arrangement of ____________; all minerals are ____________ solids.

B. ____________—solid with atoms arranged in orderly, repeating patterns
1. Some crystals form from ____________, hot melted rock below the Earth's surface.
a. When magma cools ____________, crystals are large.
b. When magma cools ____________, crystals are small.
2. Crystals can form from ____________ as water evaporates or if too much of a substance is dissolved in water.

C. Mineral groups are defined by their ____________.
1. Silicates contain ____________, oxygen, and one or more other elements; they include most common rock-forming minerals.
2. Silicon and oxygen are the two most ____________ elements in Earth's crust; they form the building blocks of many minerals.

Section 2 Mineral Identification

A. ____________ and ____________ are not enough to distinguish most minerals.
B. Hardness is a measure of how easily a mineral can be scratched; the ____________ scale compares mineral hardness.
C. The way a mineral reflects ____________ is its luster.
1. Can be ____________ or nonmetallic
2. Nonmetallic lusters include dull, pearly, silky, and ____________.
D. Specific ____________ is the ratio of a mineral's weight to the weight of an equal volume of water; expressed as a number.

L2

Reinforcement

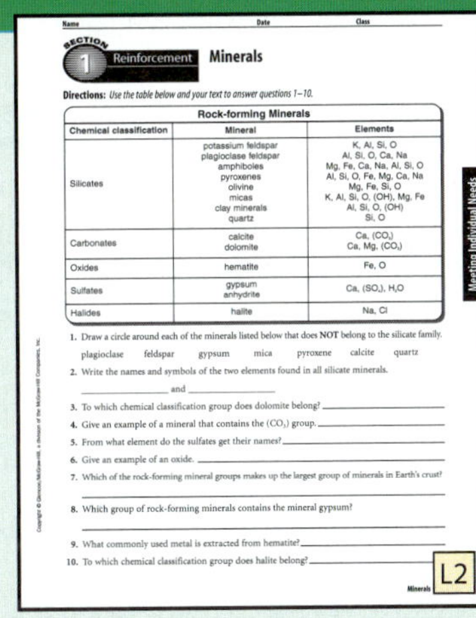
Section 1 Reinforcement — Minerals

Directions: Use the table below and your text to answer questions 1–10.

Rock-forming Minerals

Chemical classification	Mineral	Elements
Silicates	potassium feldspar, plagioclase feldspar, amphiboles, pyroxenes, olivine, micas, clay minerals, quartz	K, Al, Si, O; Al, Si, O, Ca, Na; Mg, Fe, Ca, Na, Al, Si, O; Al, Si, O, Fe, Mg, Ca, Na; Mg, Fe, Si, O; K, Al, Si, O, (OH), Mg, Fe; Al, Si, O, (OH); Si, O
Carbonates	calcite, dolomite	Ca, (CO_3); Ca, Mg, (CO_3)
Oxides	hematite	Fe, O
Sulfates	gypsum, anhydrite	Ca, (SO_4), H_2O
Halides	halite	Na, Cl

1. Draw a circle around each of the minerals listed below that does NOT belong to the silicate family. plagioclase feldspar gypsum mica pyroxene calcite quartz
2. Write the names and symbols of the two elements found in all silicate minerals.
3. To which chemical classification group does dolomite belong?
4. Give an example of a mineral that contains the (CO_3) group.
5. From what element do the sulfates get their names?
6. Give an example of an oxide.
7. Which of the rock-forming mineral groups makes up the largest group of minerals in Earth's crust?
8. Which group of rock-forming minerals contains the mineral gypsum?
9. What commonly used metal is extracted from hematite?
10. To which chemical classification group does halite belong?

L2

Enrichment

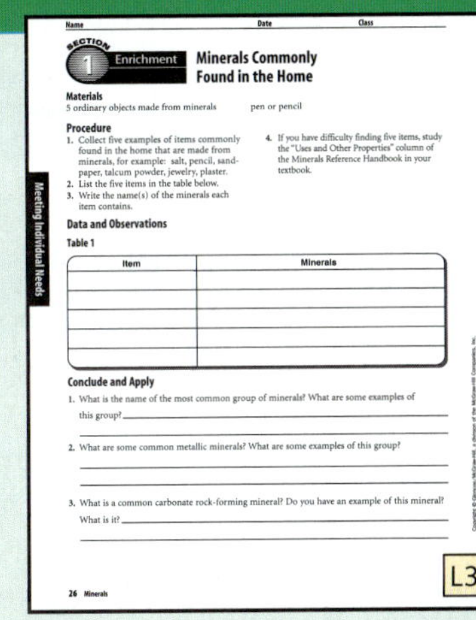
Section 1 Enrichment — Minerals Commonly Found in the Home

Materials: 5 ordinary objects made from minerals; pen or pencil

Procedure
1. Collect five examples of items commonly found in the home that are made from minerals, for example: salt, pencil, sandpaper, talcum powder, jewelry, plaster.
2. List the five items in the table below.
3. Write the name(s) of the minerals each item contains.
4. If you have difficulty finding five items, study the "Uses and Other Properties" column of the Minerals Reference Handbook in your textbook.

Data and Observations

Table 1

Item	Minerals

Conclude and Apply
1. What is the name of the most common group of minerals? What are some examples of this group?
2. What are some common metallic minerals? What are some examples of this group?
3. What is a common carbonate rock-forming mineral? Do you have an example of this mineral? What is it?

L3

Directed Reading (English/Spanish)

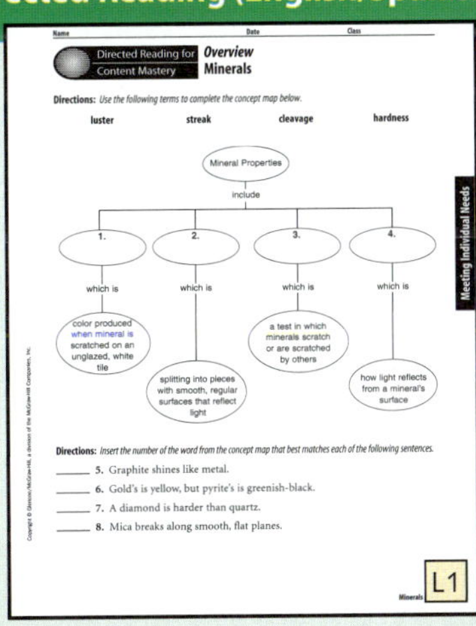
Directed Reading for Content Mastery — Overview — Minerals

Directions: Use the following terms to complete the concept map below.

luster streak cleavage hardness

Mineral Properties include 1., 2., 3., 4. — which is: color produced when mineral is scratched on an unglazed, white tile; splitting into pieces with smooth, regular surfaces that reflect light; a test in which minerals scratch or are scratched by others; how light reflects from a mineral's surface

Directions: Insert the number of the word from the concept map that best matches each of the following sentences.

____ 5. Graphite shines like metal.
____ 6. Gold's is yellow, but pyrite's is greenish-black.
____ 7. A diamond is harder than quartz.
____ 8. Mica breaks along smooth, flat planes.

L1

Study Guide

Reading Essentials

Assessment

Test Practice Workbook

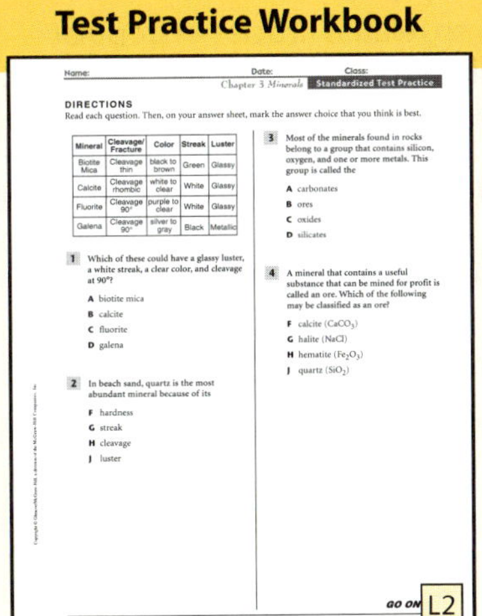
Chapter 3 Minerals — Standardized Test Practice

DIRECTIONS: Read each question. Then, on your answer sheet, mark the answer choice that you think is best.

Mineral	Cleavage/Fracture	Color	Streak	Luster
Biotite Mica	Cleavage thin	black to brown	Green	Glassy
Calcite	Cleavage rhombic	white to clear	White	Glassy
Fluorite	Cleavage 90°	purple to clear	White	Glassy
Galena	Cleavage 90°	silver to gray	Black	Metallic

1. Which of these could have a glassy luster, a white streak, a clear color, and cleavage at 90°? A biotite mica B calcite C fluorite D galena
2. In beach sand, quartz is the most abundant mineral because of its F hardness G streak H cleavage J luster
3. Most of the minerals found in rocks belong to a group that contains silicon, oxygen, and one or more metals. This group is called the A carbonates B ores C oxides D silicates
4. A mineral that contains a useful substance that can be mined for profit is called an ore. Which of the following may be classified as an ore? F calcite $(CaCO_3)$ G halite (NaCl) H hematite (Fe_2O_3) J quartz (SiO_2)

L2

Chapter Review

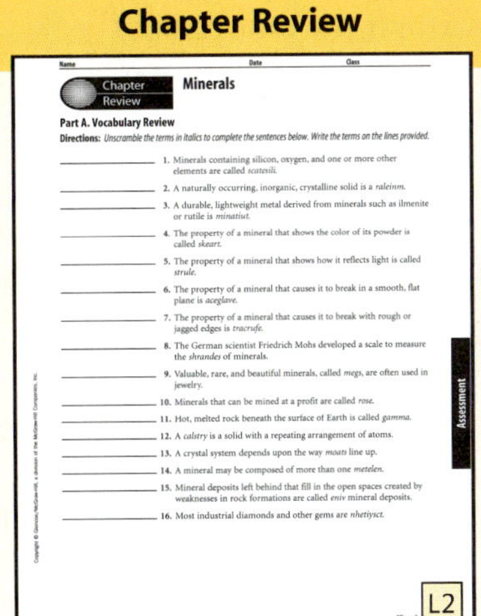
Chapter Review — Minerals

Part A. Vocabulary Review

Directions: Unscramble the terms in italics to complete the sentences below. Write the terms on the lines provided.

1. Minerals containing silicon, oxygen, and one or more other elements are called *scateiil*.
2. A naturally occurring, inorganic, crystalline solid is a *raleimn*.
3. A durable, lightweight metal derived from minerals such as ilmenite or rutile is *minattiu*.
4. The property of a mineral that shows the color of its powder is called *kaestr*.
5. The property of a mineral that shows how it reflects light is called *struel*.
6. The property of a mineral that causes it to break in a smooth, flat plane is *ucaglvee*.
7. The property of a mineral that causes it to break with rough or jagged edges is *tracrufe*.
8. The German scientist Friedrich Mohs developed a scale to measure the *dhsaenrs* of minerals.
9. Valuable, rare, and beautiful minerals, called *megs*, are often used in jewelry.
10. Minerals that can be mined at a profit are called *rose*.
11. Hot, melted rock beneath the surface of Earth is called *gamma*.
12. A *calstry* is a solid with a repeating arrangement of atoms.
13. A crystal system depends upon the way *msota* line up.
14. A mineral may be composed of more than one *metelen*.
15. Mineral deposits left behind that fill in the open spaces created by weaknesses in rock formations are called *vein* mineral deposits.
16. Most industrial diamonds and other gems are *nhetiyst*.

L2

Chapter Tests

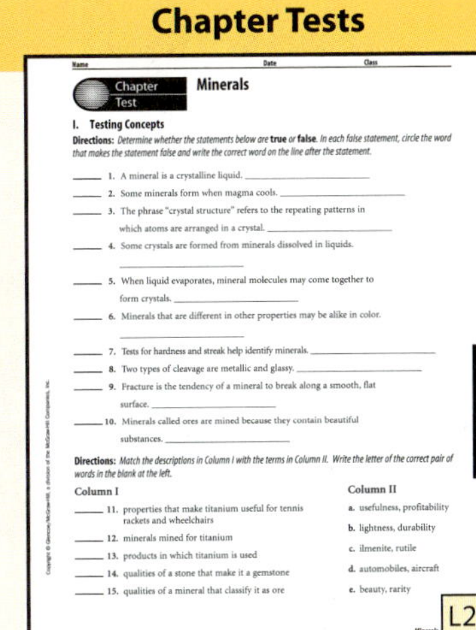
Chapter Test — Minerals

I. Testing Concepts

Directions: Determine whether the statements below are true or false. In each false statement, circle the word that makes the statement false and write the correct word on the line after the statement.

1. A mineral is a crystalline liquid.
2. Some minerals form when magma cools.
3. The phrase "crystal structure" refers to the repeating patterns in which atoms are arranged in a crystal.
4. Some crystals are formed from minerals dissolved in liquids.
5. When liquid evaporates, mineral molecules may come together to form crystals.
6. Minerals that are different in other properties may be alike in color.
7. Tests for hardness and streak help identify minerals.
8. Two types of cleavage are metallic and glassy.
9. Fracture is the tendency of a mineral to break along a smooth, flat surface.
10. Minerals called ores are mined because they contain beautiful substances.

Directions: Match the descriptions in Column I with the terms in Column II. Write the letter of the correct pair of words in the blank at the left.

Column I	Column II
11. properties that make titanium useful for tennis rackets and wheelchairs	a. usefulness, profitability
12. minerals mined for titanium	b. lightness, durability
13. products in which titanium is used	c. ilmenite, rutile
14. qualities of a stone that make it a gemstone	d. automobiles, aircraft
15. qualities of a mineral that classify it as ore	e. beauty, rarity

L2

Science Content Background

section 1 Minerals

General Characteristics

Between 40 and 50 new minerals are identified each year. Only eight elements comprise the majority of minerals. The 24 or so minerals that make up most rocks are called the rock-forming minerals. Coal, which fits many of the characteristics of minerals, is not a mineral because it comes from once-living material.

Graphite is among the softest minerals, and diamond is the hardest mineral. Both of these minerals are composed of carbon. The type of bond between the atoms of carbon and different atomic arrangements cause the differences in the properties of these minerals. Because of these differences, the two minerals have widely different uses. Diamonds are used in jewelry, as abrasives, and in cutting tools. Graphite is used as a lubricant and in pencil lead. Minerals that have identical composition but different properties are called polymorphs.

Mineral Composition

Although each type of mineral has a unique composition, the composition of individual mineral specimens can vary within limits. For example, the chemical composition of olivine ranges from Fe_2SiO_4 to Mg_2SiO_4. For this reason, the composition of olivine is often written $(Fe,Mg)_2SiO_4$. A more precise chemical formula can be written when the amounts of iron and magnesium are known, for example $(Fe_{.77}Mg_{.23})_2SiO_4$.

section 2 Mineral Identification

Hardness

Most silicate minerals do not leave a streak because they have a hardness greater than 6 on the Mohs scale. The hardness of a mineral is related to its crystal structure and to the type of atomic bonds in the mineral. The difference between the hardness of diamond (Mohs 10) and corundum (Mohs 9) is far greater than the difference in hardness between corundum and the softest mineral, talc. This means that the Mohs scale is nonlinear.

section 3 Uses of Minerals

Valuing Diamonds

The value of diamonds varies depending on whether they are tinted. The most common diamonds of gem quality have yellow or brown tints. These diamonds are less valuable than those with no tint. Diamonds with blue or pink tints are the most valuable.

chapter content resources

Internet Resources

For additional content background, visit **bookf.msscience.com** to:

- access your book online
- find references to related articles in popular science magazines
- access Web links with related content background
- access current events with science journal topics

Print Resources

A Field Guide to Rocks and Minerals (The Peterson Field Guide series), by Frederick H. Pough, Houghton Mifflin Company, 1998

Minerals of the World, by Charles A. Sorrell, Golden Press, 1973

The Mineral Kingdom, by Paul E. Desautels, The Ridge Press, Inc., 1968

Barry L. Runk/Grant Heilman Photography, Inc.

IDENTIFYING Misconceptions

Find Out What Students Think

Students may think that . . .

Minerals are rare objects that are seldom used in everyday materials.

Students are often asked to observe and classify minerals. However, some students fail to understand that these materials are part of their everyday lives.

Discussion

Ask students to raise their hands if they used minerals before coming to school this morning. Then ask how many brushed their teeth using toothpaste. Explain that when they use toothpaste, they are using a product made from minerals. Phosphate materials, aluminum oxide, and silica are used as abrasives in toothpaste. Fluoride comes from the mineral fluorite. Stannous fluoride (or tin fluoride) is also found in toothpaste. The sparkle in toothpaste comes from the mineral mica.

Have students handle and observe the minerals that furnish the ingredients for toothpaste.

Promote Understanding

Discussion

Reproduce the information shown below, which lists the amount of elements, minerals, and other earth materials the average person uses in a lifetime. Have students discuss the different ways in which minerals impact their lives.

Every American Born Will Need...

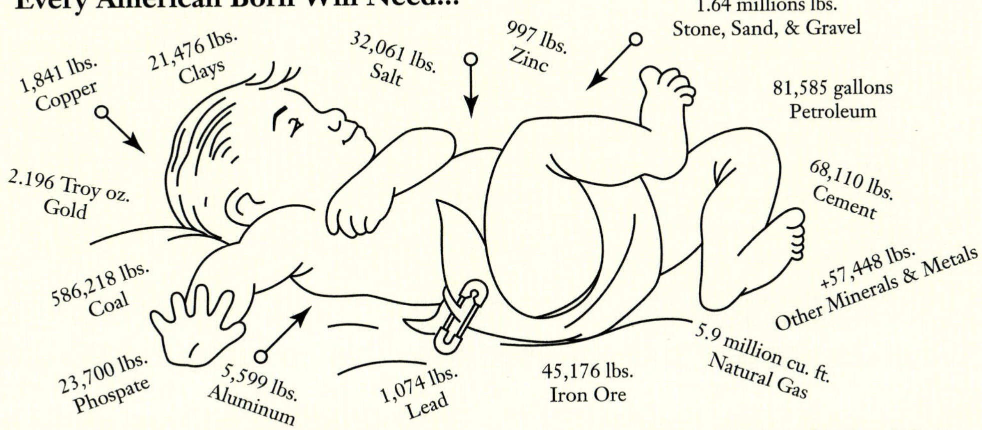

3.7 million pounds of minerals, metals, and fuels in his/her lifetime

Assess

After completing the chapter, see *Identifying Misconceptions* in the Study Guide at the end of the chapter.

chapter 1

Chapter Vocabulary

mineral, p. 8
crystal, p. 9
magma, p. 11
silicate, p. 12
hardness, p. 15
luster, p. 16
specific gravity, p. 16
streak, p. 17
cleavage, p. 17
fracture, p. 17
gem, p. 19
ore, p. 23

Science Journal Student responses will vary, but may include questions about the properties of gem-quality minerals, where they are found, and how valuable they are.

INTERACTIVE CHALKBOARD with Image Bank

PowerPoint® Presentations

This CD-ROM is an editable Microsoft® PowerPoint® presentation that includes:

- a pre-made presentation for every chapter
- interactive graphics
- animations
- audio clips
- image bank
- all new section and chapter questions
- Standardized Test Practice
- transparencies
- pre-lab questions for all labs
- Foldables directions
- links to bookf.msscience.com

chapter 1

Minerals

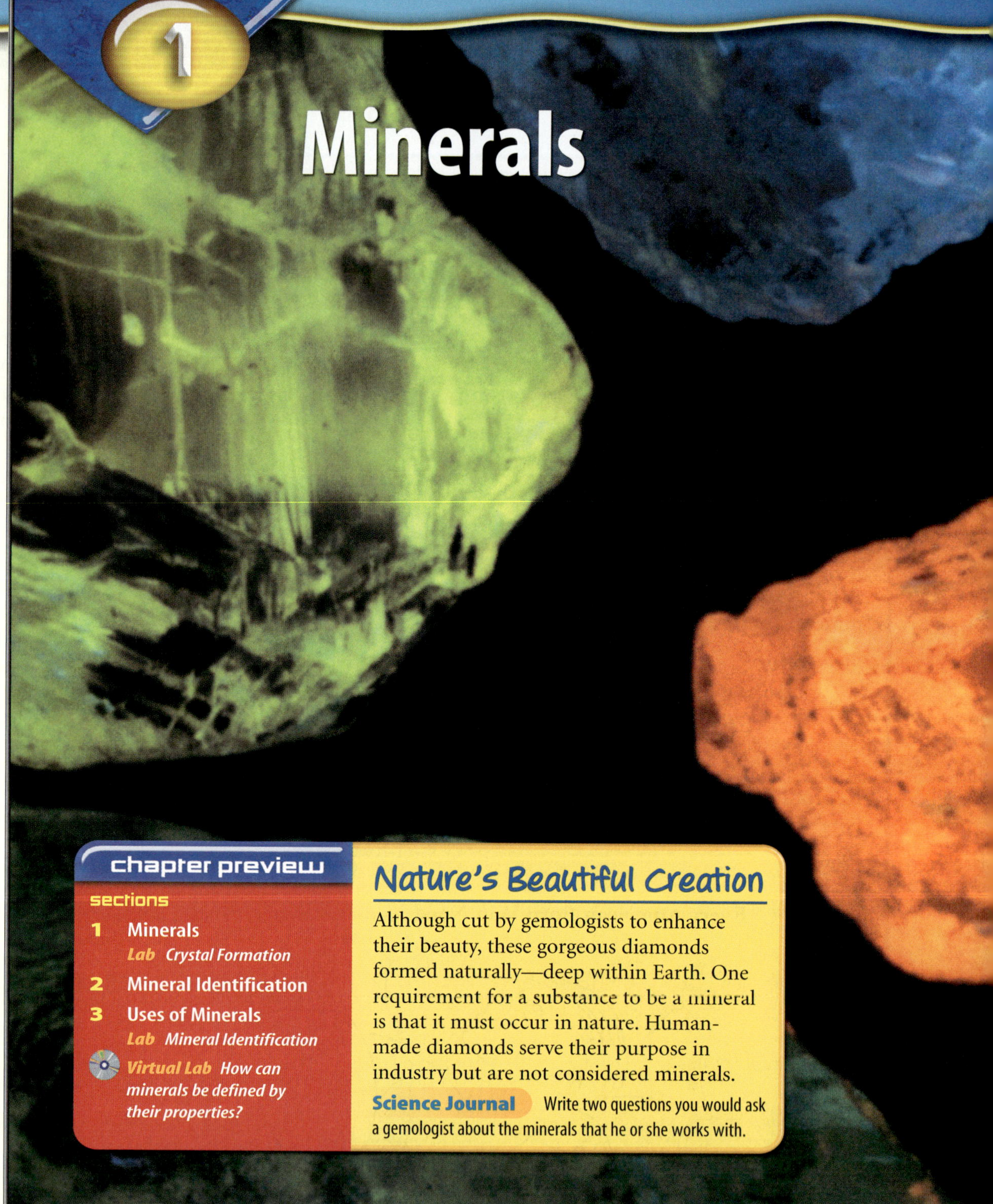

chapter preview

sections

1 Minerals
Lab *Crystal Formation*
2 Mineral Identification
3 Uses of Minerals
Lab *Mineral Identification*
Virtual Lab *How can minerals be defined by their properties?*

Nature's Beautiful Creation

Although cut by gemologists to enhance their beauty, these gorgeous diamonds formed naturally—deep within Earth. One requirement for a substance to be a mineral is that it must occur in nature. Human-made diamonds serve their purpose in industry but are not considered minerals.

Science Journal Write two questions you would ask a gemologist about the minerals that he or she works with.

Theme Connection

Scale and Structure The structure of minerals is discussed. On the atomic scale, minerals have orderly crystal structures, that is, regularly repeating patterns of atoms. Emphasizing the relationship between the internal structure of a mineral and its properties and uses will help clarify its uniqueness among these solids that make up Earth's rocks.

About the Photo

Earth's Jewels The extreme physical processes required to produce diamonds contribute to their rarity. Gem-quality diamonds are found in rocks that once were subjected to great pressures. One example is a kimberlite pipe, which forms after magma moves from great depths in Earth at high speeds, bringing diamonds closer to the surface.

Start-Up Activities

Launch LAB

Distinguish Rocks from Minerals

When examining rocks, you'll notice that many of them are made of more than one material. Some rocks are made of many different crystals of mostly the same mineral. A mineral, however, will appear more like a pure substance and will tend to look the same throughout. Can you tell a rock from a mineral?

1. Use a magnifying lens to observe a quartz crystal, salt grains, and samples of sandstone, granite, calcite, mica, and schist (SHIHST).
2. Draw a sketch of each sample.
3. Infer which samples are made of one type of material and should be classified as minerals.
4. Infer which samples should be classified as rocks.
5. **Think Critically** In your Science Journal, compile a list of descriptions for the minerals you examined and a second list of descriptions for the rocks. Compare and contrast your observations of minerals and rocks.

FOLDABLES™ Study Organizer

Minerals Make the following Foldable to help you better understand minerals.

STEP 1 **Fold** a vertical sheet of notebook paper from side to side.

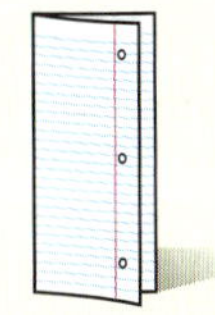

STEP 2 **Cut** along every third line of only the top layer to form tabs.

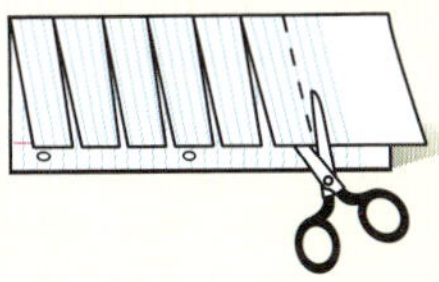

STEP 3 **Label** each tab with a question.

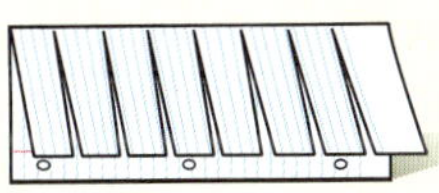

Ask Questions Before you read the chapter, write questions you have about minerals on the front of the tabs. As you read the chapter, add more questions and write answers under the appropriate tabs.

Science Online — Preview this chapter's content and activities at bookf.msscience.com

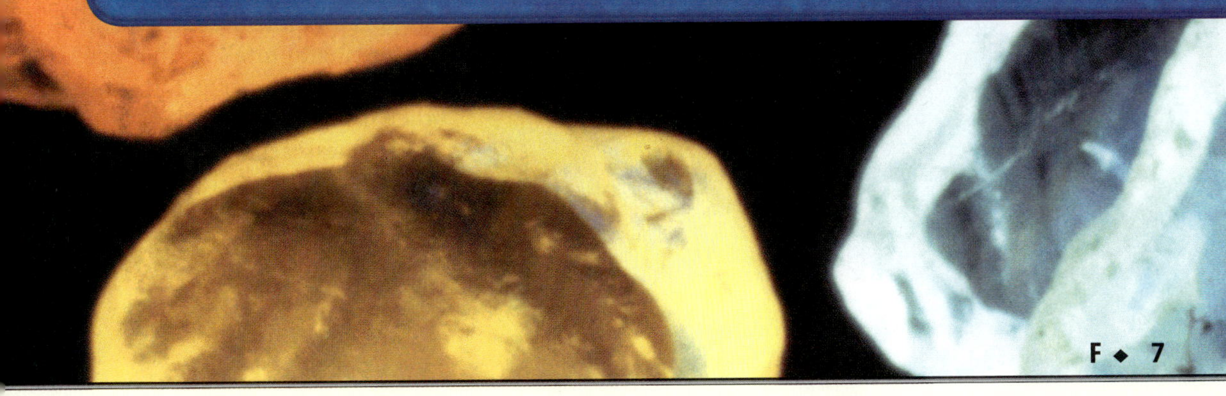

Launch LAB

Purpose Use this Launch Lab to introduce to students the differences between rocks and minerals.

Preparation Place one sample of each type of rock and mineral in a container. Each group will receive one container of samples.

Materials magnifying lens, quartz crystal, salt grains, sample of sandstone, sample of granite, sample of calcite, sample of mica, sample of schist per group

Teaching Strategy To ensure that all group members observe all the samples, have each student write a list of the samples and check off each sample as it is observed.

Safety Precautions Remind students not to taste or eat anything in the lab.

Think Critically

- Minerals: quartz, salt, calcite, mica
- Rocks: granite, sandstone, schist
- minerals are homogeneous, rocks are heterogeneous

Assessment

Process Ask students to collect a half-dozen rock or mineral samples from around the school or their homes and to classify each sample as either a rock or a mineral. Use **Performance Assessment in the Science Classroom,** p. 121. L2

FOLDABLES™ Study Organizer — Dinah Zike Study Fold

Student preparation materials for this Foldable are available in the Chapter *Fast File* Resources.

1 Motivate

INTERACTIVE CHALKBOARD
PowerPoint® Presentations

Bellringer

Section Focus Transparencies also are available on the Interactive Chalkboard CD-ROM. L2 ELL

Section 1 Section Focus Transparency **A Study in Patterns**

In a repeating pattern, the same structure occurs over and over again. Nature is full of repeating patterns similar to the pattern in the kaleidoscope image below. Crystals, for example, are patterns of repeating elements.

1. **What basic structure is repeated and how is it grouped to form the total design?**
2. **How might a kaleidoscope image relate to the structure of a mineral?** L2

Minerals

Tie to Prior Knowledge

Atom Patterns Help students recall that atoms are the building blocks of matter and that atoms can combine to form compounds. The structure of a mineral is determined by the repeating pattern of atoms in the mineral.

Caption Answer

Figure 1 Students might list parts of the building, such as cement and glass. They should not list living items, such as trees.

section 1

Minerals

as you read

What You'll Learn

- **Describe** characteristics that all minerals share.
- **Explain** how minerals form.

Why It's Important

You use minerals and products made from them every day.

Review Vocabulary
atoms: tiny particles that make up matter; composed of protons, electrons, and neutrons

New Vocabulary
- mineral
- crystal
- magma
- silicate

What is a mineral?

How important are minerals to you? Very important? You actually own or encounter many things made from minerals every day. Ceramic, metallic, and even some paper items are examples of products that are derived from or include minerals. **Figure 1** shows just a few of these things. Metal bicycle racks, bricks, and the glass in windows would not exist if it weren't for minerals. A **mineral** is a naturally occurring, inorganic solid with a definite chemical composition and an orderly arrangement of atoms. About 4,000 different minerals are found on Earth, but they all share these four characteristics.

Mineral Characteristics First, all minerals are formed by natural processes. These are processes that occur on or inside Earth with no input from humans. For example, salt formed by the natural evaporation of seawater is the mineral halite, but salt formed by evaporation of saltwater solutions in laboratories is not a mineral. Second, minerals are inorganic. This means that they aren't made by life processes. Third, every mineral is an element or compound with a definite chemical composition. For example, halite's composition, NaCl, gives it a distinctive taste that adds flavor to many foods. Fourth, minerals are crystalline solids. All solids have a definite volume and shape. Gases and liquids like air and water have no definite shape, and they aren't crystalline. Only a solid can be a mineral, but not all solids are minerals.

Figure 1 You probably use minerals or materials made from minerals every day without thinking about it.
Infer *How many objects in this picture might be made from minerals?*

Atom Patterns The word *crystalline* means that atoms are arranged in a pattern that is repeated over and over again. For example, graphite's atoms are arranged in layers. Opal, on the other hand, is not a mineral in the strictest sense because its atoms are not all arranged in a definite, repeating pattern, even though it is a naturally occurring, inorganic solid.

Section 1 Resource Manager

Chapter *FAST FILE* Resources
- Transparency Activity, p. 40
- Directed Reading for Content Mastery, pp. 15, 16
- MiniLAB, p. 3
- Enrichment, p. 26
- Lab Activity, pp. 9–10
- Lab Worksheets, pp. 5–6

Mathematics Skill Activities, p. 47
Cultural Diversity, p. 37

Figure 2 More than 200 years ago, the smooth, flat surfaces on crystals led scientists to infer that minerals had an orderly structure inside.

Even though this rose quartz looks uneven on the outside, its atoms have an orderly arrangement on the inside.

The well-formed crystal shapes exhibited by these clear quartz crystals suggest an orderly structure.

The Structure of Minerals

Do you have a favorite mineral sample or gemstone? If so, perhaps it contains well-formed crystals. A **crystal** is a solid in which the atoms are arranged in orderly, repeating patterns. You can see evidence for this orderly arrangement of atoms when you observe the smooth, flat outside surfaces of crystals. A crystal system is a group of crystals that have similar atomic arrangements and therefore similar external crystal shapes.

Reading Check *What is a crystal?*

Crystals Not all mineral crystals have smooth surfaces and regular shapes like the clear quartz crystals in **Figure 2.** The rose quartz in the smaller photo of **Figure 2** has atoms arranged in repeating patterns, but you can't see the crystal shape on the outside of the mineral. This is because the rose quartz crystals developed in a tight space, while the clear quartz crystals developed freely in an open space. The six-sided, or hexagonal crystal shape of the clear quartz crystals in **Figure 2,** and other forms of quartz can be seen in some samples of the mineral. **Figure 3** illustrates the six major crystal systems, which classify minerals according to their crystal structures. The hexagonal system to which quartz belongs is one example of a crystal system.

Crystals form by many processes. Next, you'll learn about two of these processes—crystals that form from magma and crystals that form from solutions of salts.

Mini LAB

Inferring Salt's Crystal System

Procedure

1. Use a **magnifying lens** to observe grains of common **table salt** on a dark sheet of **construction paper.** Sketch the shape of a salt grain. **WARNING:** *Do not taste or eat mineral samples. Keep hands away from your face.*
2. Compare the shapes of the salt crystals with the shapes of crystals shown in **Figure 3.**

Analysis

1. Which characteristics do all the grains have in common?
2. Research another mineral with the same crystal system as salt. What is this crystal system called?

Try at Home

2 Teach

Visual Learning

Figure 2 Both of these mineral samples are quartz. What shape do you think the transparent crystals shown in **Figure 2** might have? hexagonal

Reading Check

Answer a solid in which the atoms are arranged in orderly, repeating patterns

Mini LAB

Purpose Students will observe cubic crystals of table salt. L2
LS **Visual-Spatial**

Materials magnifying lens, construction paper, table salt, **Figure 3**

Teaching Strategy Tell students to observe crystals against a dark background.

Analysis

1. All the salt grains are cubes.
2. Other minerals of the isometric (cubic) system include fluorite, galena, and pyrite.

Assessment

Oral Have students classify samples of crystals that you provide. Use **Performance Assessment in the Science Classroom,** p. 121.

Try at Home

Active Reading

Metacognition Journal In this strategy, each student analyzes his or her own thought processes. Have students divide a sheet of paper in half. On the left, have them record what they have learned about a topic. On the right, have them record the reason they learned it. Have students write a Metacognition Journal about crystal structures. L2

NATIONAL GEOGRAPHIC

Visualizing Crystal Systems

Have students examine the pictures and read the captions. Then ask the following question.

Which of the crystal systems includes the fewest right angles? Triclinic crystals have no right angles; all angles where the crystal surfaces meet are oblique.

Inquiry Lab

Modeling Mineral Coordinate Systems

Purpose students practice working with coordinate systems and thinking in 3-D

Possible Materials, toothpicks, tongue depressors, pipe cleaners, protractor, connecting toys, glue, tape, string, ruler, **Figure 3**

Estimated Time 1 class session

Teaching Strategies

- Students can build 3-D models of the coordinate systems that describe each crystal system. L3
- Students could sketch the model coordinate systems of their choice in different orientations. L2
- Students who are having difficulties can focus on the cubic, tetragonal, and orthorhombic crystal systems. L2
- Allow students to explore other questions that arise.

For additional inquiry activities, see *Science Inquiry Labs*.

NATIONAL GEOGRAPHIC **VISUALIZING CRYSTAL SYSTEMS**

Figure 3

A crystal's shape depends on how its atoms are arranged. Crystal shapes can be organized into groups known as crystal systems—shown here in 3-D with geometric models (in blue). Knowing a mineral's crystal system helps researchers understand its atomic structure and physical properties.

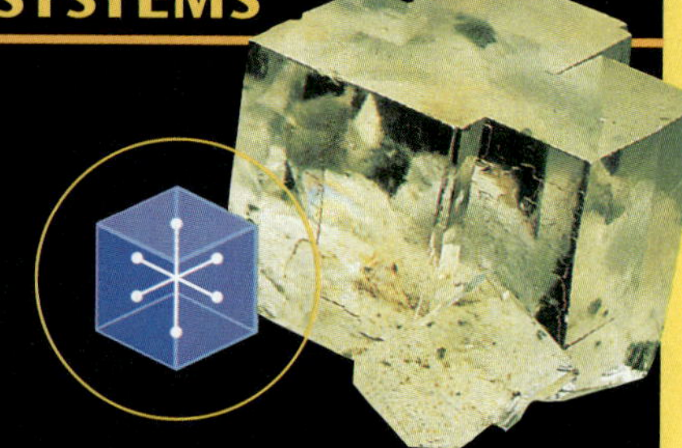

▲ CUBIC Fluorite is an example of a mineral that forms cubic crystals. Minerals in the cubic crystal system are equal in size along all three principal dimensions.

▲ HEXAGONAL (hek SA guh nul) In hexagonal crystals, horizontal distances between opposite crystal surfaces are equal. These crystal surfaces intersect to form 60° or 120° angles. The vertical length is longer or shorter than the horizontal lengths.

◀ TETRAGONAL (te TRA guh nul) Zircon crystals are tetragonal. Tetragonal crystals are much like cubic crystals, except that one of the principal dimensions is longer or shorter than the other two dimensions.

▲ ORTHORHOMBIC (awr thuh RAHM bihk) Minerals with orthorhombic structure, such as barite, have dimensions that are unequal in length, resulting in crystals with a brick-like shape.

▲ MONOCLINIC (mah nuh KLIH nihk) Minerals in the monoclinic system, such as orthoclase, also exhibit unequal dimensions in their crystal structure. Only one right angle forms where crystal surfaces meet. The other angles are oblique, which means they don't form 90° angles where they intersect.

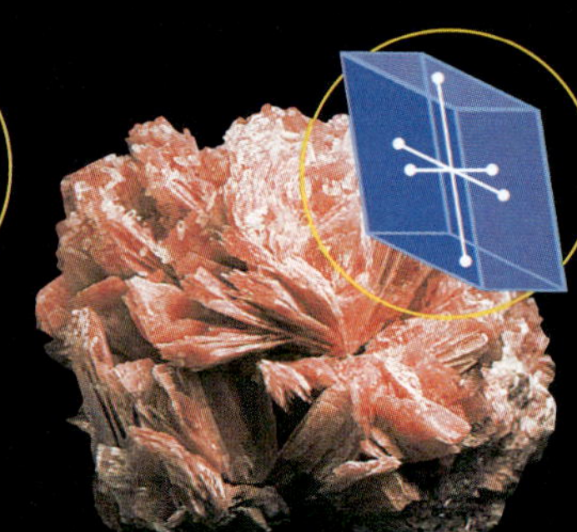

▲ TRICLINIC (tri KLIH nihk) The triclinic crystal system includes minerals exhibiting the least symmetry. Triclinic crystals, such as rhodonite (ROH dun ite), are unequal in all dimensions, and all angles where crystal surfaces meet are oblique.

10 ◆ F CHAPTER 1 Minerals

Teacher FYI

Hexagonal Crystal System The hexagonal crystal system possesses one axis of six-fold or three-fold symmetry perpendicular to, and not equal in length to, three equal axes intersecting at 120°. Crystals with six-fold symmetry are classified as hexagonal—apatite and graphite are examples. Crystals with three-fold symmetry are classified as trigonal—calcite and some varieties of quartz are examples.

Figure 4 Minerals form by many natural processes.

A This rock formed as magma cooled slowly, allowing large mineral grains to form.

Labradorite

B Some minerals form when salt water evaporates, such as these white crystals of halite in Death Valley, California.

Crystals from Magma Natural processes form minerals in many ways. For example, hot melted rock material, called **magma**, cools when it reaches Earth's surface, or even if it's trapped below the surface. As magma cools, its atoms lose heat energy, move closer together, and begin to combine into compounds. During this process, atoms of the different compounds arrange themselves into orderly, repeating patterns. The type and amount of elements present in a magma partly determine which minerals will form. Also, the size of the crystals that form depends partly on how rapidly the magma cools.

When magma cools slowly, the crystals that form are generally large enough to see with the unaided eye, as shown in **Figure 4A.** This is because the atoms have enough time to move together and form into larger crystals. When magma cools rapidly, the crystals that form will be small. In such cases, you can't easily see individual mineral crystals.

Crystals from Solution Crystals also can form from minerals dissolved in water. When water evaporates, as in a dry climate, ions that are left behind can come together to form crystals like the halite crystals in **Figure 4B.** Or, if too much of a substance is dissolved in water, ions can come together and crystals of that substance can begin to form in the solution. Minerals can form from a solution in this way without the need for evaporation.

INTEGRATE Physics

Crystal Formation Evaporites commonly form in dry climates. Research the changes that take place when a saline lake or shallow sea evaporates and halite or gypsum forms.

Activity

Mineral Materials Have students observe carefully, and write descriptions of, seven objects around the room. Encourage them to select any items they wish. Their descriptions must provide evidence of whether or not the objects are minerals or items made from minerals.
L2 LS **Linguistic** P

INTEGRATE Physics

Crystal Formation Evaporite minerals form when a saline lake or shallow saltwater sea evaporates. During this process, liquid water changes to a gas and solids that are dissolved in the water precipitate out to form layers of evaporite minerals, such as halite (rock salt), gypsum, and sylvite. The order of precipitation, from least to most soluble, is gypsum, halite, and sylite.

Research Have students research several geographic regions that contain evaporite mineral deposits. Students can prepare a report that discusses the geologic environment that contributed to the formation of the deposits and their approximate ages.

IDENTIFYING Misconceptions

Crystal Aesthetics Students may think that all minerals are as beautiful as cut gems. Have students contrast a gemstone in a piece of jewelry with mineral samples that do not have well-formed crystals. L2 LS **Visual-Spatial**

Differentiated Instruction

Challenge Have students search through mineral guides and find one mineral conforming to each of the crystal systems. Have them describe the physical properties of each mineral. Ask them to describe ways in which minerals with similar crystal systems can have similar physical properties. L2

3 Assess

DAILY INTERVENTION

Check for Understanding

Visual-Spatial Have groups of students compete to brainstorm the greatest number of ordinary, non-mineral objects that conform to the described crystal systems. Examples might include a skyscraper, tissue box, and unsharpened colored pencil. Offer a special bonus to any group that can describe an object conforming to the triclinic system. L2

Reteach

Illustrating Elements Hold up an unlabeled circle graph showing the weight percentages of elements in Earth's crust. Ask students to identify the portions of the graph that represent aluminum, oxygen, and silicon. L2

Assessment

Process Have students make a Venn diagram that compares and contrasts crystals formed from magma with crystals formed from solutions. Use **Performance Assessment in the Science Classroom,** p. 167. L2

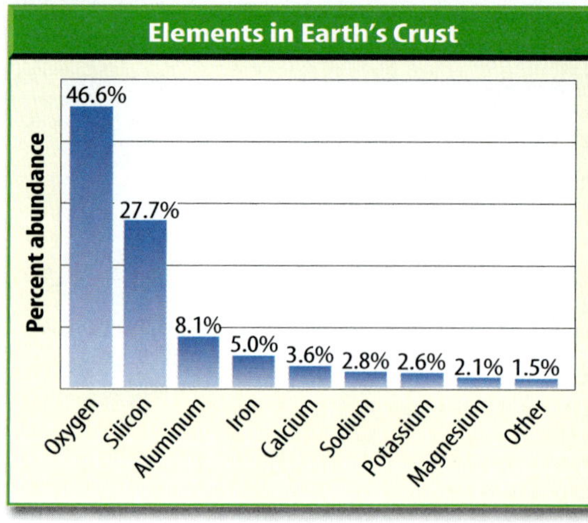

Figure 5 Most of Earth's crust is composed of eight elements.

Mineral Compositions and Groups

Ninety elements occur naturally in Earth's crust. Approximately 98 percent (by weight) of the crust is made of only eight of these elements, as shown in **Figure 5.** Of the thousands of known minerals, only a few dozen are common, and these are mostly composed of the eight most common elements in Earth's crust.

Most of the common rock-forming minerals belong to a group called the silicates. **Silicates** (SIH luh kayts) are minerals that contain silicon (Si) and oxygen (O) and usually one or more other elements. Silicon and oxygen are the two most abundant elements in Earth's crust. These two elements alone combine to form the basic building blocks of most of the minerals in Earth's crust and mantle. Feldspar and quartz, which are silicates, and calcite, which is a carbonate, are examples of common, rock-forming minerals. Other mineral groups also are defined according to their compositions.

section 1 review

Summary

What is a mineral?

- Many products used by humans are made from minerals.
- Minerals are defined by four main characteristics.

The Structure of Minerals

- The crystal shape of a mineral reflects the way in which its atoms are arranged.
- Minerals are classified according to the types of atoms in their structures and the way that the atoms are arranged.

Mineral Compositions and Groups

- Only eight elements form approximately 98 percent (by weight) of Earth's crust.
- The majority of Earth's crust is composed of silicate minerals.

Self Check

1. **List** four characteristics that all minerals share.
2. **Describe** two ways that minerals can form from solution.
3. **Explain** whether diamonds made in the laboratory are considered to be minerals.
4. **Describe** how crystals of minerals are classified.
5. **Think Critically** The mineral dolomite, a rock-forming mineral, contains oxygen, carbon, magnesium, and calcium. Is dolomite a silicate? Explain.

Applying Skills

6. **Graph** Make a graph of your own design that shows the relative percentages of the eight most common elements in Earth's crust. Then determine the approximate percentage of the crust that is made up of iron and aluminum. If one is available, you may use an electronic spreadsheet program to make your graph and perform the calculation.

Science Online bookf.msscience.com/self_check_quiz

section 1 review

1. A mineral must be a naturally occurring, inorganic, crystalline solid with a definite chemical composition.
2. Minerals can precipitate out of a solution where too much of a substance is dissolved, or form as a solution evaporates.
3. No; because the diamonds are made in the lab, they are not naturally occurring.
4. Crystals are classified by the shapes produced by the internal arrangement of their atoms.
5. No; this rock-forming mineral is a carbonate. Silicate rock-forming minerals contain silicon and oxygen, both of which are very common. Silicates are one type of rock-forming mineral.
6. Student's graphs should include the information from **Figure 5.** They can be bar graphs, shaded-block graphs, or circle graphs; 13.1 percent.

LAB

Crystal Formation

In this lab, you'll have a chance to learn how crystals form from solutions.

Real-World Question

How do crystals form from solution?

Goals

- **Compare and contrast** the crystals that form from salt and sugar solutions.
- **Observe** crystals and infer how they formed.

Materials

250-mL beakers (2)
cardboard
large paper clip
table salt
flat wooden stick
granulated sugar
cotton string
hot plate
magnifying lens
thermal mitt
shallow pan
spoon

Safety Precautions

WARNING: *Never taste or eat any lab materials.*

Procedure

1. Gently mix separate solutions of salt in water and sugar in water in the two beakers. Keep stirring the solutions as you add salt or sugar to the water. Stop mixing when no more salt or sugar will dissolve in the solutions. Label each beaker.
2. Place the sugar solution beaker on a hot plate. Use the hot plate to heat the sugar solution gently. **WARNING:** *Do not touch the hot beaker without protecting your hands.*
3. Tie one end of the thread to the middle of the wooden stick. Tie a large paper clip to the free end of the string for weight. Place the stick across the opening of the sugar beaker so the thread dangles in the sugar solution.
4. Remove the beaker from the hot plate and cover it with cardboard. Place it in a location where it won't be disturbed.
5. Pour a thin layer of the salt solution into the shallow pan.
6. Leave the beaker and the shallow pan undisturbed for at least one week.
7. After one week, examine each solution with a magnifying lens to see whether crystals have formed.

Conclude and Apply

1. **Compare and contrast** the crystals that formed from the salt and the sugar solutions. How do they compare with samples of table salt and sugar?
2. **Describe** what happened to the saltwater solution in the shallow pan.
3. Did this same process occur in the sugar solution? Explain.

Communicating Your Data

Make a poster that describes your methods of growing salt and sugar crystals. Present your results to your class.

BENCH TESTED

Real-World Question

Purpose Students demonstrate and describe two methods of growing crystals. L2 ELL COOP LEARN LS **Visual-Spatial**

Process Skills observe and infer, communicate, experiment, classify, compare and contrast

Time Required 20 minutes on each of two days, one week apart

Procedure

Safety Precautions Caution students not to handle beakers without protecting their hands. Also, caution students to wear safety goggles and not to touch the hot plate.

Teaching Strategy The beakers need to remain undisturbed for at least one week. The more gradually the solution is cooled, the larger the crystals tend to grow.

Troubleshooting Add seed crystals on the thread to help precipitation.

Conclude and Apply

1. salt crystals—cubic; sugar crystals—orthorhombic; sample crystals—same structure, probably smaller
2. The water evaporated, leaving the salt behind.
3. No; the liquid in the beaker cooled, and sugar crystals precipitated from a supersaturated solution.

Assessment

Content Encourage students to write a story that explains what happens to the atoms of a material to allow them to gradually form crystals. Have them work in groups of four to complete their stories. Use **Performance Assessment in the Science Classroom**, p. 159. L2

Communicating Your Data

Have students draw illustrations of the different crystals produced in this lab. L2

section 2

Mineral Identification

1 Motivate

Bellringer

INTERACTIVE CHALKBOARD PowerPoint® Presentations

Section Focus Transparencies also are available on the Interactive Chalkboard CD-ROM. L2 ELL

Section 2 Section Focus Transparency **Sorting It Out**

About 4,000 different kinds of minerals can be found on Earth. Some minerals exist in a variety of forms. You'll need skill and knowledge to identify that interesting rock you found at the beach. What can you tell about the three mineral samples shown here?

1. What are some similarities and differences in appearance among the three samples?
2. Do you think that the photos show the same mineral or different ones? Explain.
3. What other traits might help you identify different minerals?

L2

Tie to Prior Knowledge

Physical Properties Remind students that the physical properties of a substance can be helpful for identifying the substance. Show students two or three different minerals, and have them list several physical properties of each. L1 LS **Visual-Spatial**

as you read

***What* You'll Learn**

- **Describe** physical properties used to identify minerals.
- **Identify** minerals using physical properties such as hardness and streak.

***Why* It's Important**

Identifying minerals helps you recognize valuable mineral resources.

Review Vocabulary
physical property: any characteristic of a material that you can observe without changing the identity of the material

New Vocabulary
- hardness
- luster
- specific gravity
- streak
- cleavage
- fracture

Physical Properties

Why can you recognize a classmate when you see him or her in a crowd away from school? A person's height or the shape of his or her face helps you tell that person from the rest of your class. Height and facial shape are two properties unique to individuals. Individual minerals also have unique properties that distinguish them.

Mineral Appearance Just like height and facial characteristics help you recognize someone, mineral properties can help you recognize and distinguish minerals. Color and appearance are two obvious clues that can be used to identify minerals.

However, these clues alone aren't enough to recognize most minerals. The minerals pyrite and gold are gold in color and can appear similar, as shown in **Figure 6.** As a matter of fact, pyrite often is called fool's gold. Gold is worth a lot of money, whereas pyrite has little value. You need to look at other properties of minerals to tell them apart. Some other properties to study include how hard a mineral is, how it breaks, and its color when crushed into a powder. Every property you observe in a mineral is a clue to its identity.

Figure 6 The general appearance of a mineral often is not enough to identify it.

Gold

Pyrite

Using only color, observers can be fooled when trying to distinguish between pyrite and gold.

Azurite

The mineral azurite is identified readily by its striking blue color.

Section 2 Resource Manager

Chapter *FAST FILE* Resources
Transparency Activity, pp. 41, 43–44
Directed Reading for Content Mastery, p. 17
Enrichment, p. 27
MiniLAB, p. 4
Reinforcement, p. 24

Science Inquiry Lab, pp. 31–32

Hardness A measure of how easily a mineral can be scratched is its **hardness.** The mineral talc is so soft you can scratch it loose with your fingernail. Talcum powder is made from this soft mineral. Diamonds, on the other hand, are the hardest mineral. Some diamonds are used as cutting tools, as shown in **Figure 7.** A diamond can be scratched only by another diamond. Diamonds can be broken, however.

Reading Check *Why is hardness sometimes referred to as scratchability?*

Figure 7 Some saw blades have diamonds embedded in them to help slice through materials, such as this limestone. Blades are kept cool by running water over them.

Sometimes the concept of hardness is confused with whether or not a mineral will break. It is important to understand that even though a diamond is extremely hard, it can shatter if given a hard enough blow in the right direction along the crystal.

Mohs Scale In 1824, the Austrian scientist Friedrich Mohs developed a list of common minerals to compare their hardnesses. This list is called Mohs scale of hardness, as seen in **Table 1.** The scale lists the hardness of ten minerals. Talc, the softest mineral, has a hardness value of one, and diamond, the hardest mineral, has a value of ten.

Here's how the scale works. Imagine that you have a clear or whitish-colored mineral that you know is either fluorite or quartz. You try to scratch it with your fingernail and then with an iron nail. You can't scratch it with your fingernail but you can scratch it with the iron nail. Because the hardness of your fingernail is 2.5 and that of the iron nail is 4.5, you can determine the unknown mineral's hardness to be somewhere around 3 or 4. Because it is known that quartz has a hardness of 7 and fluorite has a hardness of 4, the mystery mineral must be fluorite.

Some minerals have a hardness range rather than a single hardness value. This is because atoms are arranged differently in different directions in their crystal structures.

Table 1 Mineral Hardness

Mohs Scale	Hardness	Hardness of Common Objects
Talc (softest)	1	
Gypsum	2	fingernail (2.5)
Calcite	3	piece of copper (2.5 to 3.0)
Fluorite	4	iron nail (4.5)
Apatite	5	glass (5.5)
Feldspar	6	steel file (6.5)
Quartz	7	streak plate (7.0)
Topaz	8	
Corundum	9	
Diamond (hardest)	10	

2 Teach

Activity

Synthetic Diamonds Diamonds that are used in cutting tools can be made synthetically. Have students research how synthetic diamonds are produced. Have them write a one-page report on their findings. Synthetic diamonds can be made by heating graphite at high pressure. L2 LS **Linguistic** P

Reading Check

Answer Hardness is a measure of how easily a mineral can be scratched. It is tested by attempting to scratch one mineral with a common object or another mineral of known hardness.

Discussion

Fake Diamonds Tell students that someone has placed an ad in the newspaper that reads, "Real gems for sale! We'll prove these are genuine diamonds by cutting glass with them before your very eyes!" Have students describe why cutting glass wouldn't prove the stones were diamonds. Possible answer: Many minerals other than diamonds cut glass. LS **Linguistic**

Curriculum Connection

Auto Mechanics Some motor oil is dark gray in color. This is caused by the presence of graphite in the oil. Have students find out why graphite is added to motor oil. Because of its softness and atomic arrangement, graphite can be used to lubricate the moving parts of a motor. Graphite can also be used to lubricate moving parts in door hinges and locks. L2

SECTION 2 Mineral Identification F ◆ 15

Visual Learning

Figure 8 The two minerals shown have metallic and glassy lusters. Have students relate the way that light reflects off different minerals to the way it reflects off a piece of chrome, a stainless-steel butter knife, brushed aluminum framing, a chalkboard, and a piece of plastic. The lusters of the pieces of metal are all metallic. However, some are shiny and others are not. The luster of the plastic is shiny, but not metallic. L2 ELL LS **Visual-Spatial**

Applying Science

Answers

1. Hardness test: Hematite would scratch an iron nail; copper would not scratch an iron nail.
2. streak test; unglazed porcelain tile
3. Apply other tests; observe luster, specific gravity, etc.

Figure 8 Luster is an important physical property that is used to distinguish minerals. Graphite has a metallic luster. Fluorite has a nonmetallic, glassy luster.

Luster The way a mineral reflects light is known as **luster.** Luster can be metallic or nonmetallic. Minerals with a metallic luster, like the graphite shown in **Figure 8,** shine like metal. Metallic luster can be compared to the shine of a metal belt buckle, the shiny chrome trim on some cars, or the shine of metallic cooking utensils. When a mineral does not shine like metal, its luster is nonmetallic. Examples of terms for nonmetallic luster include dull, pearly, silky, and glassy. Common examples of minerals with glassy luster are quartz, calcite, halite, and fluorite.

Specific Gravity Minerals also can be distinguished by comparing the weights of equal-sized samples. The **specific gravity** of a mineral is the ratio of its weight compared with the weight of an equal volume of water. Like hardness, specific gravity is expressed as a number. If you were to research the specific gravities of gold and pyrite, you'd find that gold's specific gravity is about 19, and pyrite's is 5. This means that gold is about 19 times heavier than water and pyrite is 5 times heavier than water. You could experience this by comparing equal-sized samples of gold and pyrite in your hands—the pyrite would feel much lighter. The term *heft* is sometimes used to describe how heavy a mineral sample feels.

Applying Science

How can you identify minerals?

Properties of Minerals

Mineral	Hardness	Streak
Copper	2.5–3	copper-red
Galena	2.5	dark gray
Gold	2.5–3	yellow
Hematite	5.5–6.5	red to brown
Magnetite	6–6.5	black
Silver	2.5–3	silver-white

You have learned that minerals are identified by their physical properties, such as streak, hardness, cleavage, and color. Use your knowledge of mineral properties and your ability to read a table to solve the following problems.

Identifying the Problem

The table includes hardnesses and streak colors for several minerals. How can you use these data to distinguish minerals?

Solving the Problem

1. What test would you perform to distinguish hematite from copper? How would you carry out this test?
2. How could you distinguish copper from galena? What tool would you use?
3. What would you do if two minerals had the same hardness and the same streak color?

LAB DEMONSTRATION

Purpose to show that minerals can be identified by unique physical properties

Materials sulfur, calcite, magnetite

Preparation Wear safety goggles and an apron.

Procedure Show students the bright yellow color of sulfur. Drop dilute HCl on calcite to show its fizzing reaction. Show the magnetic property of magnetite by using it to attract a magnet.

Expected Outcome Students will see that color can help identify certain minerals, and that unique properties might be the only clue needed to identify certain minerals.

Assessment

What would happen if HCl were dropped on other minerals in the same group as calcite? The minerals would probably fizz (although not all carbonates will readily react to acid).

Streak When a mineral is rubbed across a piece of unglazed porcelain tile, as in **Figure 9,** a streak of powdered mineral is left behind. **Streak** is the color of a mineral when it is in a powdered form. The streak test works only for minerals that are softer than the streak plate. Gold and pyrite can be distinguished by a streak test. Gold has a yellow streak and pyrite has a greenish-black or brownish-black streak.

Some soft minerals will leave a streak even on paper. The last time you used a pencil to write on paper, you left a streak of the mineral graphite. One reason that graphite is used in pencil lead is because it is soft enough to leave a streak on paper.

Figure 9 Streak is more useful for mineral identification than is mineral color. Hematite, for example, can be dark red, gray, or silver in color. However, its streak is always dark reddish-brown.

Reading Check *Why do gold and pyrite leave a streak, but quartz does not?*

Cleavage and Fracture The way a mineral breaks is another clue to its identity. Minerals that break along smooth, flat surfaces have **cleavage** (KLEE vihj). Cleavage, like hardness, is determined partly by the arrangement of the mineral's atoms. Mica is a mineral that has one perfect cleavage. **Figure 10** shows how mica breaks along smooth, flat planes. If you were to take a layer cake and separate its layers, you would show that the cake has cleavage. Not all minerals have cleavage. Minerals that break with uneven, rough, or jagged surfaces have **fracture.** Quartz is a mineral with fracture. If you were to grab a chunk out of the side of that cake, it would be like breaking a mineral that has fracture.

Mica

Figure 10 Weak or fewer bonds within the structures of mica and halite allow them to be broken along smooth, flat cleavage planes.
Infer *If you broke quartz, would it look the same?*

Reading Check

Answer Both gold and pyrite are softer than the streak plate; quartz is harder than the streak plate.

Use an Analogy

Cleavage Explain that some minerals show perfect cleavage, breaking into thin sheets that can be compared to sheets of paper. Relatively large samples of muscovite or biotite are called "books," because thin sheets can be separated along one cleavage plane, similar to separating the pages of a book.

Quick Demo

Pyrite Properties

Materials pyrite, glass plate, streak plate

Estimated Time 5 minutes

Procedure Demonstrate how pyrite can be distinguished from gold with two simple tests. Rub the sample on a streak plate and show students the greenish streak. Place the glass plate flat on a tabletop and firmly scrape the pyrite against the glass. Have students feel for the scratch left in the glass.

Caption Answer

Figure 10 No; quartz breaks with uneven, often conchoidal, fracture surfaces.

Virtual Labs

Mineral Properties *How can minerals be defined by their properties?*

Differentiated Instruction

Challenge Have students research the atomic structure of a mica, such as muscovite. Have them write about the chemical bonds that exist between structural layers, and relate this to cleavage in micas. L3

Teacher FYI

Mica Uses Sheets of white mica once were used to make the "isinglass" heatproof windows of stoves and ranges. It was also used in Russia for windowpanes and became known as "Muscovy glass." This later became "muscovite," the current name for white mica.

Mini LAB

Purpose to observe how properties can be used to identify minerals LS **Visual-Spatial**

Teaching Strategy Place a clear calcite cleavage fragment over a dot on an overhead projector transparency. Rotate the calcite so that one dot revolves around the other.

Analysis

1. calcite
2. reaction with HCl

Assessment

Oral Have students describe the double image of a dot through different thicknesses of calcite. The greater the distance between cleavage planes, the farther apart the dots appear. Use **PASC**, p. 89. L2

3 Assess

DAILY INTERVENTION

Check for Understanding

Kinesthetic Have students hypothesize whether mineral samples contain iron. Iron content will make a mineral feel heavy, give it a dark color, and possibly make it magnetic.

Reteach

Observing Properties Obtain samples of magnetite. Have students determine its properties. magnetic, has a metallic luster, hardness greater than glass, black in color and streak, exhibits uneven fracture

Assessment

Performance Have students make a table classifying familiar minerals according to types of cleavage. Use **PASC**, p. 109. L2

Mini LAB

Observing Mineral Properties

Procedure

1. Obtain samples of some of the following clear minerals: **gypsum, muscovite mica, halite,** and **calcite.**
2. Place each sample over the print on this page and observe the letters.

Analysis

1. Which mineral can be identified by observing the print's double image?
2. What other special property is used to identify this mineral?

Figure 11 Some minerals are natural magnets, such as this lodestone, which is a variety of magnetite.

Other Properties Some minerals have unique properties. Magnetite, as you can guess by its name, is attracted to magnets. Lodestone, a form of magnetite, will pick up iron filings like a magnet, as shown in **Figure 11.** Light forms two separate rays when it passes through calcite, causing you to see a double image when viewed through transparent specimens. Calcite also can be identified because it fizzes when hydrochloric acid is put on it.

Now you know that you sometimes need more information than color and appearance to identify a mineral. You also might need to test its streak, hardness, luster, and cleavage or fracture. Although the overall appearance of a mineral can be different from sample to sample, its physical properties remain the same.

section 2 review

Summary

Physical Properties

- Minerals are identified by observing their physical properties.
- Hardness is a measure of how easily a mineral can be scratched.
- Luster describes how a mineral reflects light.
- Specific gravity is the ratio of the weight of a mineral sample compared to the weight of an equal volume of water.
- Streak is the color of a powdered mineral.
- Minerals with cleavage break along smooth, flat surfaces in one or more directions.
- Fracture describes any uneven manner in which a mineral breaks.
- Some minerals react readily with acid, form a double image, or are magnetic.

Self Check

1. **Compare and contrast** a mineral fragment that has one cleavage direction with one that has only fracture.
2. **Explain** how an unglazed porcelain tile can be used to identify a mineral.
3. **Explain** why streak often is more useful for mineral identification than color.
4. **Determine** What hardness does a mineral have if it does not scratch glass but it scratches an iron nail?
5. **Think Critically** What does the presence of cleavage planes within a mineral tell you about the chemical bonds that hold the mineral together?

Applying Skills

6. **Draw Conclusions** A large piece of the mineral halite is broken repeatedly into several perfect cubes. How can this be explained?

Science Online bookf.msscience.com/self_check_quiz

section 2 review

1. A mineral with cleavage breaks along smooth, flat planes. One that fractures breaks with uneven or jagged surfaces.
2. A mineral softer than an unglazed porcelain tile will leave a powdered streak on the tile. The streak color can help identify the mineral.
3. Several different samples of one mineral may have different colors, but they usually will have the same streak.
4. between 4.5 and 5.5
5. Cleavage planes occur where relatively weak bonds exist within the mineral's structure.
6. Halite has three directions of cleavage oriented at 90° to each other.

section 3

Uses of Minerals

Gems

Walking past the window of a jewelry store, you notice a large selection of beautiful jewelry—a watch sparkling with diamonds, a necklace holding a brilliant red ruby, and a gold ring. For thousands of years, people have worn and prized minerals in their jewelry. What makes some minerals special? What unusual properties do they have that make them so valuable?

Properties of Gems As you can see in **Figure 12, gems** or gemstones are highly prized minerals because they are rare and beautiful. Most gems are special varieties of a particular mineral. They are clearer, brighter, or more colorful than common samples of that mineral. The difference between a gem and the common form of the same mineral can be slight. Amethyst is a gem form of quartz that contains just traces of iron in its structure. This small amount of iron gives amethyst a desirable purple color. Sometimes a gem has a crystal structure that allows it to be cut and polished to a higher quality than that of a nongem mineral. **Table 2** lists popular gems and some locations where they have been collected.

as you read

What You'll Learn

- **Describe** characteristics of gems that make them more valuable than other minerals.
- **Identify** useful elements that are contained in minerals.

Why It's Important

Minerals are necessary materials for decorative items and many manufactured products.

Review Vocabulary
metal: element that typically is a shiny, malleable solid that conducts heat and electricity well

New Vocabulary
- gem
- ore

Figure 12 It is easy to see why gems are prized for their beauty and rarity. Shown here is The Imperial State Crown, made for Queen Victoria of England in 1838. It contains thousands of jewels, including diamonds, rubies, sapphires, and emeralds.

Section 3 Resource Manager

Chapter FAST FILE Resources
- Transparency Activity, p. 42
- Directed Reading for Content Mastery, pp. 17, 18
- Enrichment, p. 28
- Lab Worksheets, pp. 7–8
- Lab Activity, pp. 11–12
- Reinforcement, p. 25

Cultural Diversity, p. 35
Home and Community Involvement, p. 34
Lab Management and Safety, pp. 54–57

1 Motivate

INTERACTIVE CHALKBOARD
PowerPoint® Presentations

Bellringer

Section Focus Transparencies also are available on the Interactive Chalkboard CD-ROM. L2 ELL

Section 3 Section Focus Transparency **Golden Days**

Gold is a valuable metal that occurs in small amounts in all igneous rocks. It's unusual, however, to find rock that is very rich in gold. In that case, the rock is called gold ore. The gold ore must be hauled to the surface and refined before it looks like the metal you see in jewelry.

1. Why do you think some minerals and metals are considered valuable, while others are not?
2. Describe the gold vein pictured above. Does the gold look the same as it does in jewelry?
3. What other metals can you name that are mined like gold? L2

Minerals

Tie to Prior Knowledge

Jewelry Have students recall the last time they saw jewelry in a jewelry store or department store. Have students describe the jewelry. Most will say it was shiny, colorful, or looked brilliant under the lights in the store. Explain that jewelry is one common use of minerals. L1 LS **Visual-Spatial**

SECTION 3 Uses of Minerals F ◆ 19

2 Teach

Use Science Words

Word Usage Have students read advertisements and use **Table 2** to discover what gems are commonly used in jewelry. Ask them also to find out what minerals the gems are set in. They also could discover where the minerals were found. Have them write their findings in their Science Journals. Possible answer: gems used in jewelry—emeralds, rubies, tanzanite, topaz, peridot, almandine, pyrope, amethyst, sapphires, diamonds; gem settings—gold, silver, platinum; where minerals found—quartz veins and placer deposits, among others L2 ELL LS **Visual-Spatial and Linguistic** P

Make a Model

Crystal Systems Have students research the crystal systems of the gemstones listed in **Table 2.** Once the crystal system has been identified, have students construct models of each crystal system. Students should use **Figure 3** and a book on mineralogy to assist in planning their models. Student models should include the following crystal systems: cubic—spinel and garnet; hexagonal—beryl, quartz, and corundum; orthorhombic—zoisite, topaz, and olivine. L3 ELL LS **Kinesthetic**

Table 2 Minerals and Their Gems

Fun Facts	Mineral	Gem Example	Some Important Locations
Beryl is named for the element beryllium, which it contains. Some crystals reach several meters in length.	Beryl 	Emerald 	Colombia, Brazil, South Africa, North Carolina
A red spinel in the British crown jewels has a mass of 352 carats. A carat is 0.2 g.	Spinel 	Ruby spinel 	Sri Lanka, Thailand, Myanmar (Burma)
Purplish-blue examples of zoisite were discovered in 1967 near Arusha, Tanzania.	Zoisite 	Tanzanite 	Tanzania
The most valuable examples are yellow, pink, and blue varieties.	Topaz (uncut) 	Topaz (gem) 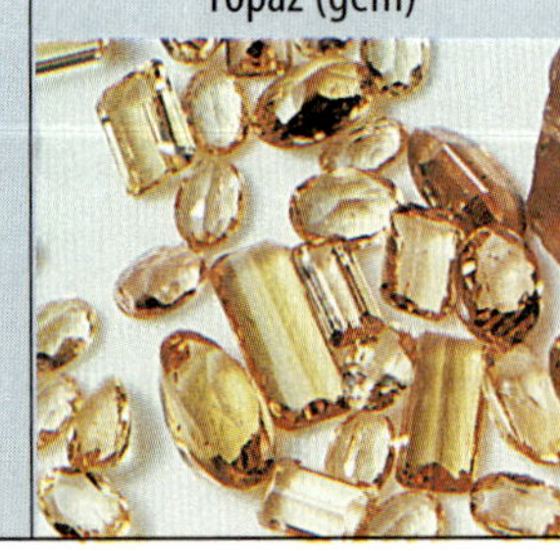	Siberia, Germany, Japan, Mexico, Brazil, Colorado, Utah, Texas, California, Maine, Virginia, South Carolina

Cultural Diversity

Quartz The mineral quartz has been recognized for its hardness and utility since prehistoric times. Quartz tools that are more than 350,000 years old have been found at sites such as Zhoukoudianzhen, China. Quartz has also been esteemed by many cultures for its beauty. The Japanese and Chinese carved clear rock quartz into art objects and spheres, as did the Greeks. Large transparent crystals are found in Madagascar, Japan, Burma, and Brazil. Quartz occurs in many colors and has been used for ornamentation. Purple amethyst comes from Mexico and Brazil. Cairngorm, a brown quartz, was named after the Cairngorm Mountains of Scotland. It also can be found in Switzerland, Japan, and Colorado. Have students research modern uses of quartz. Possible answer: timepieces and jewelry L2

Fun Facts	Mineral	Gem Example	Some Important Locations
Olivine composes a large part of Earth's upper mantle. It is also present in moon rocks.	Olivine 	Peridot 	Myanmar (Burma), Zebirget (Saint John's Island, located in the Red Sea), Arizona, New Mexico
Garnet is a common mineral found in a wide variety of rock types. The red color of the variety almandine is caused by iron in its crystal structure.	Garnet 	Almandine 	Ural Mountains, Italy, Madagascar, Czech Republic, India, Sri Lanka, Brazil, North Carolina, Arizona, New Mexico
Quartz makes up about 30 percent of Earth's continental crust.	Quartz 	Amethyst 	Colorless varieties in Hot Springs, Arkansas; Amethyst in Brazil, Uruguay, Madagascar, Montana, North Carolina, California, Maine
The blue color of sapphire is caused by iron or titanium in corundum. Chromium in corundum produces the red color of ruby.	Corundum 	Blue sapphire 	Thailand, Cambodia, Sri Lanka, Kashmir

Activity

Separating Crystals Have students suppose that they have a liquid that has a density of 4.0 g/cm^3. Ask students to design an experiment that would enable them to separate several gems into two piles; one of diamonds and the other of cubic zirconium crystals. Possible answer: The density of diamond is 3.52 g/cm^3 and the density of cubic zirconium is 4.7 g/cm^3. Place the gems in the liquid. The diamonds will float and the cubic zirconium crystals will sink. L3 LS **Logical-Mathematical**

Discussion

Gem Color Ruby and sapphire are varieties of the mineral corundum. Why are rubies red and some sapphires blue? Blue sapphire is colored by iron and titanium impurities, while the red color of ruby is caused by the presence of minor chromium.

Differentiated Instruction

Challenge Have students research a method used to make diamond-edged surgical scalpels, razor blades, dental drills, and diamond-coated computer parts. Possible answer: Microwaves are used to strip the hydrogen atoms away from the methane molecule. The carbon atoms then link together on the surface of the instrument being coated, forming tiny rows of diamonds. L3

SECTION 3 Uses of Minerals F ◆ 21

Visual Learning

Figure 13 Two famous and very valuable diamonds are shown here. What makes a diamond valuable? The value of diamonds depends partly on whether or not they are tinted. The most common diamonds of gem quality have yellow or brown tints and are less valuable than diamonds with no tints. Diamonds with blue or pink tints are more valuable than those with no tints. Why might these stones be exceptionally valuable? Possible answer: They are rare and have large carat weights. The Hope diamond is tinted blue.

Science Online

Topic: Gemstone Data

Visit bookf.msscience.com for Web links to information about gems at the Smithsonian Museum of Natural History.

Activity List three important examples of gems other than those described on this page. Prepare a data table with the heads *Gem Name/Type, Weight (carats/grams), Mineral,* and *Location.* Fill in the table entries for the gemstones you selected.

Important Gems All gems are prized, but some are truly spectacular and have played an important role in history. For example, the Cullinan diamond, found in South Africa in 1905, was the largest uncut diamond ever discovered. Its mass was 3,106.75 carats (about 621 g). The Cullinan diamond was cut into 9 main stones and 96 smaller ones. The largest of these is called the Cullinan 1 or Great Star of Africa. Its mass is 530.20 carats (about 106 g), and it is now part of the British monarchy's crown jewels, shown in **Figure 13A.**

Another well-known diamond is the blue Hope diamond, shown in **Figure 13B.** This is perhaps the most notorious of all diamonds. It was purchased by Henry Philip Hope around 1830, after whom it is named. Because his entire family as well as a later owner suffered misfortune, the Hope diamond has gained a reputation for bringing its owner bad luck. The Hope diamond's mass is 45.52 carats (about 9 g). Currently it is displayed in the Smithsonian Institution in Washington, D.C.

Useful Gems In addition to their beauty, some gems serve useful purposes. You learned earlier that diamonds have a hardness of 10 on Mohs scale. They can scratch almost any material—a property that makes them useful as industrial abrasives and cutting tools. Other useful gems include rubies, which are used to produce specific types of laser light. Quartz crystals are used in electronics and as timepieces. When subjected to an electric field, quartz vibrates steadily, which helps control frequencies in electronic devices and allows for accurate timekeeping.

Most industrial diamonds and other gems are synthetic, which means that humans make them. However, the study of natural gems led to their synthesis, allowing the synthetic varieties to be used by humans readily.

Figure 13 These gems are among the most famous examples of precious stones.

A The Great Star of Africa is part of a sceptre in the collection of British crown jewels.

B Beginning in 1668, the Hope diamond was part of the French crown jewels. Then known as the French Blue, it was stolen in 1792 and later surfaced in London, England in 1812.

Curriculum Connection

Art Ask an art teacher to bring jewelry-making supplies to class. Materials might include copper, brass, and silver, as well as precut and polished stones. Students could even make jewelry from stones they find near their homes. Have samples of the mineral ores that copper and other materials come from available for students to observe. L2 ELL LS **Visual-Spatial**

Differentiated Instruction

Challenge Students can investigate the areas of the world that produce most of the world's gems. Have them make a collage of what they discover. Their collages could include images of the gems and the places where they are found. L2 LS **Visual-Spatial** P

22 ◆ F CHAPTER 1 Minerals

Bauxite

Figure 14 Bauxite, an ore of aluminum, is processed to make pure aluminum metal for useful products.

Useful Elements in Minerals

Gemstones are perhaps the best-known use of minerals, but they are not the most important. Look around your home. How many things made from minerals can you name? Can you find anything made from iron?

Ores Iron, used in everything from frying pans to ships, is obtained from its ore, hematite. A mineral or rock is an **ore** if it contains a useful substance that can be mined at a profit. Magnetite is another mineral that contains iron.

Reading Check *When is a mineral also an ore?*

INTEGRATE Chemistry

Aluminum sometimes is refined, or purified, from the ore bauxite, shown in **Figure 14.** In the process of refining aluminum, aluminum oxide powder is separated from unwanted materials that are present in the original bauxite. After this, the aluminum oxide powder is converted to molten aluminum by a process called smelting.

During smelting, a substance is melted to separate it from any unwanted materials that may remain. Aluminum can be made into useful products like bicycles, soft-drink cans, foil, and lightweight parts for airplanes and cars. The plane flown by the Wright brothers during the first flight at Kitty Hawk had an engine made partly of aluminum.

INTEGRATE Social Studies

Historical Mineralogy An early scientific description of minerals was published by Georgius Agricola in 1556. Use print and online resources to research the mining techniques discussed by Agricola in his work *De Re Metallica*.

Reading Check

Answer when it contains a useful substance that can be mined at a profit

Quick Demo

Smelting

Materials black copper oxide, powdered charcoal, test tube, beaker with water, bunsen burner

Estimated Time 15 minutes

Procedure Demonstrate a smelting process used for separating metals from an ore by mixing one teaspoon each of black copper oxide and powdered charcoal in the bottom of a test tube. Carefully heat the mixture for a few minutes. Pour the hot mixture into a beaker of water. The copper will settle to the bottom, and the waste will float on the water. **WARNING:** *Wear goggles, gloves, and an apron.*

LS **Visual-Spatial**

INTEGRATE Social Studies

Historical Mineralogy Agricola described many mining practices, including trenching to prospect for ore, hand cranks to remove ore, methods to remove water from mines, and methods to protect workers such as using fans and bellows to ventilate mine shafts.

Teacher FYI

Graphite Scribes in ancient Rome often wrote with thin metal rods on papyrus, an early form of paper. Some of these rods were made of lead. Most still refer to the graphite core in pencils as lead. Graphite was discovered during the sixteenth century in England. This mystery mineral was called plumbago, or black lead.

Fun Fact

Titanium does not activate the body's immune system. Artificial bone made of titanium can be made in such a way that natural bone will grow over and around it.

Reading Check

Answer Fluids travel through weaknesses in rocks, such as natural fractures or cracks, faults, and surfaces between rock layers.

IDENTIFYING Misconceptions

Rarity of Minerals Many students think that minerals are rare objects that are seldom used in everyday materials. Refer to page 6F for teaching strategies that address this misconception.

Figure 15 The mineral sphalerite (greenish when nearly pure) is an important source of zinc. Iron often is coated with zinc to prevent rust in a process called galvanization.

Vein Minerals Under certain conditions, metallic elements can dissolve in fluids. These fluids then travel through weaknesses in rocks and form mineral deposits. Weaknesses in rocks include natural fractures or cracks, faults, and surfaces between layered rock formations. Mineral deposits left behind that fill in the open spaces created by the weaknesses are called vein mineral deposits.

Reading Check *How do fluids move through rocks?*

Sometimes vein mineral deposits fill in the empty spaces after rocks collapse. An example of a mineral that can form in this way is shown in **Figure 15.** This is the shiny mineral sphalerite, a source of the element zinc, which is used in batteries. Sphalerite sometimes fills spaces in collapsed limestone.

Minerals Containing Titanium You might own golf clubs with titanium shafts or a racing bicycle containing titanium. Perhaps you know someone who has a titanium hip or knee replacement. Titanium is a durable, lightweight, metallic element derived from minerals that contain this metal in their crystal structures. Two minerals that are sources of the element titanium are ilmenite (IHL muh nite) and rutile (rew TEEL), shown in **Figure 16.** Ilmenite and rutile are common in rocks that form when magma cools and solidifies. They also occur as vein mineral deposits and in beach sands.

Figure 16 Rutile and ilmenite are common ore minerals of the element titanium.

Rutile Ilmenite

Science Journal

Copper Wire Have students research the property of copper that makes it useful in electrical wire. Ask them to determine what types of chemical elements have this property. Possible answer: The electrical conductivity of copper makes it useful in electrical wire. Usually, metals are the types of chemical elements that conduct electricity well. L2

Uses for Titanium Titanium is used in automobile body parts, such as connecting rods, valves, and suspension springs. Low density and durability make it useful in the manufacture of aircraft, eyeglass frames, and sports equipment such as tennis rackets and bicycles. Wheelchairs used by people who want to race or play basketball often are made from titanium, as shown in **Figure 17.** Titanium is one of many examples of useful materials that come from minerals and that enrich humans' lives.

Figure 17 Wheelchairs used for racing and playing basketball often have parts made from titanium.

section 3 review

Summary

Gems

- Gems are highly prized mineral specimens often used as decorative pieces in jewelry or other items.
- Some gems, especially synthetic ones, have industrial uses.

Useful Elements in Minerals

- Economically important quantities of useful elements or compounds are present in ores.
- Ores generally must be processed to extract the desired material.
- Iron, aluminum, zinc, and titanium are common metals that are extracted from minerals.

Self Check

1. **Explain** why the Cullinan diamond is an important gem.
2. **Identify** Examine **Table 2.** What do rubies and sapphires have in common?
3. **Describe** how vein minerals form.
4. **Explain** why bauxite is considered to be a useful rock.
5. **Think Critically** Titanium is nontoxic. Why is this important in the manufacture of artificial body parts?

Applying Skills

6. **Use Percentages** Earth's average continental crust contains 5 percent iron and 0.007 percent zinc. How many times more iron than zinc is present in average continental crust?

Science Online bookf.msscience.com/self_check_quiz

3 Assess

DAILY INTERVENTION

Check for Understanding

Visual-Spatial Show students a photograph of a rock that contains a vein mineral deposit that formed within it. Ask them to infer how the vein mineral formed. Possible answer: As a fluid solution flowed through cracks in the host rock, conditions changed to favor the precipitation of the vein mineral in the cracks. L2

Reteach

Mineral Resources Ask students to describe how a mineral deposit containing a useful element might not be classified as an ore. Possible answers: The mineral deposit cannot be mined at a profit because the need for the mineral has decreased.

Assessment

Content What must be true for a mineral or rock to be classified as an ore? It must contain a useful substance that can be mined at a profit. L2

section 3 review

1. It was the largest uncut diamond ever found. Part of it is now among the British Monarchy's crown jewels.
2. Both are forms of corundum.
3. Vein minerals form when minerals precipitate from fluids moving along faults or fractures.
4. Bauxite is an ore of aluminum.
5. Body parts manufactured from titanium are less likely to damage body tissues.
6. 0.05/0.00007 = 714; average continental crust contains about 714 times more iron than zinc.

BENCH TESTED

Real-World Question

Purpose Students investigate how physical properties are used to identify unknown mineral samples. L2 LS **Interpersonal**

Process Skills compare and contrast, hypothesize, interpret data, observe and infer

Time Required 60–80 minutes

Materials Mineral testing kits can be purchased from scientific supply houses.

Procedure

Safety Precaution Students should never hold a glass plate in their hands while testing hardness because the plate could break.

Teaching Strategies

- Spread out all minerals on a lab bench so students can study their appearance and specific gravity.
- Begin by having students determine the hardness of each mineral and use this property to attempt mineral identification.
- Narrow possibilities or confirm ID by subjecting the sample to further tests.

Troubleshooting Encourage students to test the same physical properties of each specimen several times.

LAB

Mineral Identification

Goals

- **Hypothesize** which properties of each mineral are most useful for identification purposes.
- **Test** your hypothesis as you attempt to identify unknown mineral samples.

Materials

mineral samples
magnifying lens
pan balance
graduated cylinder
water
piece of copper
**copper penny*
glass plate
small iron nail
steel file
streak plate
5% HCl with dropper
Mohs scale of hardness
Minerals Appendix
**minerals field guide*
safety goggles
**Alternate materials*

Safety Precautions

WARNING: *If an HCl spill occurs, notify your teacher and rinse with cool water until you are told to stop. Do not taste, eat, or drink any lab materials.*

Real-World Question

Although certain minerals can be identified by observing only one property, others require testing several properties to identify them. How can you identify unknown minerals?

Procedure

1. Copy the data table into your Science Journal. Obtain a set of unknown minerals.
2. Observe a numbered mineral specimen carefully. Write a star in the table entry that represents what you hypothesize is an important physical property. Choose one or two properties that you think will help most in identifying the sample.
3. Perform tests to observe your chosen properties first.
   a. To estimate hardness:
      - Rub the sample firmly against objects of known hardness and observe whether it leaves a scratch on the objects.
      - Estimate a hardness range based on which items the mineral scratches.
   b. To estimate specific gravity: Perform a density measurement.
      - Use the pan balance to determine the sample's mass, in grams.

Alternative Inquiry Lab

Sorting Minerals To make this Lab an Inquiry Lab, ask students to divide minerals into groups according to a property of their choice before they attempt serious identification. For example, they could sort the minerals first by luster (metallic or nonmetallic) or heft (heavy, medium, lightweight). L2

Differentiated Instruction

English-Language Learners After students have completed the lab, have them play a game based on 20 Questions. One student can think of a mineral while others try to narrow the choices by asking five *yes* or *no* questions related to luster, hardness, cleavage or fracture, and so on. This will encourage students to use good questioning strategies and classification and memory skills. L1 ELL LS **Interpersonal**

Using Scientific Methods

- Measure its volume using a graduated cylinder partially filled with water. The amount of water displaced by the immersed sample, in mL, is an estimate of its volume in cm^3.
- Divide mass by volume to determine density. This number, without units, is comparable to specific gravity.

4. With the help of the Mineral Appendix or a field guide, attempt to identify the sample using the properties from step 2. Perform more physical property observations until you can identify the sample. Repeat steps 2 through 4 for each unknown.

Physical Properties of Minerals

Sample Number	Hardness	Cleavage or Fracture	Color	Specific Gravity	Luster and Streak	Crystal Shape	Other Properties	Mineral Name
1								
2			Answers will vary					
etc.								

Analyze Your Data

1. Which properties were most useful in identifying your samples? Which properties were least useful?
2. **Compare** the properties that worked best for you with those that worked best for other students.

Conclude and Apply

1. **Determine** two properties that distinguish clear, transparent quartz from clear, transparent calcite. Explain your choice of properties.
2. Which physical properties would be easiest to determine if you found a mineral specimen in the field?

Communicating Your Data

For three minerals, list physical properties that were important for their identification. **For more help, refer to the Science Skill Handbook.**

Analyze Your Data

Expected Outcome Students will discover that using a combination of properties is best when identifying unknown mineral samples.

Answers to Questions

1. Although answers will vary, hardness and streak will generally be the properties commonly cited as most useful. Color usually is the least helpful property.
2. Comparisons among groups will vary but should lead to discussions about different ways to test minerals.

Error Analysis Have students compare their results and explain why any differences occurred. Students who did not get the correct outcome should repeat tests or examine other physical properties of the same specimens.

Conclude and Apply

1. Students could choose hardness, cleavage/fracture, reaction to acid, or double refraction.
2. Properties easily field tested include color, luster, crystal shape, and the presence of cleavage and/or fracture.

Assessment

Performance Ask students to bring mineral samples from home or supply students with additional mineral samples. Have them repeat this lab to identify the additional samples. Use **Performance Assessment in the Science Classroom,** p. 121. L2

Communicating Your Data

Have students compare conclusions by asking each group to prepare and deliver a five-minute presentation. Have each group use their data table as a visual aid for the presentation. L2

TIME

Content Background

In 1895 Wilhelm Conrad Roentgen discovered a type of non-visible radiation, which he termed X rays because their nature was unknown.

In 1912 Max von Laue showed that X rays were part of the electromagnetic spectrum and, like visible light, could be diffracted, absorbed, and polarized. Two years later William and Lawrence Bragg developed the means to analyze the patterns created by the diffraction of X rays in crystalline substances.

Applying this knowledge to biochemistry, Dorothy Hodgkin was able to describe the atomic structure of insulin and vitamin B12. Her work led to a better understanding of how organic chemicals function and the development of synthetic versions of the same compounds.

Discussion

Atom Positions Why is the knowledge of atomic structure important? Possible answers: Knowing the atomic structure of substances helps chemists make those substances artificially and manufacture new products and medicines based on their structures.

What problems might Dr. Hodgkin have faced as a result of being excluded from her colleagues' meetings because she was a woman? Possible answers: She would not have been in a position to easily communicate the results of her work or gain the necessary support from her colleagues to acquire funding.

TIME SCIENCE AND HISTORY

SCIENCE CAN CHANGE THE COURSE OF HISTORY!

Dr. Dorothy Crowfoot Hodgkin

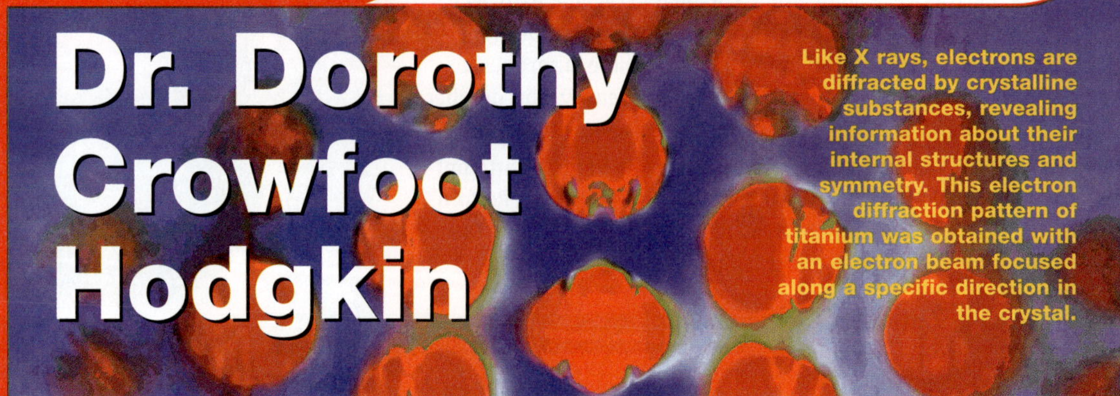

Like X rays, electrons are diffracted by crystalline substances, revealing information about their internal structures and symmetry. This electron diffraction pattern of titanium was obtained with an electron beam focused along a specific direction in the crystal.

Trailblazing scientist and humanitarian

What contributions did Dorothy Crowfoot Hodgkin make to science?

Dr. Hodgkin used a method called X-ray crystallography (kris tuh LAH gruh fee) to figure out the structures of crystalline substances, including vitamin B^{12}, vitamin D, penicillin, and insulin.

What's X-ray crystallography?

Scientists expose a crystalline sample to X rays. As X rays travel through a crystal, the crystal diffracts, or scatters, the X rays into a regular pattern. Like an individual's fingerprints, each crystalline substance has a unique diffraction pattern. Crystallography has applications in the life, Earth, and physical sciences. For example, geologists use X-ray crystallography to identify and study minerals found in rocks.

1910–1994

What were some obstacles Hodgkin overcame?

During the 1930s, there were few women scientists. Hodgkin was not even allowed to attend meetings of the chemistry faculty where she taught because she was a woman. Eventually, she won over her colleagues with her intelligence and tenacity.

How does Hodgkin's research help people today?

Dr. Hodgkin's discovery of the structure of insulin helped scientists learn how to control diabetes, a disease that affects more than 15 million Americans. Diabetics' bodies are unable to process sugar efficiently. Diabetes can be fatal. Fortunately, Dr. Hodgkin's research with insulin has saved many lives.

Research Look in reference books or go to the Glencoe Science Web site for information on how X-ray crystallography is used to study minerals. Write your findings and share them with your class.

Science Online
For more information, visit bookf.msscience.com/time

Resources for Teachers and Students

X ray: The First Hundred Years, edited by Alan Michette and Slawka Pfauntsch Chichester; New York, John Wiley & Sons, 1996

Women in Chemistry and Physics, a Bibliographic Sourcebook, edited by L.S. Grinstein, R.K. Rose, M.H. Rafailovich; Westport, CT; Greenwood Press, 1993

Research X-ray crystallography is routinely used to identify minerals of interest in many geological studies. It also is applied to research new minerals or structural and chemical variations in known minerals.

Reviewing Main Ideas

Section 1 Minerals

1. Much of what you use each day is made at least in some part from minerals.
2. All minerals are formed by natural processes and are inorganic solids with definite chemical compositions and orderly arrangements of atoms.
3. Minerals have crystal structures in one of six major crystal systems.

Section 2 Mineral Identification

1. Hardness is a measure of how easily a mineral can be scratched.
2. Luster describes how light reflects from a mineral's surface.
3. Streak is the color of the powder left by a mineral on an unglazed porcelain tile.
4. Minerals that break along smooth, flat surfaces have cleavage. When minerals break with rough or jagged surfaces, they are displaying fracture.
5. Some minerals have special properties that aid in identifying them. For example, magnetite is identified by its attraction to a magnet.

Section 3 Uses of Minerals

1. Gems are minerals that are more rare and beautiful than common minerals.
2. Minerals are useful for their physical properties and for the elements they contain.

Visualizing Main Ideas

Copy and complete the following concept map about minerals. Use the following words and phrases: the way a mineral breaks, the way a mineral reflects light, ore, a rare and beautiful mineral, how easily a mineral is scratched, streak, *and* a useful substance mined for profit.

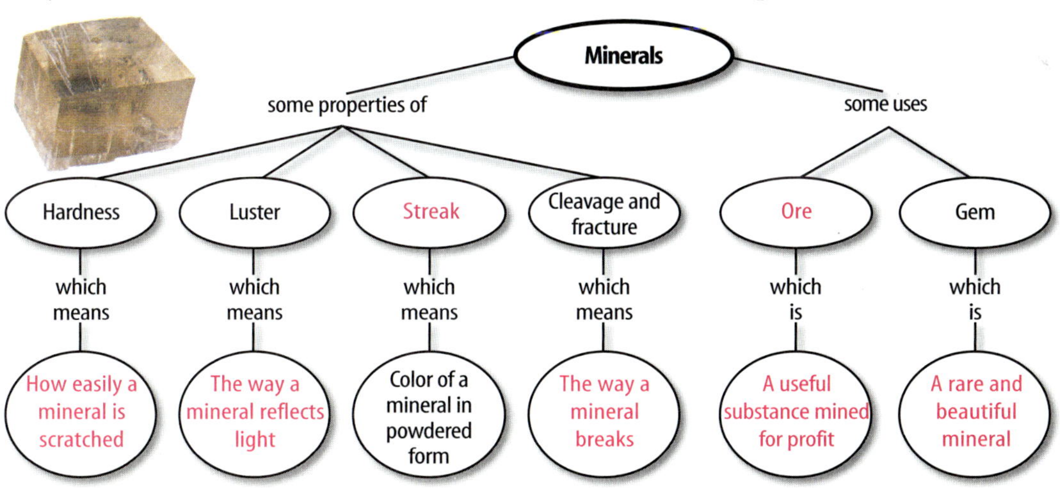

Science Online bookf.msscience.com/interactive_tutor

Identifying Misconceptions

Assess

Use the assessment as follow-up to page F at the beginning of the chapter after students have completed the chapter.

Materials mineral samples that correspond to substances found in vitamin tablets, such as compounds of calcium, iron, and zinc

Procedure Provide each small group of students with a mineral sample and the matching vitamin supplement. Samples could include the mineral halite from salt or the mineral calcite and the supplement calcium.

Expected Outcome Students will realize that they use minerals every day, even as essential parts of the foods they eat. L2

chapter 1 Study Guide

Reviewing Main Ideas

Summary statements can be used by students to review the major concepts of the chapter.

Visualizing Main Ideas

See student page.

Science Online

Visit bookf.msscience.com
/self_check_quiz
/interactive_tutor
/vocabulary_puzzlemaker
/chapter_review
/standardized_test

Assessment Transparency

For additional assessment questions, use the *Assessment Transparency* located in the transparency book.

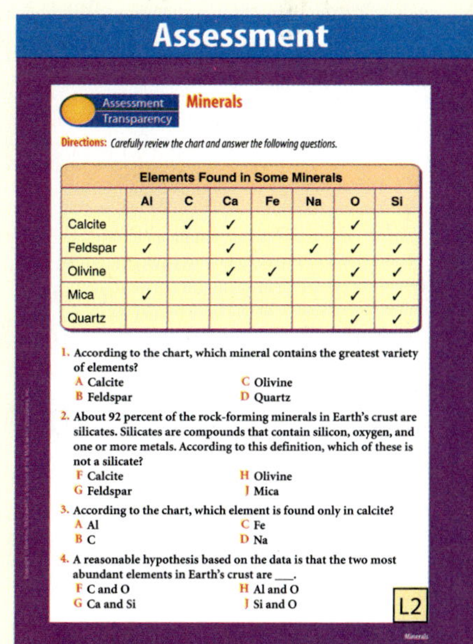

Assessment

Assessment Transparency **Minerals**

Directions: *Carefully review the chart and answer the following questions.*

Elements Found in Some Minerals	Al	C	Ca	Fe	Na	O	Si
Calcite		✓	✓			✓	
Feldspar	✓		✓		✓	✓	✓
Olivine			✓	✓		✓	✓
Mica	✓					✓	✓
Quartz						✓	✓

1. According to the chart, which mineral contains the greatest variety of elements?
   A Calcite C Olivine
   B Feldspar D Quartz
2. About 92 percent of the rock-forming minerals in Earth's crust are silicates. Silicates are compounds that contain silicon, oxygen, and one or more metals. According to this definition, which of these is not a silicate?
   F Calcite H Olivine
   G Feldspar J Mica
3. According to the chart, which element is found only in calcite?
   A Al C Fe
   B C D Na
4. A reasonable hypothesis based on the data is that the two most abundant elements in Earth's crust are ___.
   F C and O H Al and O
   G Ca and Si J Si and O

L2

chapter 1 Review

Using Vocabulary

1. Cleavage—mineral breaks along flat surfaces; Fracture—mineral breaks along uneven surfaces
2. crystal—regular, repeating pattern of atoms in a solid; mineral—a naturally occurring, inorganic crystalline solid
3. luster—how light reflects from a mineral; streak—color of the powdered mineral
4. Magma is molten rock material. Crystals form from magma as it cools and hardens.
5. hardness—how easily a mineral can be scratched; specific gravity—weight of a mineral relative to weight of an equal volume of water
6. ore—mineral or rock with useful substance that can be mined for profit; mineral—naturally occurring inorganic crystalline solid with a definite chemical composition
7. crystal—solid with an orderly internal arrangement of atoms; luster—the way a mineral reflects light
8. Minerals make up rocks. Silicate is a rock-forming mineral that contains silicon and oxygen.
9. A gem is a rare and beautiful mineral. A crystal is a solid with an orderly arrangement of atoms.
10. streak—color of the powdered form of a mineral; specific gravity—weight of a mineral relative to weight of an equal volume of water

Checking Concepts

11. D
12. B
13. B
14. B
15. C
16. D
17. C

chapter 1 Review

Using Vocabulary

cleavage p. 17	magma p. 11
crystal p. 9	mineral p. 8
fracture p. 17	ore p. 23
gem p. 19	silicate p. 12
hardness p. 15	specific gravity p. 16
luster p. 16	streak p. 17

Explain the difference between the vocabulary words in each of the following sets.

1. cleavage—fracture
2. crystal—mineral
3. luster—streak
4. magma—crystal
5. hardness—specific gravity
6. ore—mineral
7. crystal—luster
8. mineral—silicate
9. gem—crystal
10. streak—specific gravity

Checking Concepts

Choose the word or phrase that best answers the question.

11. Which is a characteristic of a mineral?
    A) It can be a liquid.
    B) It is organic.
    C) It has no crystal structure.
    D) It is inorganic.

12. What must all silicates contain?
    A) magnesium
    B) silicon and oxygen
    C) silicon and aluminum
    D) oxygen and carbon

13. What is the measure of how easily a mineral can be scratched?
    A) luster
    B) hardness
    C) cleavage
    D) fracture

Use the photo below to answer question 14.

14. Examine the photo of quartz above. In what way does quartz break?
    A) cleavage **C)** luster
    B) fracture **D)** flat planes

15. Which of the following must crystalline solids have?
    A) carbonates
    B) cubic structures
    C) orderly arrangement of atoms
    D) cleavage

16. What is the color of a powdered mineral formed when rubbing it against an unglazed porcelain tile?
    A) luster
    B) density
    C) hardness
    D) streak

17. Which is hardest on Mohs scale?
    A) talc
    B) quartz
    C) diamond
    D) feldspar

Science Online bookf.msscience.com/vocabulary_puzzlemaker

Use the ExamView® Pro Testmaker CD-ROM to:

- create multiple versions of tests
- create modified tests with one mouse click for inclusion students
- edit existing questions and add your own questions
- build tests aligned with state standards using built-in State Curriculum Tags
- change English tests to Spanish with one mouse click and vice versa

Thinking Critically

18. Classify Water is an inorganic substance that is formed by natural processes on Earth. It has a unique composition. Sometimes water is a mineral and other times it is not. Explain.

19. Determine how many sides a perfect salt crystal has.

20. Apply Suppose you let a sugar solution evaporate, leaving sugar crystals behind. Are these crystals minerals? Explain.

21. Predict Will a diamond leave a streak on a streak plate? Explain.

22. Collect Data Make an outline of how at least seven physical properties can be used to identify unknown minerals.

23. Explain how you would use **Table 1** to determine the hardness of any mineral.

24. Concept Map Copy and complete the concept map below, which includes two crystal systems and two examples from each system. Use the following words and phrases: *hexagonal, corundum, halite, fluorite,* and *quartz.*

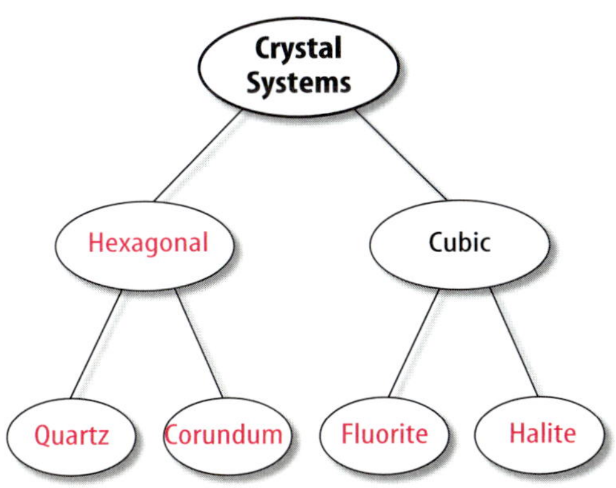

Performance Activities

25. Display Make a display that shows the six crystal systems of minerals. Research the crystal systems of minerals and give three examples for each crystal system. Indicate whether any of the minerals are found in your state. Describe any important uses of these minerals. Present your display to the class.

Applying Math

26. Mineral Volume Recall that 1 mL = 1 cm^3. Suppose that the volume of water in a graduated cylinder is 107.5 mL. A specimen of quartz, tied to a piece of string, is immersed in the water. The new water level reads 186 mL. What is the volume, in cm^3, of the piece of quartz?

Use the graph below to answer questions 27 and 28.

U.S. Slab Zinc Consumption 1978–1998

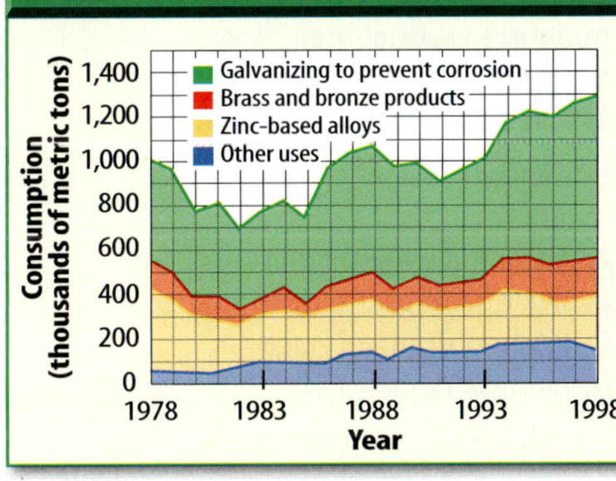

27. Zinc Use According to the graph above, what was the main use of zinc consumed in the United States between 1978 and 1998?

28. Metal Products According to the graph, approximately how many thousand metric tons of zinc were used to make brass and bronze products in 1998?

chapter 1 Review

Thinking Critically

18. On Earth, water can exist as a solid, a liquid, or a gas. Water is a mineral only when it is a solid that has formed naturally such as glacial ice.

19. Each perfect cubic salt crystal has six sides.

20. No; sugar, although it is a solid formed in nature with a definite composition and internal structure, is not a mineral because it is an organic compound.

21. No; a diamond is harder than a streak plate and will scratch the plate.

22. Accept all reasonable outlines.

23. Using common objects and the minerals on the Mohs hardness scale allows you to determine the relative hardness of any mineral. Minerals are harder than other minerals or objects they scratch and softer than minerals or objects that scratch them.

24. See student page.

Performance Activities

25. Displays should correctly show the six crystal systems with three examples listed for each one. Be sure students answer the directives concerning their minerals. Use **Performance Assessment in the Science Classroom,** p. 145.

Applying Math

National Math Standards

1, 5, 6

26. 186 mL − 107.5 mL = 78.5 mL = 78.5 cm^3

27. galvanizing

28. about 175 thousand metric tons

Assessment Resources

Reproducible Masters

Chapter *Fast File* Resources
- Chapter Review, pp. 33–34
- Chapter Tests, pp. 35–38
- Assessment Transparency Activity, p. 45

Glencoe Science Web site
- Chapter Review Test
- Standardized Test Practice

Glencoe Technology

- Assessment Transparency
- Exam*View*® Pro Testmaker
- MindJogger Videoquiz
- Interactive Chalkboard

FAST FILE

Answer Sheet A practice answer sheet can be found at bookf.msscience.com/answer_sheet.

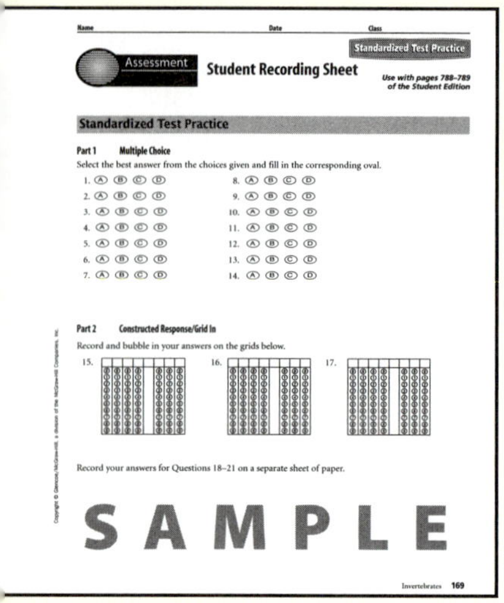
Assessment Student Recording Sheet
Standardized Test Practice
Use with pages 788–789 of the Student Edition
Standardized Test Practice
Part 1 Multiple Choice
Select the best answer from the choices given and fill in the corresponding oval.
Part 2 Constructed Response/Grid In
Record and bubble in your answers on the grids below.
Record your answers for Questions 18–21 on a separate sheet of paper.
SAMPLE

Part 1 Multiple Choice

1. B
2. C
3. A
4. A
5. B
6. C
7. D
8. C
9. A
10. B

chapter 1

Standardized Test Practice

Part 1 Multiple Choice

Record your answers on the answer sheet provided by your teacher or on a sheet of paper.

Use the photo below to answer question 1.

1. To which crystal system does the crystal shown above belong?
   A. hexagonal **C.** triclinic
   B. cubic **D.** monoclinic

2. Which of the following is a common rock-forming mineral?
   A. azurite **C.** quartz
   B. gold **D.** diamond

3. Which term refers to the resistance of a mineral to scratching?
   A. hardness **C.** luster
   B. specific gravity **D.** fracture

4. Which is a special property of the mineral magnetite?
   A. attracted by a magnet
   B. fizzes with dilute hydrochloric acid
   C. forms a double image
   D. has a salty taste

5. Which causes some minerals to break along smooth, flat surfaces?
   A. streak **C.** luster
   B. cleavage **D.** fracture

Test-Taking Tip

If you are taking a timed test, keep track of time during the test. If you find that you're spending too much time on a multiple-choice question, mark your best guess and move on.

6. Which of these forms in cracks or along faults?
   A. bauxite
   B. silicates
   C. vein minerals
   D. rock-forming minerals

7. Which is the most abundant element in Earth's crust?
   A. silicon **C.** iron
   B. manganese **D.** oxygen

Use the table below to answer questions 8–10.

Mineral	Hardness
Talc	1
Gypsum	2
Calcite	3
Fluorite	4
Apatite	5
Feldspar	6
Quartz	7
Topaz	8
Corundum	9
Diamond	10

8. Which mineral in the table is softest?
   A. diamond **C.** talc
   B. feldspar **D.** gypsum

9. Which mineral will scratch feldspar but not topaz?
   A. quartz **C.** apatite
   B. calcite **D.** diamond

10. After whom is the scale shown above named?
    A. Neil Armstrong
    B. Friedrich Mohs
    C. Alfred Wegener
    D. Isaac Newton

Part 2 Short Response/Grid In

11. A mineral is a naturally occurring, inorganic solid with a definite chemical composition and an orderly arrangement of atoms.
12. Gems are valuable because they are rare, beautiful, and usually durable. They have properties that are well suited for making jewelry.
13. Cleavage occurs when minerals break along planar surfaces representing planes of weakness in their structures. Fracture occurs when minerals break along random or nonplanar surfaces.
14. Many different minerals can have the same color, and some minerals occur in a wide variety of colors.
15. 8.08 g
16. 34.5 carats
17. 4,335,986 g
18. They are synthetic, which means that humans make them.
19. Minerals are useful because most products could not be made without them. Society depends on minerals for metals, chemicals, building materials, and more.

Part 2 Short Response/Grid In

Record your answers on the answer sheet provided by your teacher or on a sheet of paper.

11. What is the definition of a mineral?

12. Why are gems valuable?

13. Explain the difference between fracture and cleavage.

14. Why is mineral color sometimes not helpful for identifying minerals?

Use the conversion factor and table below to answer questions 15–17.

1.0 carat = 0.2 grams

Diamond	Carats	Grams
Uncle Sam: largest diamond found in United States	40.4	?
Punch Jones: second largest U.S. diamond; named after boy who discovered it	?	6.89
Theresa: discovered in Wisconsin in 1888	21.5	4.3
2001 diamond production from western Australia	21,679,930	?

15. How many grams is the *Uncle Sam* diamond?

16. How many carats is the *Punch Jones* diamond?

17. How many grams of diamond were produced in western Australia in 2001?

18. What is the source of most of the diamonds that are used for industrial purposes?

19. Explain how minerals are useful to society. Describe some of their uses.

Part 3 Open Ended

Record your answers on a sheet of paper.

Use the photo below to answer question 20.

20. The mineral crystals in the rock above formed when magma cooled and are visible with the unaided eye. Hypothesize about how fast the magma cooled.

21. What is a crystal system? Why is it useful to classify mineral crystals this way?

22. How can a mineral be identified using its physical properties?

23. What is a crystal? Do all crystals have smooth crystal faces? Explain.

24. Are gases that are given off by volcanoes minerals? Why or why not?

25. What is the most abundant mineral group in Earth's crust? What elements always are found in the minerals included in this group?

26. Several layers are peeled from a piece of muscovite mica? What property of minerals does this illustrate? Describe this property in mica.

Science Online bookf.msscience.com/standardized_test

Part 3 Open Ended

20. The magma cooled slowly beneath Earth's surface.

21. All minerals in a crystal system have characteristic similarities in shape caused by their atomic arrangements.

22. A mineral can be identified by performing tests to characterize its physical properties. These are then compared to characteristics listed in mineral identification tables.

23. A crystal is a solid in which all of the atoms are arranged in orderly repeating patterns. Many crystals do not have smooth faces, but all crystals have orderly atomic structures.

24. No, minerals are not gases or liquids.

25. silicates; Silicates always contain silicon and oxygen.

26. cleavage; Mica has one perfect cleavage.

Rubrics

The following rubrics are sample scoring devices for short response and open-ended questions.

Short Response

Points	Description
2	The student demonstrates a thorough understanding of the science of the task. The response may contain minor flaws that do not detract from the demonstration of a thorough understanding.
1	The student has provided a response that is only partially correct.
0	The student has provided a completely incorrect solution or no response at all.

Open Ended

Points	Description
4	The student demonstrates a thorough understanding of the science of the task. The response may contain minor flaws that do not detract from the demonstration of a thorough understanding.
3	The student demonstrates an understanding of the science of the task. The response is essentially correct and demonstrates an essential but less than thorough understanding of the science.
2	The student demonstrates only a partial understanding of the science of the task. Although the student may have used the correct approach to a solution or may have provided a correct solution, the work lacks an essential understanding of the underlying science concepts.
1	The student demonstrates a very limited understanding of the science of the task. The response is incomplete and exhibits many flaws.
0	The student provides a completely incorrect solution or no response at all.

chapter 2 Organizer

Section/Objectives	Standards: National	Standards: State/Local	Labs/Features
Chapter Opener	See 9T–10T for a Key to Standards.		**Launch Lab:** Observe and Describe Rocks, p. 35 **Foldables,** p. 35
Section 1 The Rock Cycle 2 sessions 1 block 1. **Distinguish** between a rock and a mineral. 2. **Describe** the rock cycle and some changes that a rock could undergo.	National Content Standards: UCP.1–3, UCP.5, A.1, A.2, D.1, D.2		**MiniLAB:** Modeling Rock, p. 37 **Visualizing the Rock Cycle,** p. 38
Section 2 Igneous Rocks 2 sessions 1 block 3. **Recognize** magma and lava as the materials that cool to form igneous rocks. 4. **Contrast** the formation of intrusive and extrusive igneous rocks. 5. **Contrast** granitic and basaltic igneous rocks.	National Content Standards: UCP.1–3, UCP.5, A.1, A.2, D.1, D.2		**Science Online,** p. 42 **Integrate Chemistry,** p. 43 **Lab:** Igneous Rock Clues, p. 44
Section 3 Metamorphic Rocks 2 sessions 1 block 6. **Describe** the conditions in Earth that cause metamorphic rocks to form. 7. **Classify** metamorphic rocks as foliated or nonfoliated.	National Content Standards: UCP.1–3, UCP.5, A.1, A.2, D.1, D.2		**Science Online,** p. 46
Section 4 Sedimentary Rocks 3 sessions 1.5 blocks 8. **Explain** how sedimentary rocks form from sediments. 9. **Classify** sedimentary rocks as detrital, chemical, or organic in origin.	National Content Standards: UCP.1–3, UCP.5, A.1, A.2, D.1, D.2		**MiniLAB:** Classifying Sediments, p. 50 **Integrate Career,** p. 52 **Applying Math:** Coal Formation, p. 54 **Lab:** Sedimentary Rocks pp. 56–57 **Science and Society:** Australia's Controversial Rock Star, p. 58

Glencoe Exclusive! **TeacherWorks™** All-In-One Planner and Resource Center

Lab Materials	Reproducible Resources	Section Assessment	Technology
Launch Lab: 3 rock samples, magnifying lens	**Chapter *Fast File* Resources** Foldables Worksheet, p. 13 Directed Reading Overview, p. 15 Note-taking Worksheets, pp. 31–33	Glencoe's Assessment Advantage	TeacherWorks includes: • Interactive Teacher Edition • Lesson Planner with calendar • Access to all program blacklines • Correlations to standards • Web links
MiniLAB: white glue (about 10 mL), dirt or sand (about 7 g), small paper cup	**Chapter *Fast File* Resources** Transparency Activity, p. 42 MiniLAB, p. 3 Enrichment, p. 27 Reinforcement, p. 23 Directed Reading, p. 16 Lab Activity, pp. 9–10 Transparency Activity, pp. 47–48	**Portfolio** Curriculum Connection, p. 38 **Performance** MiniLAB, p. 37 Applying Skills, p. 39 **Content** Section Review, p. 39	Section Focus Transparency Teaching Transparency Virtual Labs CD-ROM Guided Reading Audio Program Interactive Chalkboard CD-ROM
Lab: rhyolite, basalt, vesicular basalt, pumice, granite, obsidian, gabbro, magnifying lens	**Chapter *Fast File* Resources** Transparency Activity, p. 43 Lab Worksheets, pp. 5–6 Enrichment, p. 28 Reinforcement, p. 24 Directed Reading, p. 16 **Reading and Writing Skill Activities,** p. 27	**Portfolio** Reteach, p. 43 **Performance** Applying Skills, p. 43 **Content** Section Review, p. 43	Section Focus Transparency Virtual Labs CD-ROM Guided Reading Audio Program Interactive Chalkboard CD-ROM
	Chapter *Fast File* Resources Transparency Activity, p. 44 Enrichment, p. 29 Reinforcement, p. 25 Directed Reading, p. 17 Lab Activity, pp. 11–12 **Science Inquiry Labs,** p. 31–32	**Portfolio** **Performance** Applying Skills, p. 48 **Content** Section Review, p. 48	Section Focus Transparency Virtual Labs CD-ROM Guided Reading Audio Program Interactive Chalkboard CD-ROM
MiniLAB: paper, samples of sediment, tweezers, magnifying lens **Lab:** sedimentary rocks, marking pen, 5% hydrochloric acid or vinegar, dropper, paper towels, magnifying lens, metric ruler	**Chapter *Fast File* Resources** Transparency Activity, p. 45 MiniLAB, p. 4 Enrichment, p. 30 Reinforcement, p. 26 Directed Reading, pp. 17, 18 Lab Worksheets, pp. 7–8	**Portfolio** Science Journal, p. 51 **Performance** MiniLAB, p. 50 Applying Math, p. 54 Applying Math, p. 55 **Content** Section Review, p. 55	Section Focus Transparency Virtual Labs CD-ROM Guided Reading Audio Program Interactive Chalkboard CD-ROM Video Lab

Need materials? Contact Science Kit at 1-800-828-7777 or www.sciencekit.com on the Internet.

End of Chapter Assessment

Blackline Masters	Technology	Professional Series
Chapter *Fast File* Resources Chapter Review, pp. 35–36 Chapter Tests, pp. 37–40 **Standardized Test Practice,** pp. 11–14	MindJogger Videoquiz Virtual Labs CD-ROM Exam*View*® Pro Testmaker TeacherWorks CD-ROM Interactive Chalkboard CD-ROM	**Performance Assessment in the Science Classroom (PASC)**

Transparencies

Section Focus

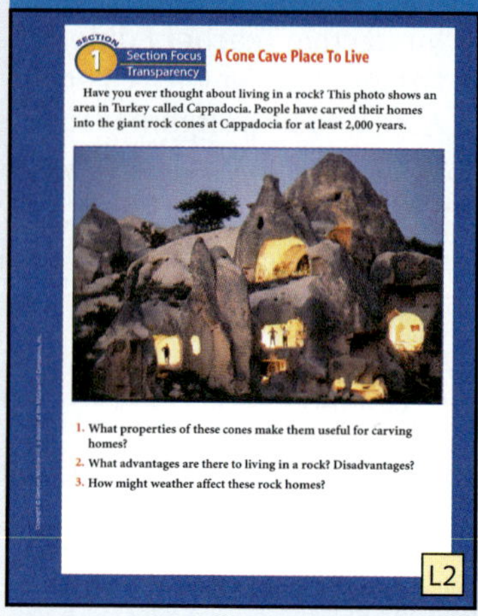

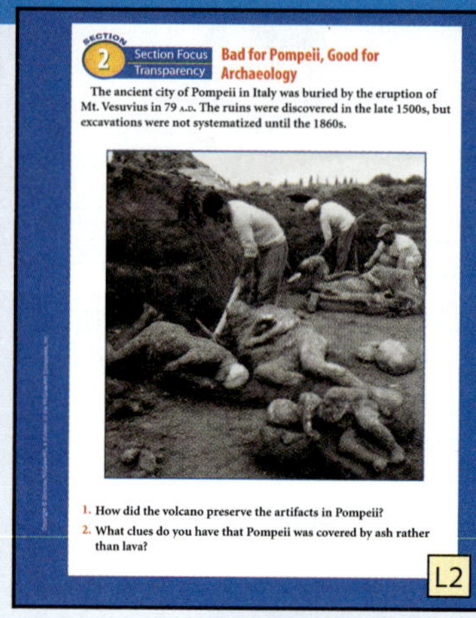

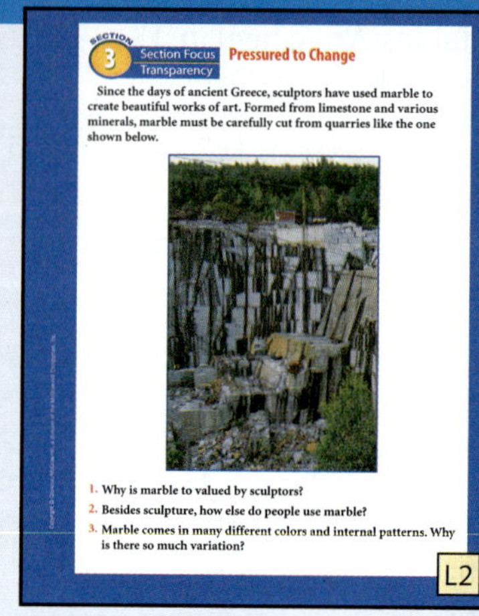

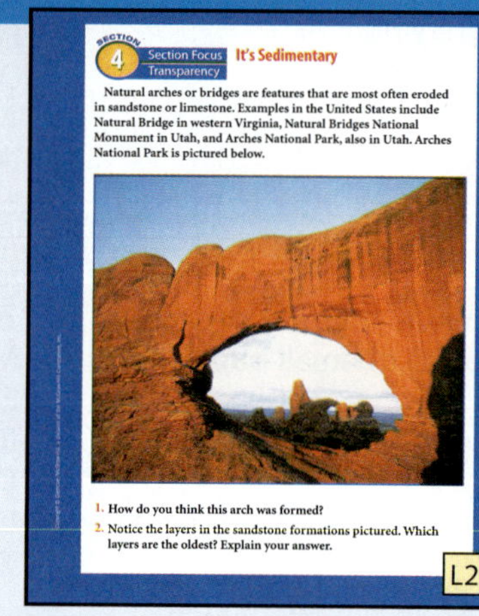

This is a representation of key blackline masters available in the Teacher Classroom Resources. See Resource Manager boxes within the chapter for additional information.

Assessment

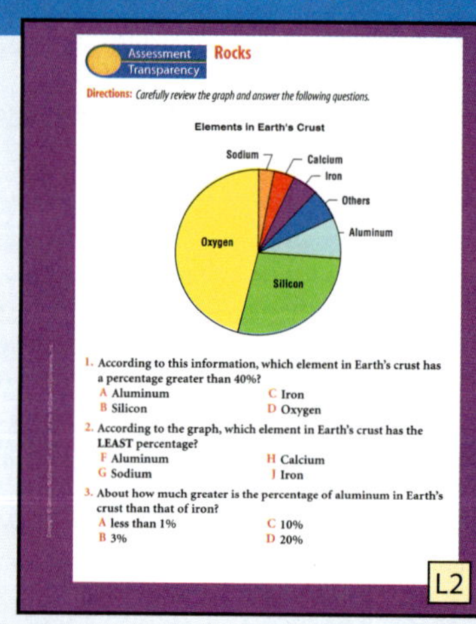

Teaching

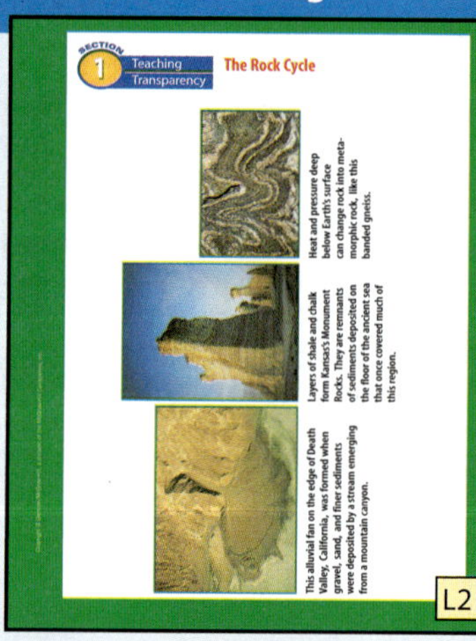

Key to Teaching Strategies

The following designations will help you decide which activities are appropriate for your students.

L1 Level 1 activities should be appropriate for students with learning difficulties.

L2 Level 2 activities should be within the ability range of all students.

L3 Level 3 activities are designed for above-average students.

ELL ELL activities should be within the ability range of English Language Learners.

COOP LEARN Cooperative Learning activities are designed for small group work.

LS Multiple Learning Styles logos, as described on page 6T, are used throughout to indicate strategies that address different learning styles.

P These strategies represent student products that can be placed into a best-work portfolio.

PBL Problem-Based Learning activities apply real-world situations to learning.

Hands-on Activities

Student Text Lab Worksheet

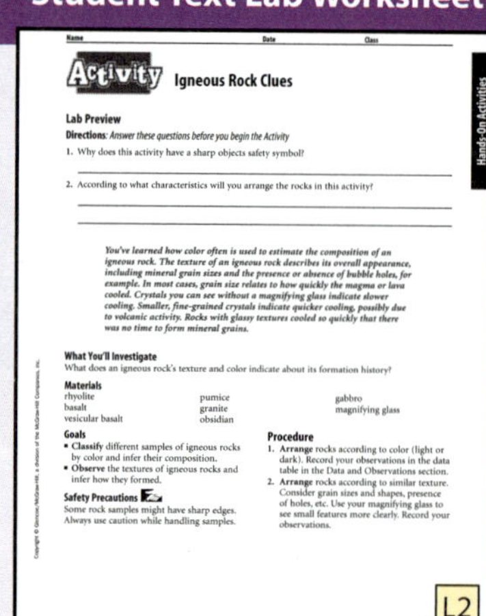

Laboratory Activities

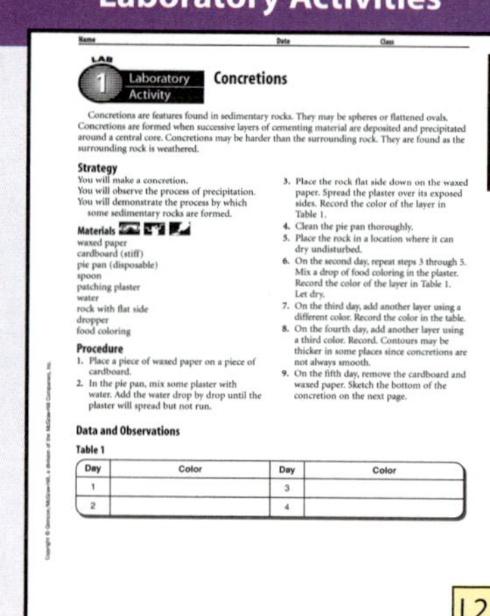

Meeting Different Ability Levels

Content Outline

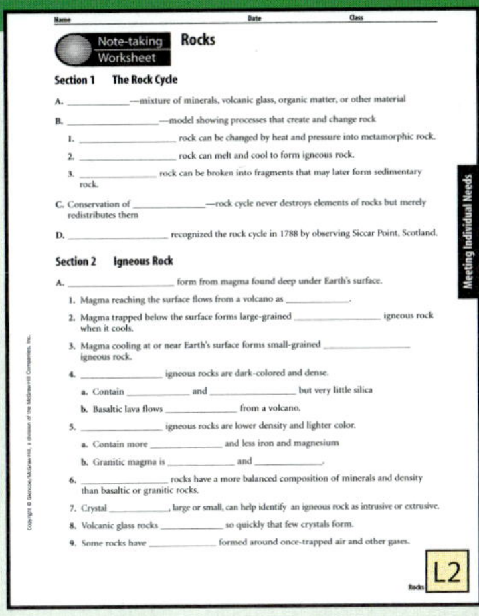

Reinforcement

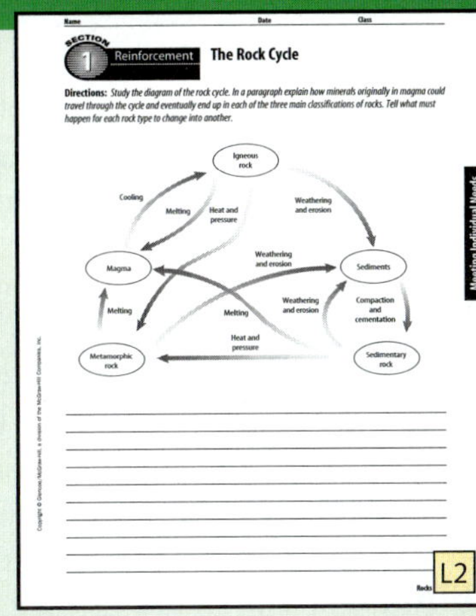

Enrichment

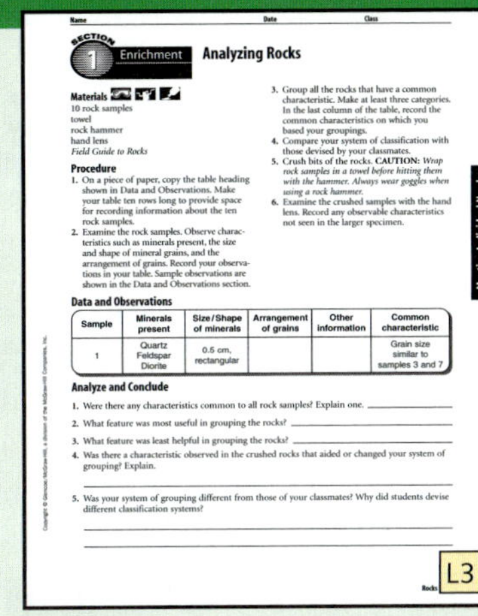

Directed Reading (English/Spanish)

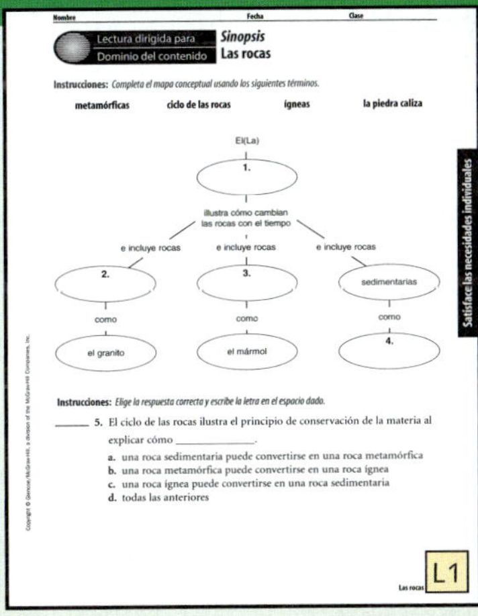

Study Guide

Reading Essentials

Assessment

Test Practice Workbook

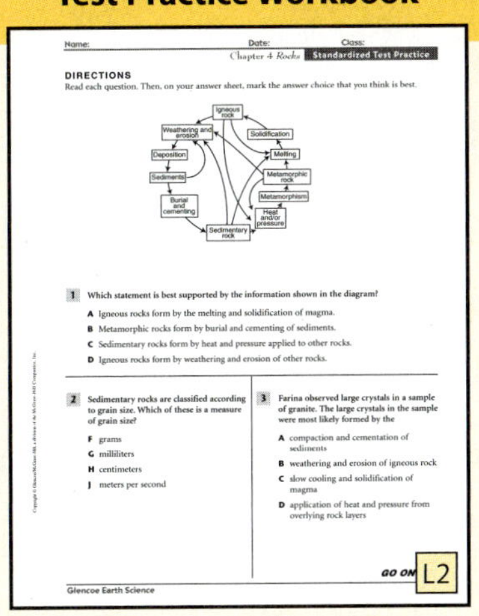

Chapter Review

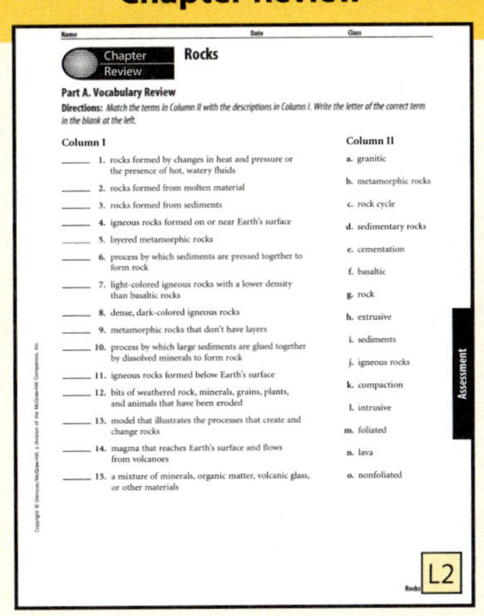

Chapter Tests

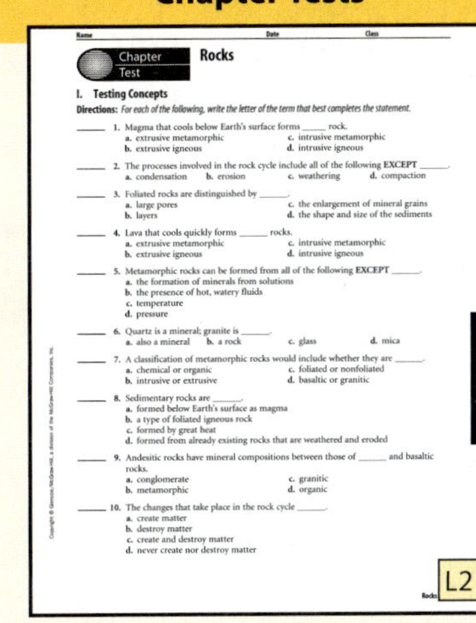

Science Content Background

section 1 The Rock Cycle

Many Paths

There is no set path through the rock cycle. A rock from any of the three groups of rocks can be changed into a rock belonging to another group, or it can be changed into a rock belonging to its original group. A study of the rock cycle reveals the relationships among the three groups of rocks.

section 2 Igneous Rocks

Texture and Grain Size

The size and arrangement of mineral grains determine igneous rock textures. Some rocks are coarse grained, and individual grains can be seen with the unaided eye. Other rocks are fine grained, and individual minerals are not visible with the unaided eye. In rocks with glassy textures, few or no mineral crystals are present. The sizes of mineral grains in most igneous rocks are determined by cooling rates. The faster the magma or lava cooled during an igneous rock's formation, the smaller the grain size is.

Magma

As magma cools, different minerals will crystallize at different temperatures. This relationship is known as Bowen's reaction series. The first silicate minerals to crystallize from magma are lower in silica and higher in iron and magnesium. As crystallization progresses, minerals higher in silica and lower in iron and magnesium form.

Olivine is one of the first minerals to crystallize. Next, pyroxenes such as augite form, which have a different composition and structure. This reaction continues through amphiboles such as hornblende, and biotite mica. One of the last minerals to crystallize is potassium feldspar. Quartz (SiO_2) crystallizes when most of the remaining melt consists of silicon and oxygen.

Teacher to Teacher

Steve Federman
Loveland Middle School
Loveland, Ohio

"I have students review characteristics of sedimentary, igneous, and metamorphic rocks by working with a partner to create a dichotomous key. Students consider characteristics such as grain size, texture, cement type, sediment type, foliated or nonfoliated, and so on, then write *if, then* type statements that apply."

Steve Federman

D. Cavagnaro/DRK Photo

As minerals continue to crystallize from a melt, igneous rocks of different composition form, from ultramafic through mafic (basaltic) and intermediate (andesitic) to felsic (granitic).

section 3 Metamorphic Rocks

How Rocks Change

The overall chemical compositions of metamorphic rocks can be similar to the compositions of the rocks from which they form. However, new minerals may form in response to changes in temperature and pressure.

Heat, Pressure, and Fluids

Changes in temperature, pressure, and fluids can have several causes. One common cause is mountain building. Processes of mountain building cause metamorphism of rock over a large area. This type of metamorphism, called regional metamorphism, produces large amounts of metamorphic rock. Another type of metamorphism, called contact metamorphism, occurs when magma comes in contact with surrounding rocks. A zone of metamorphic rock called an aureole forms around the magma. No melting occurs to form this metamorphic aureole, but hot fluids from the magma do pass into and react with the surrounding rock.

section 4 Sedimentary Rocks

Formation

Understanding the processes by which sedimentary rocks form allows one to discern the history of the rock: the source of the sediment, the means by which the sediment was transported, and the environment that existed where the sediment was deposited.

Sedimentary rocks form under two main sets of conditions. These are referred to as environments of deposition and are either marine (in the sea) or terrestrial (on continents). Rocks that form in a shore zone are considered to form under conditions that are transitional between the marine and terrestrial environments.

Sedimentary Rock Layers

Three of the most important principles used by geologists to interpret sedimentary rock layers were formulated by Nicholas Steno in 1669 and are referred to as Steno's principles. The principle of superposition states that the oldest rock layers lie at the bottom of an undisturbed sequence of sedimentary rock. The principle of original horizontality states that tilted sedimentary rock layers must have been disturbed after they were deposited. The principle of original lateral continuity states that sedimentary rock layers originally extended until they thinned to zero thickness or butted into the edge of the area in which they were deposited. This last principle allows geologists to conclude that a rock layer now cut by a canyon was once continuous across it.

chapter content resources

Internet Resources

For additional content background, visit bookf.msscience.com to:

- access your book online
- find references to related articles in popular science magazines
- access Web links with related content background
- access current events with science journal topics

Print Resources

Peterson First Guide to Rocks and Minerals, by Frederick H. Pough, Houghton Mifflin Company, 1998

Encyclopedia of Rocks, Minerals and Gemstones, by Chris Pellant and Henry Russell, Thunder Bay Press, 2001

How the Earth Works, by John Farndon, Dorling Kindersley Limited, 1992

chapter 2

Chapter Vocabulary

rock, p. 36
rock cycle, p. 37
igneous rock, p. 40
lava, p. 40
intrusive, p. 41
extrusive, p. 41
basaltic, p. 43
granitic, p. 43
metamorphic rock, p. 45
foliated, p. 47
nonfoliated, p. 48
sediment, p. 49
sedimentary rock, p. 49
compaction, p. 50
cementation, p. 51

Science Journal Students might discuss the rock's color or shape, or where it was found; descriptions of rocks in the photo should be detailed and vivid.

INTERACTIVE CHALKBOARD with Image Bank

PowerPoint® Presentations

This CD-ROM is an editable Microsoft® PowerPoint® presentation that includes:

- a pre-made presentation for every chapter
- interactive graphics
- animations
- audio clips
- image bank
- all new section and chapter questions
- Standardized Test Practice
- transparencies
- pre-lab questions for all labs
- Foldables directions
- links to bookf.msscience.com

chapter 2

Rocks

chapter preview

sections

1. The Rock Cycle
2. Igneous Rocks
   Lab Igneous Rock Clues
3. Metamorphic Rocks
4. Sedimentary Rocks
   Lab Sedimentary Rocks

Virtual Lab How are rocks classified?

How did it get there?

The giant rocky peak of El Capitan towers majestically in Yosemite National Park. Surrounded by flat landscape, it seems out of place. How did this expanse of granite rock come to be?

Science Journal Are you a rock collector? If so, write two sentences about your favorite rock. If not, describe the rocks you see in the photo in enough detail that a nonsighted person could visualize them.

Theme Connection

Systems and Interactions Physical conditions, such as heat and pressure inside Earth and weathering and erosion on Earth's surface, interact with Earth materials, constantly changing rock through processes of the rock cycle.

About the Photo

Exposed Rock El Capitan is a huge granite batholith that rises straight up from the valley floor in Yosemite National Park, California. It is igneous rock that formed at high temperatures and pressures. Exposed to Earth's surface today, it is unstable, gradually weathering and shedding material as wind and water come in contact with it.

Start-Up Activities

Launch LAB

Observe and Describe Rocks

Some rocks are made of small mineral grains that lock together, like pieces of a puzzle. Others are grains of sand tightly held together or solidified lava that once flowed from a volcano. If you examine rocks closely, you sometimes can tell what they are made of.

1. Collect three different rock samples near your home or school.
2. Draw a picture of the details you see in each rock.
3. Use a magnifying lens to look for different types of materials within the same rock.
4. Describe the characteristics of each rock. Compare your drawings and descriptions with photos, drawings, and descriptions in a rocks and minerals field guide.
5. Use the field guide to try to identify each rock.
6. **Think Critically** Decide whether you think your rocks are mixtures. If so, infer or suggest what these mixtures might contain. Write your explanations in your Science Journal.

FOLDABLES™ Study Organizer

Major Rock Types Make the following Foldable to help you organize facts about types of rocks.

STEP 1 **Fold** a sheet of paper in half lengthwise. Make the back edge about 5 cm longer than the front edge.

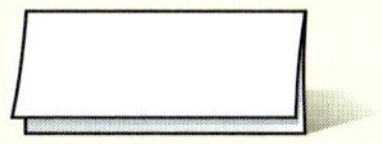

STEP 2 **Turn** the paper so the fold is on the bottom. Then fold it into thirds.

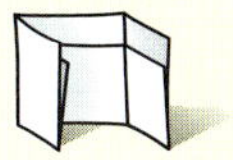

STEP 3 **Unfold and cut** only the top layer along both folds to make three tabs.

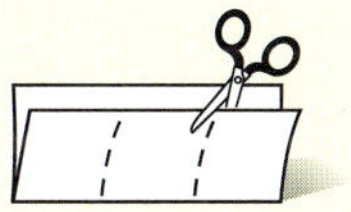

STEP 4 **Label** the Foldable as shown.

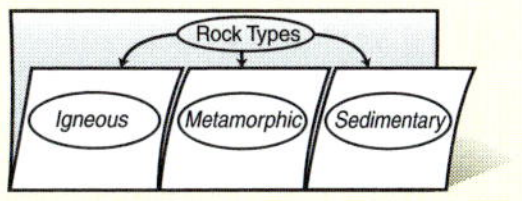

Make an Organizational Study Fold As you read the chapter, write and illustrate what you learn about the three main types of rocks in your study fold.

Science Online — Preview this chapter's content and activities at bookf.msscience.com

Launch LAB

Purpose Use the Launch Lab to introduce students to the composition of different rocks. L2 ELL COOP LEARN LS **Visual-Spatial**

Preparation Have each student bring in several rocks.

Materials magnifying lens, rocks

Teaching Strategy Have each group appoint one student to report the group's observations to the rest of the class.

Think Critically

Students might notice grains of materials of different colors or textures that are part of the rock. Rocks are mixtures of minerals, rock fragments, glass and organic matter. These materials are present in a wide variety of proportions in natural samples.

Assessment

Performance Have groups exchange rocks and repeat the activity. Have them compare and contrast observations. Use **Performance Assessment in the Science Classroom,** p. 35.

FOLDABLES™ Study Organizer — Dinah Zike Study Fold

Student preparation materials for this Foldable are available in the Chapter *Fast File* Resources.

Section 1 The Rock Cycle

as you read

What **You'll Learn**

- **Distinguish** between a rock and a mineral.
- **Describe** the rock cycle and some changes that a rock could undergo.

Why **It's Important**

Rocks exist everywhere, from under deep oceans and in high mountain ranges, to the landscape beneath your feet.

Review Vocabulary

mineral: a naturally occurring, inorganic solid with a definite chemical composition and an orderly arrangement of atoms

New Vocabulary

- rock
- rock cycle

What is a rock?

Imagine you and some friends are exploring a creek. Your eye catches a glint from a piece of rock at the edge of the water. As you wander over to pick up the rock, you notice that it is made of different-colored materials. Some of the colors reflect light, while others are dull. You put the rock in your pocket for closer inspection in science lab.

Common Rocks The next time you walk past a large building or monument, stop and take a close look at it. Chances are that it is made out of common rock. In fact, most rock used for building stone contains one or more common minerals, called rock-forming minerals, such as quartz, feldspar, mica, or calcite. When you look closely, the sparkles you see are individual crystals of minerals. A **rock** is a mixture of such minerals, rock fragments, volcanic glass, organic matter, or other natural materials. **Figure 1** shows minerals mixed together to form the rock granite. You might even find granite near your home.

Figure 1 Mount Rushmore, in South Dakota, is made of granite. Granite is a mixture of feldspar, quartz, mica, hornblende, and other minerals.

1 Motivate

INTERACTIVE CHALKBOARD PowerPoint® Presentations

Bellringer

Section Focus Transparencies also are available on the Interactive Chalkboard CD-ROM. L2 ELL

SECTION 1 Section Focus Transparency **A Cone Cave Place To Live**

Have you ever thought about living in a rock? This photo shows an area in Turkey called Cappadocia. People have carved their homes into the giant rock cones at Cappadocia for at least 2,000 years.

1. What properties of these cones make them useful for carving homes?
2. What advantages are there to living in a rock? Disadvantages?
3. How might weather affect these rock homes?

L2

Rocks

Tie to Prior Knowledge

Rock Formation Ask students to recall rocks they have seen. Do all rocks look the same? no What differences can you recall? Possible answers: color, texture, density, luster (shininess). Tell students that the composition of rocks, and the way they look, are clues to how they formed. Let students know that they will learn how rocks form and change in this section.

Section 1 Resource Manager

Chapter *FAST FILE* Resources

- Transparency Activity, pp. 42, 47–48
- Directed Reading for Content Mastery, pp. 15, 16
- MiniLAB, p. 3
- Lab Activity, pp. 9–10
- Enrichment, p. 27
- Reinforcement, p. 23
- Note-taking Worksheets, pp. 31–33

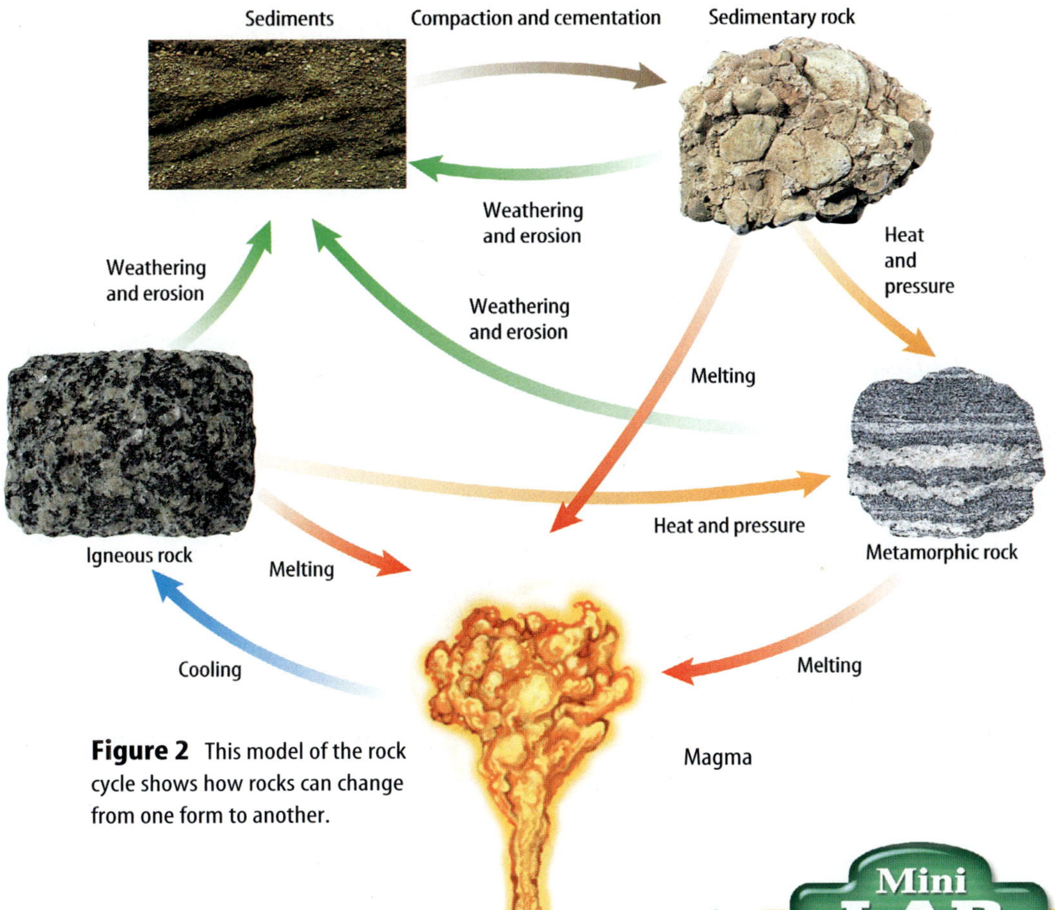

Figure 2 This model of the rock cycle shows how rocks can change from one form to another.

The Rock Cycle

To show how rocks slowly change through time, scientists have created a model called the **rock cycle,** shown in **Figure 2.** It illustrates the processes that create and change rocks. The rock cycle shows the three types of rock—igneous, metamorphic, and sedimentary—and the processes that form them.

Look at the rock cycle and notice that rocks change by many processes. For example, a sedimentary rock can change by heat and pressure to form a metamorphic rock. The metamorphic rock then can melt and later cool to form an igneous rock. The igneous rock then could be broken into fragments by weathering and erode away. The fragments might later compact and cement together to form another sedimentary rock. Any given rock can change into any of the three major rock types. A rock even can transform into another rock of the same type.

Reading Check *What is illustrated by the rock cycle?*

Mini LAB

Modeling Rock

Procedure

1. Mix about 10 mL of **white glue** with about 7 g of **dirt** or **sand** in a **small paper cup.**
2. Stir the mixture and then allow it to harden overnight.
3. Tear away the paper cup carefully from your mixture.

Analysis

1. Which rock type is similar to your hardened mixture?
2. Which part of the rock cycle did you model?

Try at Home

2 Teach

Visual Learning

Figure 2 What are some of the processes that can change rock from one form to another? melting, cooling, weathering and erosion, compaction and cementation, and heat and pressure

Mini LAB

Purpose Students will model sedimentary rock. L2 ELL LS **Kinesthetic**

Materials glue, sand or dirt, paper cups

Teaching Strategy Explain to students that sedimentary rocks forming under natural conditions may require thousands to millions of years.

Analysis

1. sedimentary rock
2. the cementation of mineral and rock fragments into sedimentary rock.

Assessment

Process Ask students to model sedimentary rock using a variety of different materials such as gravel, ground-up shells, and different-colored sand. Make a display of student rock models in the classroom. Use **Performance Assessment in the Science Classroom,** p. 123.

Try at Home

Reading Check

Answer the three types of rocks, how they form, and the processes that change one type into another

Differentiated Instruction

Learning Disabled Place large, labeled samples of basalt, granite, gneiss, slate, shale, sandstone, and limestone in a science center. Ask students in groups of three to study the samples and to take turns selecting a rock and telling its type and the processes that formed it. L1 COOP LEARN LS **Logical-Mathematical and Naturalist**

NATIONAL GEOGRAPHIC

Visualizing the Rock Cycle

Have students examine the pictures and read the captions. Then ask the following questions.

What are some possible ways the black beach sand could become rock? It could be compacted and cemented to become sedimentary rock; it could be melted and cooled to once again form igneous rock.

What processes formed Kansas's Monument Rocks? Layers of sediments that formed on the ancient seafloor were compacted and/or cemented to form the sedimentary rocks.

Activity

Travel Brochure Assign small groups a rock structure, such as Garden of the Gods in Colorado, the Palisades in New York, the White Cliffs of Dover in England, the Cathedral Spires of South Dakota's Black Hills, or Devil's Postpile in California. Ask each group to prepare a travel brochure that shows a map of the feature's location and a description of how it was formed. L2 ELL LS **Visual-Spatial and Linguistic**

Discussion

Metamorphic Rock What could happen to rock to cause the pressure and temperature conditions necessary to metamorphose it? It could become deeply buried by many younger rock formations situated above it.

NATIONAL GEOGRAPHIC **VISUALIZING THE ROCK CYCLE**

Figure 3

Rocks continuously form and transform in a process that geologists call the rock cycle. For example, molten rock—from volcanoes such as Washington's Mount Rainier, background—cools and solidifies to form igneous rock. It slowly breaks down when exposed to air and water to form sediments. These sediments are compacted or cemented into sedimentary rock. Heat and pressure might transform sedimentary rock into metamorphic rock. When metamorphic rock melts and hardens, igneous rock forms again. There is no distinct beginning, nor is there an end, to the rock cycle.

▲ The black sand beach of this Polynesian island is sediment weathered and eroded from the igneous rock of a volcano nearby.

▲ This alluvial fan on the edge of Death Valley, California, was formed when gravel, sand, and finer sediments were deposited by a stream emerging from a mountain canyon.

▲ Layers of shale and chalk form Kansas's Monument Rocks. They are remnants of sediments deposited on the floor of the ancient sea that once covered much of this region.

▲ Heat and pressure deep below Earth's surface can change rock into metamorphic rock, like this banded gneiss.

38 ◆ F CHAPTER 2 Rocks

Curriculum Connection

Language Arts The word igneous comes from the Latin word *ignis*, which means "fire." Ask students to write a paragraph describing where the words *metamorphic* and *sedimentary* come from. *Metamorphic*, which means "change of form," comes from the Greek roots *meta*, meaning "after" and *morphe*, meaning "form." *Sedimentary* comes from the word "sediment." L2 P

INTEGRATE Physics

Matter and the Rock Cycle

The rock cycle, illustrated in **Figure 3,** shows how rock can be weathered to small rock and mineral grains. This material then can be eroded and carried away by wind, water, or ice. When you think of erosion, it might seem that the material is somehow destroyed and lost from the cycle. This is not the case. The chemical elements that make up minerals and rocks are not destroyed. This fact illustrates the principle of conservation of matter. The changes that take place in the rock cycle never destroy or create matter. The elements are just redistributed in other forms.

Figure 4 The rock formations at Siccar Point, Scotland, show that rocks undergo constant change.

Reading Check *What is the principle of conservation of matter?*

Discovering the Rock Cycle James Hutton, a Scottish physician and naturalist, first recognized in 1788 that rocks undergo profound changes. Hutton noticed, among other things, that some layers of solid rock in Siccar Point, shown in **Figure 4,** had been altered since they formed. Instead of showing a continuous pattern of horizontal layering, some of the rock layers at Siccar Point are tilted and partly eroded. However, the younger rocks above them are nearly horizontal.

Hutton published these and other observations, which proved that rocks are subject to constant change. Hutton's early recognition of the rock cycle continues to influence geologists.

section 1 review

Summary

What is a rock?

- Rocks are mixtures of minerals, rock fragments, organic matter, volcanic glass, and other materials found in nature.

The Rock Cycle

- The three major types of rock are igneous, metamorphic, and sedimentary.
- Rock cycle processes do not create or destroy matter.
- Processes that are part of the rock cycle change rocks slowly over time.
- In the late eighteenth century, James Hutton recognized some rock cycle processes by observing rocks in the field.
- Some of Hutton's ideas continue to influence geologic thinking today.

Self Check

1. **Explain** how rocks differ from minerals.
2. **Compare and contrast** igneous and metamorphic rock formation.
3. **Describe** the major processes of the rock cycle.
4. **Explain** one way that the rock cycle can illustrate the principle of conservation of matter.
5. **Think Critically** How would you define magma based on the illustration in **Figure 2?** How would you define sediment and sedimentary rock?

Applying Skills

6. **Communicate** Review the model of the rock cycle in **Figure 2.** In your Science Journal, write a story or poem that explains what can happen to a sedimentary rock as it changes throughout the rock cycle.

Science Online bookf.msscience.com/self_check_quiz

Teacher FYI

Rock Cycle The rock cycle did not happen to form the rocks that exist today and then stop. It is always occurring as various processes in different parts of the crust and on Earth's surface.

Reading Check

Answer Matter is neither created nor destroyed.

3 Assess

DAILY INTERVENTION

Check For Understanding

Interpersonal Have students write a list of possible paths that a crystal in granite bedrock might take through the rock cycle. L2 LS

Reteach

Mineral Mixtures Demonstrate that rocks are mixtures by showing a sample of granite. Point out to students that granite is a mixture of the minerals feldspar, quartz, hornblende, and mica.

Assessment

Content Have students research the word *cycle* and then write a short paragraph explaining why the rock cycle truly is a cycle. Use **Performance Assessment in the Science Classroom,** p. 159.

section 1 review

1. Minerals are naturally occurring crystalline solids with definite chemical compositions. Rocks most often are mixtures of more than one mineral or other natural materials.
2. Both form at relatively high temperatures. Igneous rocks solidify from melted rock material; metamorphic rocks begin and remain in the solid state as they form.
3. Weathering breaks down rock; erosion and deposition transport and deposit the sediment, which is compacted and cemented into new rock; pressure and heat inside Earth change rock from one form to another. Magma from melting of preexisting rock solidifies to form igneous rock.
4. The changes never destroy or create matter. For example, weathering of existing rock produces sediment that is eroded and deposited to form new rock. Accept any reasonable answers.
5. Magma is melted rock material. Sediments are loose mineral and rock fragments. Sedimentary rock is formed from compacted and cemented sediment.
6. Stories or poems can include any logical pathways through the rock cycle.

1 Motivate

INTERACTIVE CHALKBOARD
PowerPoint® Presentations

Bellringer

Section Focus Transparencies also are available on the Interactive Chalkboard CD-ROM. L2 ELL

SECTION 2 Section Focus Transparency **Bad for Pompeii, Good for Archaeology**

The ancient city of Pompeii in Italy was buried by the eruption of Mt. Vesuvius in 79 A.D. The ruins were discovered in the late 1500s, but excavations were not systematized until the 1860s.

1. How did the volcano preserve the artifacts in Pompeii?
2. What clues do you have that Pompeii was covered by ash rather than lava?

L2

Rocks

Tie to Prior Knowledge

Lava Have students recall a volcanic eruption they've seen in person or on television. What comes out of erupting volcanoes? lava, ash, rock, steam Describe lava. They might describe it as hot material that flows out of a volcano. Inform students that lava is a basic ingredient for a certain type of igneous rock and that in this section they will learn how lava becomes rock.

section 2

Igneous Rocks

as you read

What You'll Learn

- **Recognize** magma and lava as the materials that cool to form igneous rocks.
- **Contrast** the formation of intrusive and extrusive igneous rocks.
- **Contrast** granitic and basaltic igneous rocks.

Why It's Important

Igneous rocks are the most abundant kind of rock in Earth's crust. They contain many valuable resources.

Review Vocabulary

element: substance made of one type of atom that cannot be broken down by ordinary chemical or physical means

New Vocabulary

- igneous rock
- lava
- intrusive
- extrusive
- basaltic
- granitic

Formation of Igneous Rocks

Perhaps you've heard of recent volcanic eruptions in the news. When some volcanoes erupt, they eject a flow of molten rock material, as shown in **Figure 5.** Molten rock material, called magma, flows when it is hot and becomes solid when it cools. When hot magma cools and hardens, it forms **igneous** (IHG nee us) **rock.** Why do volcanoes erupt, and where does the molten material come from?

Magma In certain places within Earth, the temperature and pressure are just right for rocks to melt and form magma. Most magmas come from deep below Earth's surface. Magma is located at depths ranging from near the surface to about 150 km below the surface. Temperatures of magmas range from about 650°C to 1,200°C, depending on their chemical compositions and pressures exerted on them.

The heat that melts rocks comes from sources within Earth's interior. One source is the decay of radioactive elements within Earth. Some heat is left over from the formation of the planet, which originally was molten. Radioactive decay of elements contained in rocks balances some heat loss as Earth continues to cool.

Because magma is less dense than surrounding solid rock, it is forced upward toward the surface, as shown in **Figure 6.** When magma reaches Earth's surface and flows from volcanoes, it is called **lava.**

Figure 5 Some lava is highly fluid and free-flowing, as shown by this spectacular lava fall in Volcano National Park, East Rift, Kilauea, Hawaii.

Section 2 Resource Manager

Chapter FAST FILE Resources

Transparency Activity, p. 43

Directed Reading for Content Mastery, p. 16

Enrichment, p. 28

Reinforcement, p. 24

Lab Worksheets, pp. 5–6

Reading and Writing Skill Activities, p. 27

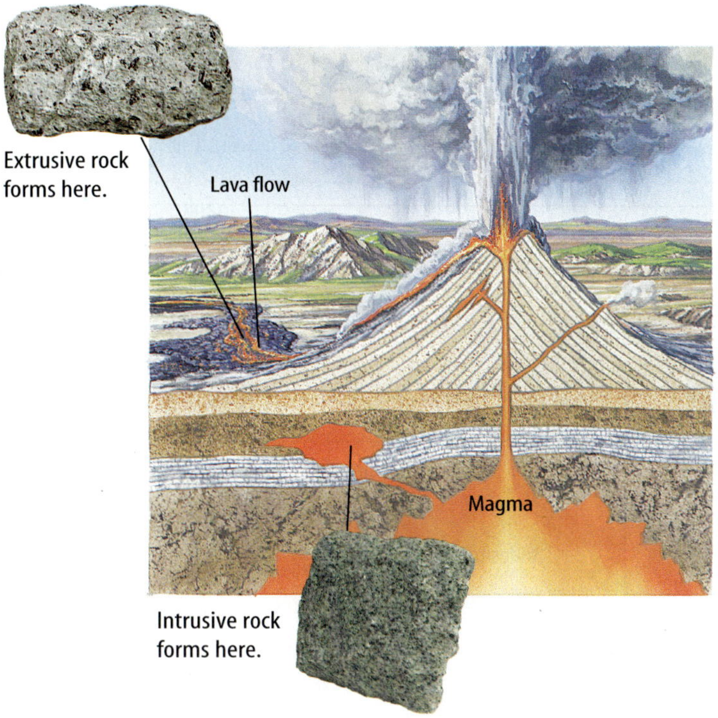

Figure 6 Intrusive rocks form from magma trapped below Earth's surface. Extrusive rocks form from lava flowing at the surface.

Intrusive Rocks Magma is melted rock material composed of common elements and fluids. As magma cools, atoms and compounds in the liquid rearrange themselves into new crystals called mineral grains. Rocks form as these mineral grains grow together. Rocks that form from magma below the surface, as illustrated in **Figure 6,** are called **intrusive** igneous rocks. Intrusive rocks are found at the surface only after the layers of rock and soil that once covered them have been removed by erosion. Erosion occurs when the rocks are pushed up by forces within Earth. Because intrusive rocks form at depth and they are surrounded by other rocks, it takes a long time for them to cool. Slowly cooled magma produces individual mineral grains that are large enough to be observed with the unaided eye.

Extrusive Rocks **Extrusive** igneous rocks are formed as lava cools on the surface of Earth. When lava flows on the surface, as illustrated in **Figure 6,** it is exposed to air and water. Lava, such as the basaltic lava shown in **Figure 5,** cools quickly under these conditions. The quick cooling rate keeps mineral grains from growing large, because the atoms in the liquid don't have the time to arrange into large crystals. Therefore, extrusive igneous rocks are fine grained.

Reading Check *What controls the grain size of an igneous rock?*

2 Teach

Visual Learning

Figure 6 Ask students to contrast magma and lava. Magma is molten rock material that is beneath Earth's surface. Magma becomes lava upon reaching the surface.

Use Science Words

Word Origin Have students use dictionaries to compare the origins of the words *intrusive* and *extrusive* with the origins of *interior* and *exterior.* L2 LS **Linguistic**

Quick Demo

Porous Rocks

Materials pumice, scoria, plastic container of water

Estimated Time 5 to 10 minutes

Procedure Place pumice and scoria in the container of water. Students should observe that pumice floats in the water, but scoria does not. Ask them to make inferences based on their observations. Both samples contain holes from trapped gases, but scoria is darker in color and denser. Scoria contains a higher proportion of heavier elements, such as iron, than pumice does.

Fun Fact

Pumice is an extrusive igneous rock that forms during explosive volcanic eruptions. Composed of glass and mineral fragments interrupted by bubble holes, pumice forms as pressure is suddenly removed from magma, allowing gases to escape.

Differentiated Instruction

English-Language Learners Provide students with one extrusive and one intrusive igneous rock sample. Have them make sketches of the rocks and then compare and contrast the textures of each in their own words. L2

Reading Check

Answer the rate at which magma or lava cools

Discussion

Obsidian Properties Obsidian is a solid, dense volcanic glass. Examine the photo of obsidian in **Table 2.** Why do you think it has been used as a cutting tool? It can be sculpted into an object with sharp, thin edges.

Virtual Labs

Rock Classification *How are rocks classified?*

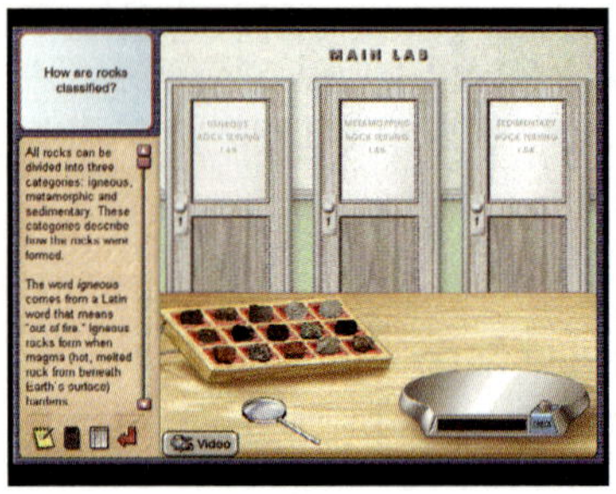

Differentiated Instruction

Challenge Invite interested students to collect rocks from various parts of your community and then use a rock and mineral guide to identify any igneous rocks they find among them. Have students label the rocks and display them in class. Invite other students to do the same with sedimentary and metamorphic rocks in upcoming sections. L2 ELL COOP LEARN IS
Visual-Spatial and Naturalist

Reading Check

Answer whether they form above or below the ground, and the type of magma from which they form

Table 1 Common Igneous Rocks

Magma Type	Basaltic	Andesitic	Granitic
Intrusive	Gabbro 	Diorite 	Granite 
Extrusive	Basalt  Scoria 	Andesite 	Rhyolite  Pumice  Obsidian 

Science Online

Topic: Rock Formation
Visit bookf.msscience.com for Web links to information about intrusive and extrusive rocks.

Activity List several geographic settings where intrusive or extrusive rocks are found. Select one setting for intrusive rocks, and one for extrusive rocks. Describe how igneous rocks form in the two settings, and locate an example of each on a map.

Volcanic Glass Pumice, obsidian, and scoria are examples of volcanic glass. These rocks cooled so quickly that few or no mineral grains formed. Most of the atoms in these rocks are not arranged in orderly patterns, and few crystals are present.

In the case of pumice and scoria, gases become trapped in the gooey molten material as it cools. Some of these gases eventually escape, but holes are left behind where the rock formed around the pockets of gas.

Classifying Igneous Rocks

Igneous rocks are intrusive or extrusive depending on how they are formed. A way to further classify these rocks is by the magma from which they form. As shown in **Table 1,** an igneous rock can form from basaltic, andesitic, or granitic magma. The type of magma that cools to form an igneous rock determines important chemical and physical properties of that rock. These include mineral composition, density, color, and melting temperature.

Reading Check *Name two ways igneous rocks are classified.*

Active Reading

Buddy Interviews Have students interview one another to find out what helps them understand what they are reading, how they find answers, and how they assimilate new vocabulary terms. Have students use Buddy Interviews to help them master the differences among igneous rocks.

Teacher FYI

Rock Textures Some intrusive igneous rocks have textures similar to extrusive rocks. These rocks may form as dikes and sills, and are intruded into relatively shallow levels of Earth's crust.

Basaltic Rocks **Basaltic** (buh SAWL tihk) igneous rocks are dense, dark-colored rocks. They form from magma that is rich in iron and magnesium and poor in silica, which is the compound SiO_2. The presence of iron and magnesium in minerals in basalt gives basalt its dark color. Basaltic lava is fluid and flows freely from volcanoes in Hawaii, such as Kilauea. How does this explain the black beach sand common in Hawaii?

Granitic Rocks **Granitic** igneous rocks are light-colored rocks of a lower density than basaltic rocks. Granitic magma is thick and stiff and contains lots of silica but lesser amounts of iron and magnesium. Because granitic magma is stiff, it can build up a great deal of gas pressure, which is released explosively during violent volcanic eruptions.

Andesitic Rocks Andesitic igneous rocks have mineral compositions between those of basaltic and granitic rocks. Many volcanoes around the rim of the Pacific Ocean formed from andesitic magmas. Like volcanoes that erupt granitic magma, these volcanoes also can erupt violently.

Take another look at **Table 1.** Basalt forms at the surface of Earth because it is an extrusive rock. Granite forms below Earth's surface from magma with a high concentration of silica. When you identify an igneous rock, you can infer how it formed and the type of magma that it formed from.

INTEGRATE Chemistry

Melting Rock Inside Earth, materials contained in rocks can melt. In your Science Journal, describe what is happening to the atoms and molecules to cause this change of state.

section 2 review

Summary

Formation of Igneous Rocks

- When molten rock material, called magma, cools and hardens, igneous rock forms.
- Intrusive igneous rocks form as magma cools and hardens slowly, beneath Earth's surface.
- Extrusive igneous rocks form as lava cools and hardens rapidly, at or above Earth's surface.

Classifying Igneous Rocks

- Igneous rocks are further classified according to their mineral compositions.
- The violent nature of some volcanic eruptions is partly explained by the composition of the magma that feeds them.

Self Check

1. **Explain** why some types of magma form igneous rocks that are dark colored and dense.
2. **Identify** the property of magma that causes it to be forced upward toward Earth's surface.
3. **Explain** The texture of obsidian is best described as glassy. Why does obsidian contain few or no mineral grains?
4. **Think Critically** Study the photos in **Table 1.** How are granite and rhyolite similar? How are they different?

Applying Skills

5. **Make and Use Graphs** Four elements make up most of the rocks in Earth's crust. They are: *oxygen—46.6 percent, aluminum—8.1 percent, silicon—27.7 percent,* and *iron—5.0 percent.* Make a bar graph of these data. What might you infer from the low amount of iron?

Science Online bookf.msscience.com/self_check_quiz

INTEGRATE Chemistry

Melting Rock As thermal energy increases, atoms move faster, and the material changes from solid to liquid.

Text Question Answer

The black beach sand is sediment derived from erosion of black basalt.

3 Assess

DAILY INTERVENTION

Check for Understanding

Kinesthetic Have students compare granitic and basaltic rocks of similar size. Have one student close his or her eyes while another student places the rocks in the first student's hands. Have students hypothesize which rocks are basaltic and which are granitic based on the heft of the rock. L2 LS

Reteach

Magma Composition Ask students to make a concept map that shows basaltic, andesitic, and granitic magmas and the rocks that form from them. L2 LS **Visual-Spatial** P

Assessment

Oral Have students write questions on the material in the section and then take turns quizzing each other aloud. Use **Performance Assessment in the Science Classroom,** p. 91.

section 2 review

1. The magma or lava is rich in iron and magnesium, which are contained in relatively dark and dense minerals.
2. Magma is less dense than surrounding rocks and is forced upward.
3. Obsidian solidifies so quickly that there is no time for mineral grains to form.
4. Granite and rhyolite both are granitic igneous rocks composed of the same minerals. Granite is intrusive—cooled slowly beneath Earth's surface; rhyolite is extrusive—cooled rapidly at the surface.
5. Students' bar graphs should be drawn to scale and properly labeled. On average, crustal rocks contain a fairly low concentration of iron as compared to oxygen, silicon, and aluminum.

BENCH TESTED

Real-World Question

Purpose Students observe the texture and color of igneous rocks to determine how they formed. L2 ELL LS **Kinesthetic**

Process Skills observe, classify, record data, make and use tables, analyze, infer

Time Required 45–50 minutes

Materials Supply magnifying lenses that students can use to observe the rocks.

Procedure

Teaching Strategy Have students work in pairs. If rock samples are in short supply, allow students to take turns using the same samples.

Conclude and Apply

1. rocks that are light in color
2. Rocks with smaller grains; quick cooling does not allow large grains (crystals) to form.
3. Pumice and vesicular basalt; holes in rock suggest that gas was escaping as it cooled.
4. Obsidian and pumice; they cool so quickly that individual grains (crystals) do not have the chance to form.
5. Answers will vary depending on samples observed. The presence of grains that all are visible to the unaided eye indicates rocks are not volcanic.

LAB

Igneous Rock Clues

You've learned how color often is used to estimate the composition of an igneous rock. The texture of an igneous rock describes its overall appearance, including mineral grain sizes and the presence or absence of bubble holes, for example. In most cases, grain size relates to how quickly the magma or lava cooled. Crystals you can see without a magnifying lens indicate slower cooling. Smaller, fine-grained crystals indicate quicker cooling, possibly due to volcanic activity. Rocks with glassy textures cooled so quickly that there was no time to form mineral grains.

Real-World Question

What does an igneous rock's texture and color indicate about its formation history?

Goals

- **Classify** different samples of igneous rocks by color and infer their composition.
- **Observe** the textures of igneous rocks and infer how they formed.

Materials

rhyolite	granite
basalt	obsidian
vesicular basalt	gabbro
pumice	magnifying lens

Safety Precautions

WARNING: *Some rock samples might have sharp edges. Always use caution while handling samples.*

Procedure

1. **Arrange** rocks according to color (light or dark). Record your observations in your Science Journal.
2. **Arrange** rocks according to similar texture. Consider grain sizes and shapes, presence of holes, etc. Use your magnifying lens to see small features more clearly. Record your observations.

Conclude and Apply

1. **Infer** which rocks are granitic based on color.
2. **Infer** which rocks cooled quickly. What observations led you to this inference?
3. **Identify** any samples that suggest gases were escaping from them as they cooled.
4. **Describe** Which samples have a glassy appearance? How did these rocks form?
5. **Infer** which samples are not volcanic. Explain.

Communicating Your Data

Research the compositions of each of your samples. Did the colors of any samples lead you to infer the wrong compositions? Communicate to your class what you learned.

Assessment

Process Encourage students to use a word processing program to make a table that lists and classifies the rocks used in the activity. Use **Performance Assessment in the Science Classroom**, p. 109.

Communicating Your Data

Some granitic rocks are dark in color, and might lead to incorrect inferences. Students should show pictures they have located in print or Internet resources to support their conclusions.

44 ◆ F CHAPTER 2 Rocks

section 3 Metamorphic Rocks

Formation of Metamorphic Rocks

Have you ever packed your lunch in the morning and not been able to recognize it at lunchtime? You might have packed a sandwich, banana, and a large bottle of water. You know you didn't smash your lunch on the way to school. However, you didn't think about how the heavy water bottle would damage your food if the bottle was allowed to rest on the food all day. The heat in your locker and the pressure from the heavy water bottle changed your sandwich. Like your lunch, rocks can be affected by changes in temperature and pressure.

Metamorphic Rocks Rocks that have changed because of changes in temperature and pressure or the presence of hot, watery fluids are called **metamorphic rocks.** Changes that occur can be in the form of the rock, shown in **Figure 7,** the composition of the rock, or both. Metamorphic rocks can form from igneous, sedimentary, or other metamorphic rocks. What Earth processes can change these rocks?

as you read

What You'll Learn

- **Describe** the conditions in Earth that cause metamorphic rocks to form.
- **Classify** metamorphic rocks as foliated or nonfoliated.

Why It's Important

Metamorphic rocks are useful because of their unique properties.

Review Vocabulary
pressure: the amount of force exerted per unit of area

New Vocabulary
- metamorphic rock
- foliated
- nonfoliated

Figure 7 The mineral grains in granite are flattened and aligned when heat and pressure are applied to them. As a result, gneiss is formed.
Describe *other conditions that can cause metamorphic rocks to form.*

SECTION 3 Metamorphic Rocks **F** ◆ **45**

Section 3 Resource Manager

Chapter FAST FILE Resources
- Transparency Activity, p. 44
- Directed Reading for Content Mastery, p. 17
- Enrichment, p. 29
- Lab Activity, pp. 11–12
- Reinforcement, p. 25
- MiniLAB, p. 4

Cultural Diversity, p. 27
Science Inquiry Labs, pp. 31–32

1 Motivate

Bellringer

INTERACTIVE CHALKBOARD PowerPoint® Presentations

Section Focus Transparencies also are available on the Interactive Chalkboard CD-ROM. L2 ELL

Section 3 Section Focus Transparency **Pressured to Change**

Since the days of ancient Greece, sculptors have used marble to create beautiful works of art. Formed from limestone and various minerals, marble must be carefully cut from quarries like the one shown below.

1. Why is marble so valued by sculptors?
2. Besides sculpture, how else do people use marble?
3. Marble comes in many different colors and internal patterns. Why is there so much variation?

L2

Tie to Prior Knowledge

Metamorphic Help students recall the meaning of the word *metamorphosis.* Have them use this knowledge to speculate about the nature of metamorphic rocks.

Text Question Answer

high temperatures and pressures deep inside Earth

Caption Answer

Figure 7 a change in temperature or pressure and the presence of hot fluids

2 Teach

Use an Analogy

Contact Metamorphism Use this analogy to explain how heat generated from hot magma can cause an adjacent area of rock to change. Remind students what happens to beef as it is seared in a pan. The outside of the meat is cooked as it comes in contact with the pan. If removed from the heat soon enough, the inside of the meat remains unchanged. If the cooking continues, the center of the meat will eventually change. Hot magma changes rock in contact with its surface. The longer it continues, the more the rock body in contact with the magma will be changed. **LS Visual-Spatial**

Reading Check

Answer Different types of metamorphic rock form depending on the temperature, pressure, and fluid conditions.

Visual Learning

Figure 8 What is the source of the heat in the fluids? The heat source is the melted rock material, or magma.

Discussion

Reactive Fluids Why might fluids at high temperatures be more effective at changing rock than low-temperature fluids? High temperature fluids would transport more thermal energy to initiate chemical reactions within rock.

Science Online

Topic: Shale Metamorphism

Visit bookf.msscience.com for Web links to information about the metamorphism of shale. Communicate to your class what you learn.

Activity Make a table with headings that are major rock types that form from shale metamorphism. Under each rock heading, make a list of minerals that can occur in the rock.

Heat and Pressure Rocks beneath Earth's surface are under great pressure from rock layers above them. Temperature also increases with depth in Earth. In some places, the heat and pressure are just right to cause rocks to melt and magma to form. In other areas where melting doesn't occur, some mineral grains can change by dissolving and recrystallizing—especially in the presence of fluids. Sometimes, under these conditions, minerals exchange atoms with surrounding minerals and new, bigger minerals form.

Depending upon the amount of pressure and temperature applied, one type of rock can change into several different metamorphic rocks, and each type of metamorphic rock can come from several kinds of parent rocks. For example, the sedimentary rock shale will change into slate. As increasing pressure and temperature are applied, the slate can change into phyllite, then schist, and eventually gneiss. Schist also can form when basalt is metamorphosed, or changed, and gneiss can come from granite.

Reading Check *How can one type of rock change into several different metamorphic rocks?*

Hot Fluids Did you know that fluids can move through rock? These fluids, which are mostly water with dissolved elements and compounds, can react chemically with a rock and change its composition, especially when the fluids are hot. That's what happens when rock surrounding a hot magma body reacts with hot fluids from the magma, as shown in **Figure 8.** Most fluids that transform rocks during metamorphic processes are hot and mainly are comprised of water and carbon dioxide.

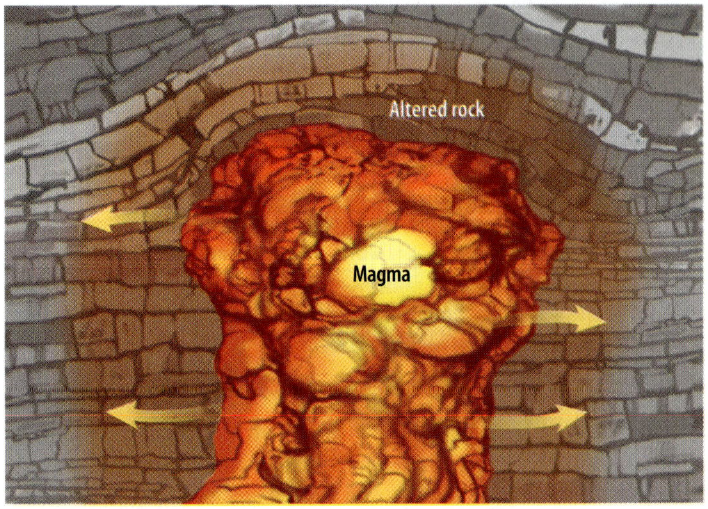

Figure 8 In the presence of hot, water-rich fluids, solid rock can change in mineral composition without having to melt.

Active Reading

Jigsaw In this collaborative learning strategy, individuals become experts on a portion of a text and share their expertise with their "home" group. Everyone shares responsibility for learning the assigned reading. Assign each person in each home group an "expert" number (1 through 5, for example). Have students gather into the expert groups that correspond to the number they were assigned. There, have them read, discuss, and master chapter concepts and determine how best to teach them to their home groups. Have students return to home groups and share the content they learned in their expert groups. Have students use the Jigsaw strategy to learn about conditions in Earth that cause metamorphic rocks to form.

Classifying Metamorphic Rocks

Metamorphic rocks form from igneous, sedimentary, or other metamorphic rocks. Heat, pressure, and hot fluids trigger the changes. Each resulting rock can be classified according to its composition and texture.

Foliated Rocks When mineral grains line up in parallel layers, the metamorphic rock is said to have a **foliated** texture. Two examples of foliated rocks are slate and gneiss. Slate forms from the sedimentary rock shale. The minerals in shale arrange into layers when they are exposed to heat and pressure. As **Figure 9** shows, slate separates easily along these foliation layers.

The minerals in slate are pressed together so tightly that water can't pass between them easily. Because it's watertight, slate is ideal for paving around pools and patios. The naturally flat nature of slate and the fact that it splits easily make it useful for roofing and tiling many surfaces.

Gneiss (NISE), another foliated rock, forms when granite and other rocks are changed. Foliation in gneiss shows up as alternating light and dark bands. Movement of atoms has separated the dark minerals, such as biotite mica, from the light minerals, which are mainly quartz and feldspar.

Figure 9 Slate often is used as a building or landscaping material.
Identify *the properties that make slate so useful for these purposes.*

Reading Check *What type of metamorphic rock is composed of mineral grains arranged in parallel layers?*

Inquiry Lab

Designing with Rock

Purpose Have students select rocks as home-building materials and design a simple home.

Possible Materials rock samples, rocks and minerals field guide, print and online resources on architecture and home building, paper, pencil

Estimated Time 1 class session

Teaching Strategies

- Ask students to imagine that they are going to build a house using three different rock types.
- Have them sketch their homes, and write a paragraph describing which rocks they selected and why.
- Students should support their choices and designs by explaining rock properties and home-building guidelines. L2

For additional inquiry activities, see *Science Inquiry Labs*.

Caption Answer

Figure 9 Slate is watertight and easily split into layers.

Activity

Pressure Direction Obtain samples of slate, schist, and gneiss. Using **Figure 7**, have students determine in which direction the greatest pressure was applied to change each rock. Possible answer: It would have been applied perpendicular to the alignment of mineral grains. L2 LS **Visual-Spatial**

Reading Check

Answer foliated metamorphic rocks

Teacher FYI

Hydrothermal Metamorphism Metamorphism that occurs when hot, mineral-rich fluids come into contact with rock deep inside Earth (hydrothermal metamorphism) is thought in some cases to help produce ores of copper, zinc, iron, and gold.

Differentiated Instruction

Challenge Ask students to search their community for uses of metamorphic rock. Have them describe the metamorphic rocks they notice in enough detail so that the rest of the class can identify them. L3

SECTION 3 Metamorphic Rocks F ◆ 47

3 Assess

DAILY INTERVENTION

Check for Understanding

Visual-Spatial Have students sketch foliated and nonfoliated metamorphic rocks. Call on individual students to describe what their sketches show. Encourage students to obtain samples of the rocks they have sketched. L2 LS

Reteach

Rock Facts Have students take turns brainstorming facts they now know about metamorphic rocks. Write their contributions on the board. Use this information to stimulate discussion about the origins and classification of metamorphic rocks. L1 LS **Interpersonal**

Assessment

Performance Provide students with a variety of metamorphic rocks. Have them infer a possible parent rock for each sample. Use **Performance Assessment in the Science Classroom,** p. 121.

Figure 10 This exhibit in Vermont shows the beauty of carved marble.

Nonfoliated Rocks In some metamorphic rocks, layering does not occur. The mineral grains grow and rearrange, but they don't form layers. This process produces a **nonfoliated** texture.

Sandstone is a sedimentary rock that's often composed mostly of quartz grains. When sandstone is heated under a lot of pressure, the grains of quartz grow in size and become interlocking, like the pieces of a jigsaw puzzle. The resulting rock is called quartzite.

Marble is another nonfoliated metamorphic rock. Marble forms from the sedimentary rock limestone, which is composed of the mineral calcite. Usually, marble contains several other minerals besides calcite. For example, hornblende and serpentine give marble a black or greenish tone, whereas hematite makes it red. As **Figure 10** shows, marble is a popular material for artists to sculpt because it is not as hard as other rocks.

So far, you've investigated only a portion of the rock cycle. You still haven't observed how sedimentary rocks are formed and how igneous and metamorphic rocks evolve from them. The next section will complete your investigation of the rock cycle.

section 3 review

Summary

Formation of Metamorphic Rocks

- Changes in pressure, temperature, or the presence of fluids can cause metamorphic rocks to form.
- Rock, altered by metamorphic processes at high temperatures and pressures, changes in the solid state without melting.
- Hot fluids that move through and react with preexisting rock are composed mainly of water and carbon dioxide.
- One source of hot, watery fluids is magma bodies close to the changing rock.
- Any parent rock type—igneous, metamorphic, or sedimentary—can become a metamorphic rock.

Classifying Metamorphic Rocks

- Texture and mineral composition determine how a metamorphic rock is classified.
- Physical properties of metamorphic rocks, such as the watertight nature of slate, make them useful for many purposes.

Self Check

1. **Explain** what role fluids play in rock metamorphism.
2. **Describe** how metamorphic rocks are classified. What are the characteristics of rocks in each of these classifications?
3. **Identify** Give an example of a foliated and a nonfoliated metamorphic rock. Name one of their possible parent rocks.
4. **Think Critically** Marble is a common material used to make sculptures, but not just because it's a beautiful stone. What properties of marble make it useful for this purpose?

Applying Skills

5. **Concept Map** Put the following events in an events-chain concept map that explains how a metamorphic rock might form from an igneous rock. *Hint: Start with "Igneous Rock Forms."* Use each event just once.

   Events: *sedimentary rock forms, weathering occurs, heat and pressure are applied, igneous rock forms, metamorphic rock forms, erosion occurs, sediments are formed, deposition occurs*

Science Online bookf.msscience.com/self_check_quiz

section 3 review

1. Hot fluids chemically react with rocks with which they are in contact, changing their compositions.
2. Foliated or nonfoliated; foliated rocks have layers of mineral grains; nonfoliated rocks have little or no layering.
3. foliated: slate; parent rock: shale; nonfoliated: marble; parent rock: limestone
4. It also is relatively homogeneous and easy to shape. It is composed mainly of the mineral calcite, which has a hardness of only 3 on the Mohs scale.
5. Igneous rock forms. Weathering occurs. Sediments are formed. Erosion occurs. Deposition occurs. Sedimentary rock forms. Heat and pressure are applied. Metamorphic rock forms.

section 4

Sedimentary Rocks

Formation of Sedimentary Rocks

Igneous rocks are the most common rocks on Earth, but because most of them exist below the surface, you might not have seen too many of them. That's because 75 percent of the rocks exposed at the surface are sedimentary rocks.

Sediments are loose materials such as rock fragments, mineral grains, and bits of shell that have been moved by wind, water, ice, or gravity. If you look at the model of the rock cycle, you will see that sediments come from already-existing rocks that are weathered and eroded. **Sedimentary rock** forms when sediments are pressed and cemented together, or when minerals form from solutions.

Stacked Rocks Sedimentary rocks often form as layers. The older layers are on the bottom because they were deposited first. Sedimentary rock layers are a lot like the books and papers in your locker. Last week's homework is on the bottom, and today's notes will be deposited on top of the stack. However, if you disturb the stack, the order in which the books and papers are stacked will change, as shown in **Figure 11.** Sometimes, forces within Earth overturn layers of rock, and the oldest are no longer on the bottom.

as you read

What You'll Learn

- **Explain** how sedimentary rocks form from sediments.
- **Classify** sedimentary rocks as detrital, chemical, or organic in origin.
- **Summarize** the rock cycle.

Why It's Important

Some sedimentary rocks, like coal, are important sources of energy.

Review Vocabulary
weathering: surface processes that work to break down rock mechanically or chemically

New Vocabulary
- sediment
- sedimentary rock
- compaction
- cementation

Figure 11 Like sedimentary rock layers, the oldest paper is at the bottom of the stack. If the stack is disturbed, then it is no longer in order.

1 Motivate

INTERACTIVE CHALKBOARD PowerPoint® Presentations

Bellringer

Section Focus Transparencies also are available on the Interactive Chalkboard CD-ROM. L2 ELL

Tie to Prior Knowledge

Sedimentary Rock Formation Have students describe what happens when sugar is stirred into a glass of cold water. What happens? Some grains dissolve, others drop to the bottom of the glass. Explain that some rocks form from sediment that drops to the bottom of water bodies. Others form from chemicals dissolved in water. Tell students they will learn in this section how this happens.

Section 4 Resource Manager

Chapter *FAST FILE* Resources
- Transparency Activity, p. 45
- Directed Reading for Content Mastery, pp. 17, 18
- MiniLAB, p. 4
- Enrichment, p. 30
- Reinforcement, p. 26
- Lab Worksheets, pp. 7–8

Reading and Writing Skill Activities, p. 21
Mathematics Skill Activities, p. 29
Lab Management and Safety, p. 70

2 Teach

Mini LAB

Purpose Students classify sediments according to grain size. L2 ELL COOP LEARN LS Kinesthetic

Materials paper, samples of silt, sand, and gravel-sized sediment grains (with both rounded and sharp edges), tweezers or dissecting probe, magnifying lens

Teaching Strategy Review with students the appearance of rounded grains as opposed to grains with sharp edges. For example, grains need not be spherical in shape to be considered well-rounded.

Safety Precautions Caution students to handle sharp instruments with care and to wear goggles while handling samples.

Analysis

1. Students should describe grains by size and the smoothness or angularity of their edges.
2. Conglomerate from rounded gravel, breccia from angular gravel; If other sediments are used: shale from clay, siltstone from silt, sandstone from sand.

Assessment

Performance Have students paste silt, sand, and gravel-sized sediment onto a note card. Then use the cards to determine the grain sizes of several sedimentary rock samples. Use **Performance Assessment in the Science Classroom,** p. 121.

Reading Check

Answer As layers of sediment build up, pressure from upper layers on lower layers compacts lower layers into rock.

Mini LAB

Classifying Sediments

Procedure

WARNING: *Use care when handling sharp objects.*

1. Collect different samples of **sediment.**
2. Spread them on a sheet of **paper.**
3. Use **Table 2** to determine the size range of gravel-sized sediment.
4. Use **tweezers or a dissecting probe** and a **magnifying lens** to separate the gravel-sized sediments.
5. Separate the gravel into piles—rounded or angular.

Analysis

1. Describe the grains in both piles.
2. Determine what rock could form from each type of sediment you have.

Classifying Sedimentary Rocks

Sedimentary rocks can be made of just about any material found in nature. Sediments come from weathered and eroded igneous, metamorphic, and sedimentary rocks. Sediments also come from the remains of some organisms. The composition of a sedimentary rock depends upon the composition of the sediments from which it formed.

Like igneous and metamorphic rocks, sedimentary rocks are classified by their composition and by the manner in which they formed. Sedimentary rocks usually are classified as detrital, chemical, or organic.

Detrital Sedimentary Rocks

The word *detrital* (dih TRI tul) comes from the Latin word *detritus,* which means "to wear away." Detrital sedimentary rocks, such as those shown in **Table 2,** are made from the broken fragments of other rocks. These loose sediments are compacted and cemented together to form solid rock.

Weathering and Erosion When rock is exposed to air, water, or ice, it is unstable and breaks down chemically and mechanically. This process, which breaks rocks into smaller pieces, is called weathering. **Table 2** shows how these pieces are classified by size. The movement of weathered material is called erosion.

Compaction Erosion moves sediments to a new location, where they then are deposited. Here, layer upon layer of sediment builds up. Pressure from the upper layers pushes down on the lower layers. If the sediments are small, they can stick together and form solid rock. This process, shown in **Figure 12,** is called **compaction.**

Reading Check *How do rocks form through compaction?*

Figure 12 During compaction, pore space between sediments decreases, causing them to become packed together more tightly.

Differentiated Instruction

Visually Impaired Provide students with samples of rounded and angular gravel that they can feel to determine the size and shape of grains. Also provide samples of conglomerate and breccia in which the students can feel the embedded grains. L2

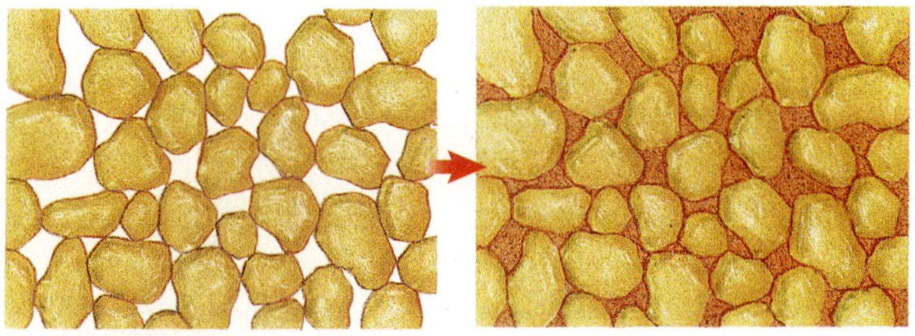

Figure 13 Sediments are cemented together as minerals crystallize between grains.

Cementation If sediments are large, like sand and pebbles, pressure alone can't make them stick together. Large sediments have to be cemented together. As water moves through soil and rock, it picks up materials released from minerals during weathering. The resulting solution of water and dissolved materials moves through open spaces between sediments. **Cementation,** which is shown in **Figure 13,** occurs when minerals such as quartz, calcite, and hematite are deposited between the pieces of sediment. These minerals, acting as natural cements, hold the sediment together like glue, making a detrital sedimentary rock.

Shape and Size of Sediments Detrital rocks have granular textures, much like granulated sugar. They are named according to the shapes and sizes of the sediments that form them. For example, conglomerate and breccia both form from large sediments, as shown in **Table 2.** If the sediments are rounded, the rock is called conglomerate. If the sediments have sharp angles, the rock is called breccia. The roundness of sediment particles depends on how far they have been moved by wind or water.

Table 2 Sediment Sizes and Detrital Rocks

Sediment	Clay	Silt	Sand	Gravel
Size Range	<0.004 mm	0.004–0.063 mm	0.063–2 mm	>2 mm
Example	Shale 	Siltstone 	Sandstone 	Conglomerate (shown) or Breccia 

Visual Learning

Figure 13 Why is a cementing material needed to hold coarse-grained sedimentary rocks together? Pressure alone can't make large grains stick together. A cementing material acts to hold them together.

Discussion

Sediment Sizes Have students imagine a swiftly-flowing river that empties into a quiet bay.

- Where in the bay would you expect to find the coarsest sediments being deposited? Explain. The coarsest sediments would be deposited nearest the mouth of the river. This is where the water loses velocity and drops the largest sediments in its load.
- Where would you expect to find the finest sediments being deposited? Explain. The finest sediments would be deposited farther from the mouth of the river. Because these sediments are lighter in weight than the coarser sediments, they stay suspended longer, even when the velocity of the water slows.
- Have students use **Table 2** to answer this question. If these sediments became cemented together, what type of rock would form in each location? Sandstone or conglomerate would form near the mouth of the river. Shale or siltstone would form farther out in the bay.

Science Journal

Observing Rock Exhibit a piece of sandstone. Have students examine the rock and write a brief geologic history that explains how the sandstone could have formed. Then have students hypothesize what might happen to the rock if it were exposed to weathering and erosion, or to heat and pressure. L2 LS **Logical-Mathematical** P

SECTION 4 Sedimentary Rocks F ◆ 51

Activity

Detrital Rocks Form groups of three students. Supply each group with containers of clay, silt, fine sand, coarse sand, and rounded and angular gravel. Make samples of shale, siltstone, sandstone, conglomerate, and breccia available to each group. Have students use a hand lens to match sediment type with the appropriate rock sample. One student in each group should present a report containing the conclusions drawn by the group. L2 COOP LEARN LS **Interpersonal**

INTEGRATE Career

Sedimentary Petrology People who study to become sedimentary petrologists focus on the formation of sedimentary rocks. They could pursue careers in industrial minerals, or in petroleum exploration, research, and development, among others. Other careers include continuing advanced research and becoming professors at colleges and universities. Graduate degrees are strongly recommended for industrial fields, and essential for academic fields.

Reading Check

Answer They form when dissolved minerals come out of solution or when liquid evaporates.

Figure 14 Although concrete strongly resembles conglomerate, concrete is not a rock because it does not occur in nature.

INTEGRATE Career

Sedimentary Petrology Research the work done by sedimentary petrologists. Include examples of careers in academia and in industry.

Materials Found in Sedimentary Rocks The gravel-sized sediments in conglomerate and breccia can consist of any type of rock or mineral. Often, they are composed of chunks of the minerals quartz and feldspar. They also can be pieces of rocks such as gneiss, granite, or limestone. The cement that holds the sediments together usually is made of quartz or calcite.

Have you ever looked at the concrete in sidewalks, driveways, and stepping stones? The concrete in **Figure 14** is made of gravel and sand grains that have been cemented together. Although the structure is similar to that of naturally occurring conglomerate, it cannot be considered a rock.

Sandstone is formed from smaller particles than conglomerates and breccias. Its sand-sized sediments can be just about any mineral, but they are usually grains of minerals such as quartz and feldspar that are resistant to weathering. Siltstone is similar to sandstone except it is made of smaller, silt-sized particles. Shale is a detrital sedimentary rock that is made mainly of clay-sized particles. Clay-sized sediments are compacted together by pressure from overlying layers.

Chemical Sedimentary Rocks

Chemical sedimentary rocks form when dissolved minerals come out of solution. You can show that salt is deposited in the bottom of a glass or pan when saltwater solution evaporates. In a similar way, minerals collect when seas or lakes evaporate. The deposits of minerals that come out of solution form sediments and rocks. For example, the sediment making up New Mexico's White Sands desert consists of pieces of a chemical sedimentary rock called rock gypsum. Chemical sedimentary rocks are different. They are not made from pieces of preexisting rocks.

Reading Check *How do chemical sedimentary rocks form?*

Cultural Diversity

Rock Art Native Americans used flint to make arrowheads, axes, and other tools that need a sharp edge. Flint is an extremely fine-grained, dense, glossy sedimentary rock made of quartz. It often is found embedded in deposits of chalk or limestone. Flint is an even-grained rock that breaks easily into smooth, curved flakes with sharp edges. This makes it ideal for the manufacture of cutting tools. Today people learn the art of flint napping as a craft. *Napping* is the method of striking a flint rock in such a way that a sharp-edged flake results. Have students find out what archaeologists have learned about the flint tools used by ancient Americans. Then have them make a classroom bulletin board showing the locations, types, and appearances of the flint artifacts. L2

Limestone Calcium carbonate is carried in solution in ocean water. When calcium carbonate ($CaCO_3$) comes out of solution as calcite and its many crystals grow together, limestone forms. Limestone also can contain other minerals and sediments, but it must be at least 50 percent calcite. Limestone usually is deposited on the bottom of lakes or shallow seas. Large areas of the central United States have limestone bedrock because seas covered much of the country for millions of years. It is hard to imagine Kansas being covered by ocean water, but it has happened several times throughout geological history.

Rock Salt When water that is rich in dissolved salt evaporates, it often deposits the mineral halite. Halite forms rock salt, shown in **Figure 15.** Rock salt deposits can range in thickness from a few meters to more than 400 m. Companies mine these deposits because rock salt is an important resource. It's used in the manufacturing of glass, paper, soap, and dairy products. The halite in rock salt is processed and used as table salt.

Figure 15 Rock salt is extracted from this mine in Germany. The same salt can be processed and used to season your favorite foods.

Organic Sedimentary Rocks

Rocks made of the remains of once-living things are called organic sedimentary rocks. One of the most common organic sedimentary rocks is fossil-rich limestone. Like chemical limestone, fossil-rich limestone is made of the mineral calcite. However, fossil-rich limestone mostly contains remains of once-living ocean organisms instead of only calcite that formed directly from ocean water.

Animals such as mussels, clams, corals, and snails make their shells from $CaCO_3$ that eventually becomes calcite. When they die, their shells accumulate on the ocean floor. When these shells are cemented together, fossil-rich limestone forms. If a rock is made completely of shell fragments that you can see, the rock is called coquina (koh KEE nuh).

Chalk Chalk is another organic sedimentary rock that is made of microscopic shells. When you write with naturally occurring chalk, you're crushing and smearing the calcite-shell remains of once-living ocean organisms.

IDENTIFYING Misconceptions

Salt Uses Some students may think that most salt is used as a food seasoning. About 50% of the salt sold in the United States is used as a winter deicer. About 15% is used in conditioning water, 15% is used by the chemical industry, 10% is used in agriculture, and about 10% is used as a food seasoning.

Quick Demo

Chemical Rocks

Materials pie pan, small beaker, salt, water

Estimated Time 5–10 minutes to set up; 5–10 minutes for observations the next day

Procedure Mix salt in a small beaker that is about 1/2 full of water to make a concentrated salt solution. Pour a thin layer of the solution in the pie pan. Leave undisturbed overnight. Ask students to predict what they will observe the next day. After they observe the pan, ask them to infer what happened. As the water evaporated, crystals of salt formed on the bottom of the pan.

Teacher FYI

Chalk Formation Many deposits of chalk formed during the Cretaceous Period, between 145 and 65 million years ago. The word *cretaceous* comes from the Latin word for chalk, which is *creta*.

SECTION 4 Sedimentary Rocks F ◆ 53

Discussion

Coal Types Display samples of peat, lignite, and bituminous and anthracite coal. Have students observe the peat to see that it is made of plant material. Demonstrate the relative ease with which you can break the lignite and the bituminous coal as opposed to the harder anthracite. Infer what happens to the hardness of coal as it changes from peat to lignite, bituminous, and anthracite. The coal becomes harder.

Applying Math

National Math Standards

Correlation to Mathematics Objectives

1, 2, 9

Teaching Strategy

Follow the steps in the example problem. Note that the plant matter required is approximately 3 times the thickness of the coal layer produced. Use this as a quick check for answers students get using the equation.

Answers to Practice Problems

1. 1.8 m of plant matter
2. 0.17 m of coal

Coal Another useful organic sedimentary rock is coal, shown in **Figure 16.** Coal forms when pieces of dead plants are buried under other sediments in swamps. These plant materials are chemically changed by microorganisms. The resulting sediments are compacted over millions of years to form coal, an important source of energy. Much of the coal in North America and Europe formed during a period of geologic time that is so named because of this important reason. The Carboniferous Period, which spans from approximately 360 to 286 million years ago, was named in Europe. So much coal formed during this interval of time that coal's composition—primarily carbon—was the basis for naming a geologic period.

Applying Math — Calculate Thickness

COAL FORMATION It took 300 million years for a layer of plant matter about 0.9 m thick to produce a bed of bituminous coal 0.3 m thick. Estimate the thickness of plant matter that produced a bed of coal 0.15 m thick.

Solution

1. *This is what you know:*
   - original thickness of plant matter = 0.9 m
   - original coal thickness = 0.3 m
   - new coal thickness = 0.15 m
2. *This is what you need to know:* thickness of plant matter needed to form 0.15 m of coal
3. *This is the equation you need to use:* (thickness of plant matter)/(new coal thickness) = (original thickness of plant matter)/(original coal thickness)
4. *Substitute the known values:* (? m plant matter)/(0.15 m coal) = (0.9 m plant matter)/(0.3 m coal)
5. *Solve the equation:* (? m plant matter) = (0.9 m plant matter)(0.15 m coal)/(0.3 m coal) = 0.45 m plant matter
6. *Check your answer:* Multiply your answer by the original coal thickness. Divide by the original plant matter thickness to get the new coal thickness.

Practice Problems

1. Estimate the thickness of plant matter that produced a bed of coal 0.6 m thick.
2. About how much coal would have been produced from a layer of plant matter 0.50 m thick?

Science Online — For more practice, visit bookf.msscience.com/math_practice

LAB DEMONSTRATION

Purpose to observe that some sedimentary rocks react with hydrochloric acid (HCl)

Materials samples of quartz sandstone, limestone, and shale; dilute HCl

Procedure Place a few drops of dilute HCl on each sample. **WARNING:** *Wear goggles and protect clothing and work surfaces.*

Expected Outcome Students will observe that the limestone sample fizzes readily.

Assessment

Which of the samples react to acid? Students should infer that the limestone sample contains a substance that reacts with HCl. Limestone reacts because it contains calcite.

Figure 16 This coal layer in Alaska is easily identified by its jet-black color, as compared with other sedimentary layers.

Another Look at the Rock Cycle

You have seen that the rock cycle has no beginning and no end. Rocks change continually from one form to another. Sediments can become so deeply buried that they eventually become metamorphic or igneous rocks. These reformed rocks later can be uplifted and exposed to the surface—possibly as mountains to be worn away again by erosion.

All of the rocks that you've learned about in this chapter formed through some process within the rock cycle. All of the rocks around you, including those used to build houses and monuments, are part of the rock cycle. Slowly, they are all changing, because the rock cycle is a continuous, dynamic process.

section 4 review

Summary

Formation of Sedimentary Rocks

- Sedimentary rocks form as layers, with older layers near the bottom of an undisturbed stack.

Classifying Sedimentary Rocks

- To classify a sedimentary rock, determine its composition and texture.

Detrital Sedimentary Rocks

- Rock and mineral fragments make up detrital rocks.

Chemical Sedimentary Rocks

- Chemical sedimentary rocks form from solutions of dissolved minerals.

Organic Sedimentary Rocks

- The remains of once-living organisms make up organic sedimentary rocks.

Self Check

1. **Identify** where sediments come from.
2. **Explain** how compaction is important in the formation of coal.
3. **Compare and contrast** detrital and chemical sedimentary rock.
4. **List** chemical sedimentary rocks that are essential to your health or that are used to make life more convenient. How is each used?
5. **Think Critically** Explain how pieces of granite and slate could both be found in the same conglomerate. How would the granite and slate pieces be held together?

Applying Math

6. **Calculate Ratios** Use information in **Table 2** to estimate how many times larger the largest grains of silt and sand are compared to the largest clay grains.

Science Online bookf.msscience.com/self_check_quiz

Make a Model

Rock Cycle Have students make a model of the rock cycle. Encourage them to be creative and to present their models in class. L2

3 Assess

DAILY INTERVENTION

Check for Understanding

Logical-Mathematical Have students explain which is older: detrital sedimentary rock or the sediments it contains. The sediments are older; they had to be present first in order for the sedimentary rock to form from them. L2

Reteach

Rock Properties Have students construct a table that summarizes the characteristics of detrital, chemical, and organic sedimentary rocks. L2

Assessment

Process Assess students' abilities to sequence events by having them make an events chain for the process of changing a metamorphic rock to a sedimentary rock. Use **Performance Assessment in the Science Classroom,** p. 163.

section 4 review

1. weathered rock and mineral material
2. Compaction concentrates transformed plant matter to form an important energy resource.
3. Detrital rocks are made from the pieces of weathered rocks; chemical rocks form as dissolved minerals come out of solution or as liquid evaporates.
4. Possible answers: salt: preservation, seasoning, manufacture of glass, paper and soap; limestone: material for buildings, roads, and statues
5. Granite and slate could be weathered in separate locations, then transported and deposited in the same location. Mineral cements would hold grains together.
6. The largest grains of silt are about 16 times larger than clay, and the largest grains of sand are about 500 times larger than clay.

BENCH TESTED

Real-World Question

Purpose Students use various characteristics to classify sedimentary rocks. L2 ELL

COOP LEARN LS **Interpersonal**

Process Skills observe and infer, communicate, classify, form operational definitions, interpret qualitative data, compare and contrast

Time Required 45 minutes

Procedure

Safety Precautions Have students wear goggles, aprons, and rubber gloves when handling acid in the lab. **WARNING:** *When mixing concentrated acid and water, always add the acid to the water (never add water to acid). Mix in a well-ventilated area.*

Teaching Strategies

- Suitable rocks include: fossil-rich limestone, rock salt, shale, siltstone, sandstone, and conglomerate.
- Dilute (5%) HCl can be made by mixing 1 part concentrated HCl with 19 parts tap water.
- Make several rock and mineral identification books available in class. The books should identify the characteristics of rocks that students encounter in class.

LAB

Sedimentary Rocks

Goals

- **Observe** sedimentary rock characteristics.
- **Compare and contrast** sedimentary rock textures.
- **Classify** sedimentary rocks as detrital, chemical, or organic.

Materials

unknown sedimentary rock samples
marking pen
5% hydrochloric acid (HCl) solution
dropper
paper towels
water
magnifying lens
metric ruler

Safety Precautions

WARNING: *HCl is an acid and can cause burns. Wear goggles and a lab apron. Rinse spills with water and wash hands afterward.*

Sedimentary rocks are formed by compaction and cementation of sediment. Because sediment is found in all shapes and sizes, do you think these characteristics could be used to classify detrital sedimentary rocks? Sedimentary rocks also can be classified as chemical or organic.

Real-World Question

How are rock characteristics used to classify sedimentary rocks as detrital, chemical, or organic?

Procedure

1. Make a Sedimentary Rock Samples chart in your Science Journal similar to the one shown on the next page.
2. **Determine** the sizes of sediments in each sample, using a magnifying lens and a metric ruler. Using **Table 2,** classify any grains of sediment in the rocks as gravel, sand, silt, or clay. In general, the sediment is silt if it is gritty and just barely visible, and clay if it is smooth and if individual grains are not visible.
3. Place a few drops of 5% HCl solution on each rock sample. Bubbling on a rock indicates the presence of calcite.
4. **Examine** each sample for fossils and describe any that are present.
5. **Determine** whether each sample has a granular or nongranular texture.

Alternative Inquiry Lab

Classification System To make this Lab an Inquiry Lab, ask students to devise a classification system of their own to categorize a second set of unknown sedimentary rocks. Then they should use their own systems to help them identify the rocks. Have them critique their classification methods according to how helpful they were in arriving at correct identifications. L3

Science Journal

Emphasizing Safety In preparation for this lab exercise, demonstrate how to check samples for reaction with HCl. Be sure to wear goggles and an apron. Ask students to write a description of the reaction in their Science Journals. Also, have students explain the need for wearing goggles and an apron while performing this activity. L2

ELL LS **Linguistic**

56 ◆ F CHAPTER 2 Rocks

Using Scientific Methods

Sedimentary Rock Samples					
Sample	Observations	Minerals or Fossils Present	Sediment Size	Detrital, Chemical, or Organic	Rock Name
A	fizzes in acid	calcite, fossils	varies	organic	fossil-rich limestone
B	feels gritty	quartz, feldspar	sand	detrital	sandstone
C	breaks in layers	kaolinite, quartz	clay	detrital	shale
D	sediment easily seen	pebbles of any rock or mineral	pebble and sand	detrital	shale
E	soft	halite	not grainy	chemical	rock salt

Analyze Your Data

1. **Classify** your samples as detrital, chemical, or organic.
2. **Identify** each rock sample.

Conclude and Apply

1. **Explain** why you tested the rocks with acid. What minerals react with acid?
2. **Compare and contrast** sedimentary rocks that have a granular texture with sedimentary rocks that have a nongranular texture.

Communicating Your Data

Compare your conclusions with those of other students in your class. **For more help, refer to the Science Skill Handbook.**

Analyze Your Data

Expected Outcome Students will identify the rock samples based on characteristics such as hardness, texture, layering, size and feel of grains, and reaction to acid. They also will classify them as detrital, chemical, or organic.

Error Analysis Rocks could be misclassified if students were not able to correctly identify grain size or the relative hardness of samples, or if they had trouble getting the samples to react in the presence of acid. Some carbonates must be powdered before they readily react with acid.

Answers to Questions

1. Classifications will vary according to available samples. As a general rule, detrital specimens have a granular, or grainy texture, organic specimens contain fossils or other remains of organisms, and chemical specimens often are not grainy and contain primarily one mineral such as limestone (calcite), rock salt (halite) and rock gypsum (gypsum).
2. Check students' identifications by comparing their observations to text and photos in a rock and mineral field guide.

Conclude and Apply

1. to determine whether calcite was present; carbonates
2. Rocks of both textures form from sedimentary processes. Rocks with a granular, or clastic texture are made of pieces of other rocks, minerals, and/or shells. Rocks with a nongranular texture are formed by chemical or organic means.

Assessment

Performance Provide students with large samples of unidentified rocks. Ask them to determine whether the rocks are sedimentary, and if so, whether they are detrital, chemical, or organic. Use **Performance Assessment in the Science Classroom,** p. 89.

Communicating Your Data

Encourage students to use a spreadsheet program to display the characteristics of sedimentary rock samples used in this activity.

TIME

Content Background

According to Anangu, Earth's surface once was featureless. Places like Uluru did not exist until ancestral beings—in the form of people, plants, and animals—started to travel across the land. As they traveled from one area to another, these ancestral beings formed the features of the landscape. The travels and activities of the ancestral beings linked places throughout the country by iwara (paths or tracks). Iwara often link places separated by hundreds of kilometers. Uluru represents one meeting point in a network of such ancestral tracks.

Anangu knowledge of their land is based on Tjukurpa, or Dream Time. This knowledge is passed on through ceremony, song, dance, and art. The four major elements of the Tjukurpa for Uluru are Kuniya (woma python), Liru (poisonous snake), Kurpany (monster dog-like creature), and Mala (hare-wallaby). These beings and the stories associated with them connect Anangu with the land around them. To the Anangu, the huge grooves in rocks of Uluru were made by Kuniya as she came and went on her search for food.

Discussion

Tourism What are the advantages and disadvantages to the Anangu of having tourists visit Uluru? Possible answer: advantages—financial rewards, opportunity to explain their beliefs to tourists; disadvantages—tourists climbing their sacred rock; tourists not respecting their beliefs and traditions.

TIME SCIENCE AND Society

SCIENCE ISSUES THAT AFFECT YOU!

Australia's controversial rock star

One of the most famous rocks in the world is causing serious problems for Australians

Uluru (yew LEW rew), also known as Ayers Rock, is one of the most popular tourist destinations in Australia. This sandstone skyscraper is more than 8 km around, over 300 m high, and extends as much as 4.8 km below the surface. One writer describes it as an iceberg in the desert. Geologists hypothesize that the mighty Uluru rock began forming 550 million years ago during Precambrian time. That's when large mountain ranges started to form in Central Australia.

For more than 25,000 years, this geological wonder has played an important role in the lives of the Aboriginal peoples, the Anangu (a NA noo). These native Australians are the original owners of the rock and have spiritual explanations for its many caves, holes, and scars.

Athlete Nova Benis-Kneebone had the honor of receiving the Olympic torch near the sacred Uluru and carried it partway to the Olympic stadium.

Tourists Take Over

In the 1980s, some 100,000 tourists visited—and many climbed—Uluru. In 2000, the rock attracted about 400,000 tourists. The Anangu take offense at anyone climbing their sacred rock. However, if climbing the rock were outlawed, tourism would be seriously hurt. That would mean less income for Australians.

To respect the Anangu's wishes, the Australian government returned Ayers Rock to the Anangu in 1985 and agreed to call it by its traditional name. The Anangu leased back the rock to the Australian government until the year 2084, when its management will return to the Anangu. Until then, the Anangu will collect 25 percent of the money people pay to visit the rock.

The Aboriginal people encourage tourists to respect their beliefs. They offer a walking tour around the rock, and they show videos about Aboriginal traditions. The Anangu sell T-shirts that say "I *didn't* climb Uluru." They hope visitors to Uluru will wear the T-shirt with pride and respect.

Write Research a natural landmark or large natural land or water formation in your area. What is the geology behind it? When was it formed? How was it formed? Write a folktale that explains its formation. Share your folktale with the class.

Science Online
For more information, visit bookf.msscience.com/time

Write Have students examine physical maps of your state to locate natural landforms and water formations. After a site is selected, have students research the geology of the formation to answer the questions in the student text before writing their folktales. Before sharing their folktales, have students explain the formation's geologic history to the class. L2

Resources for Teachers and Students

A Natural History of Australia, by Tim M. Berra, Academic Press, 1998

Australian Dreaming: 40,000 Years of Aboriginal History, edited by Jennifer Isaacs, Lansdowne Press, 1980

Reviewing Main Ideas

Section 1 The Rock Cycle

1. A rock is a mixture of one or more minerals, rock fragments, organic matter, or volcanic glass.
2. The rock cycle includes all processes by which rocks form.

Section 2 Igneous Rocks

1. Magma and lava are molten materials that harden to form igneous rocks.
2. Intrusive igneous rocks form when magma cools slowly below Earth's surface. Extrusive igneous rocks form when lava cools rapidly at the surface.
3. The compositions of most igneous rocks range from granitic to andesitic to basaltic.

Section 3 Metamorphic Rocks

1. Heat, pressure, and fluids can cause metamorphic rocks to form.
2. Slate and gneiss are examples of foliated metamorphic rocks. Quartzite and marble are examples of nonfoliated metamorphic rocks.

Section 4 Sedimentary Rocks

1. Detrital sedimentary rocks form when fragments of rocks and minerals are compacted and cemented together.
2. Chemical sedimentary rocks come out of solution or are left behind by evaporation.
3. Organic sedimentary rocks contain the remains of once-living organisms.

Visualizing Main Ideas

Copy and complete the following concept map on rocks. Use the following terms: organic, metamorphic, foliated, extrusive, igneous, *and* chemical.

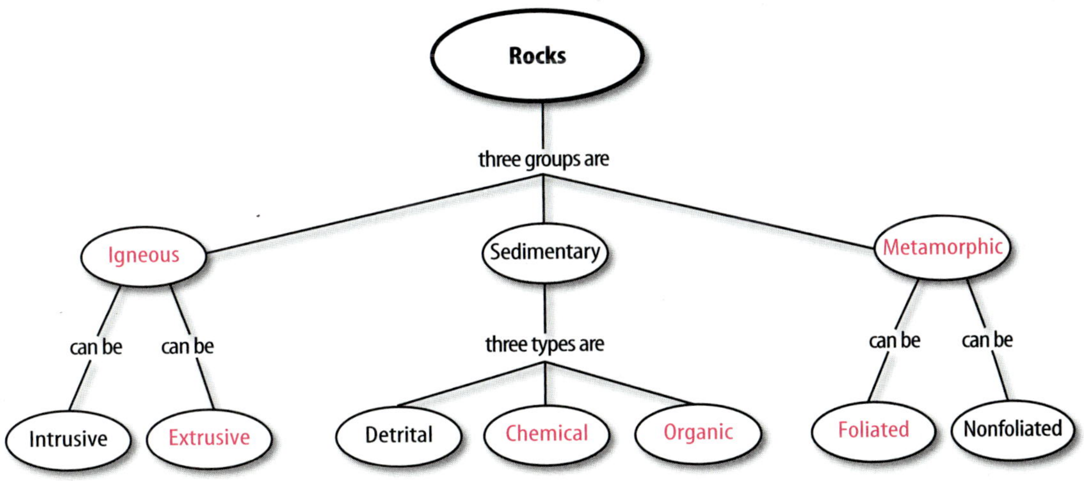

Science Online bookf.msscience.com/interactive_tutor

chapter 2 Study Guide

Reviewing Main Ideas

Summary statements can be used by students to review the major concepts of the chapter.

Visualizing Main Ideas

See student page.

Science Online

Visit bookf.msscience.com
/self_check_quiz
/interactive_tutor
/vocabulary_puzzlemaker
/chapter_review
/standardized_test

Assessment Transparency

For additional assessment questions, use the *Assessment Transparency* located in the transparency book.

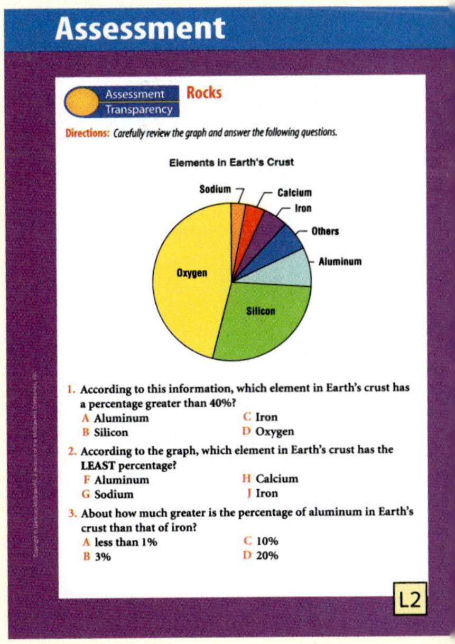
Assessment

Assessment Transparency **Rocks**

Directions: *Carefully review the graph and answer the following questions.*

1. According to this information, which element in Earth's crust has a percentage greater than 40%?
   A Aluminum C Iron
   B Silicon D Oxygen
2. According to the graph, which element in Earth's crust has the LEAST percentage?
   F Aluminum H Calcium
   G Sodium J Iron
3. About how much greater is the percentage of aluminum in Earth's crust than that of iron?
   A less than 1% C 10%
   B 3% D 20%

L2

CHAPTER STUDY GUIDE F ◆ 59

Using Vocabulary

1. Foliated metamorphic rocks are layered; nonfoliated metamorphic rocks are not layered.
2. Compaction is the pressing of mineral grains together during the formation of sedimentary rock; cementation is the gluing of grains together by dissolved mineral matter coming out of solution.
3. Sediment is weathered pieces of solid Earth materials; lava is molten rock material that reaches Earth's surface through volcanic activity.
4. Extrusive rock forms above Earth's surface; intrusive rock forms below Earth's surface.
5. A rock is a mixture of minerals, volcanic glass, organic matter, or other materials; the rock cycle describes how one type of rock can change into another.
6. Igneous rock forms when magma or lava solidifies; sedimentary rock forms from compaction and cementation of sediment or when minerals form from solution; metamorphic rock forms when any type of rock is changed by heat, pressure, or fluids inside Earth.
7. Sediment consists of grains of weathered and eroded rock; sedimentary rock is formed by the compaction and cementation of sediment or when minerals form from solution.
8. Lava is molten rock that has reached Earth's surface; igneous rock forms from magma or lava that has cooled and solidified.
9. Rock is a mixture of minerals, volcanic glass, organic matter, or other materials; sediment is composed of weathered pieces of rock and other Earth materials.
10. Basaltic magma is low in silica and high in magnesium and iron compared to granitic magma.

Checking Concepts

11. D 14. C 17. C
12. D 15. D 18. B
13. A 16. B 19. A

chapter 2 Review

Using Vocabulary

basaltic p. 43	lava p. 40
cementation p. 51	metamorphic rock p. 45
compaction p. 50	nonfoliated p. 48
extrusive p. 41	rock p. 36
foliated p. 47	rock cycle p. 37
granitic p. 43	sediment p. 49
igneous rock p. 40	sedimentary rock p. 49
intrusive p. 41	

Explain the difference between the vocabulary words in each of the following sets.

1. foliated—nonfoliated
2. cementation—compaction
3. sediment—lava
4. extrusive—intrusive
5. rock—rock cycle
6. metamorphic rock—igneous rock—sedimentary rock
7. sediment—sedimentary rock
8. lava—igneous rock
9. rock—sediment
10. basaltic—granitic

Checking Concepts

Choose the word or phrase that best answers the question.

11. Why does magma tend to rise toward Earth's surface?
    A) It is more dense than surrounding rocks.
    B) It is more massive than surrounding rocks.
    C) It is cooler than surrounding rocks.
    D) It is less dense than surrounding rocks.

12. During metamorphism of granite into gneiss, what happens to minerals?
    A) They partly melt.
    B) They become new sediments.
    C) They grow smaller.
    D) They align into layers.

13. Which rock has large mineral grains?
    A) granite **C)** obsidian
    B) basalt **D)** pumice

14. Which type of rock is shown in this photo?
    A) foliated
    B) nonfoliated
    C) intrusive
    D) extrusive

15. What do igneous rocks form from?
    A) sediments **C)** gravel
    B) mud **D)** magma

16. What sedimentary rock is made of large, angular pieces of sediments?
    A) conglomerate **C)** limestone
    B) breccia **D)** chalk

17. Which of the following is an example of a detrital sedimentary rock?
    A) limestone **C)** breccia
    B) evaporite **D)** chalk

18. What is molten material at Earth's surface called?
    A) limestone **C)** breccia
    B) lava **D)** granite

19. Which of these is an organic sedimentary rock?
    A) coquina **C)** rock salt
    B) sandstone **D)** conglomerate

Science Online bookf.msscience.com/vocabulary_puzzlemaker

Use the Exam*View*® Pro Testmaker CD-ROM to:

- create multiple versions of tests
- create modified tests with one mouse click for inclusion students
- edit existing questions and add your own questions
- build tests aligned with state standards using built-in State Curriculum Tags
- change English tests to Spanish with one mouse click and vice versa

Thinking Critically

20. Infer Granite, pumice, and scoria are igneous rocks. Why doesn't granite have airholes like the other two?

21. Infer why marble rarely contains fossils.

22. Predict Would you expect quartzite or sandstone to break more easily? Explain your answer.

23. Compare and contrast basaltic and granitic magmas.

24. Form Hypotheses A geologist was studying rocks in a mountain range. She found a layer of sedimentary rock that had formed in the ocean. Hypothesize how this could happen.

25. Concept Map Copy and complete the concept map shown below. Use the following terms and phrases: *magma, sediments, igneous rock, sedimentary rock, metamorphic rock.* Add and label any missing arrows.

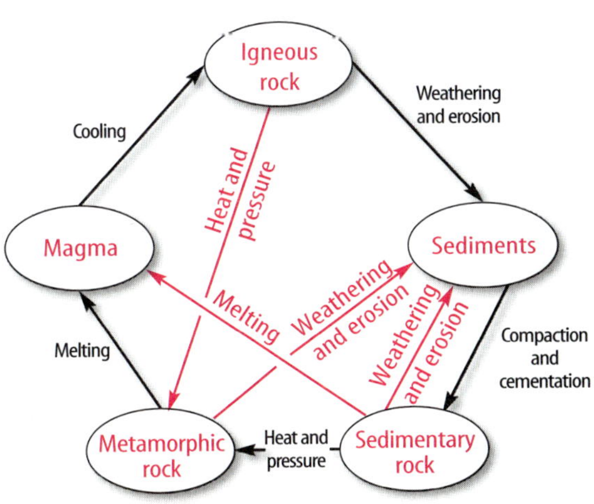

Performance Activities

26. Poster Collect a group of rocks. Make a poster that shows the classifications of rocks, and glue your rocks to the poster under the proper headings. Describe your rocks and explain where you found them.

Applying Math

27. Grain Size Assume that the conglomerate shown on the second page of the "Sedimentary Rocks" lab is one-half of its actual size. Determine the average length of the gravel in the rock.

28. Plant Matter Suppose that a 4-m layer of plant matter was compacted to form a coal layer 1 m thick. By what percent has the thickness of organic material been reduced?

Use the graph below to answer questions 29 and 30.

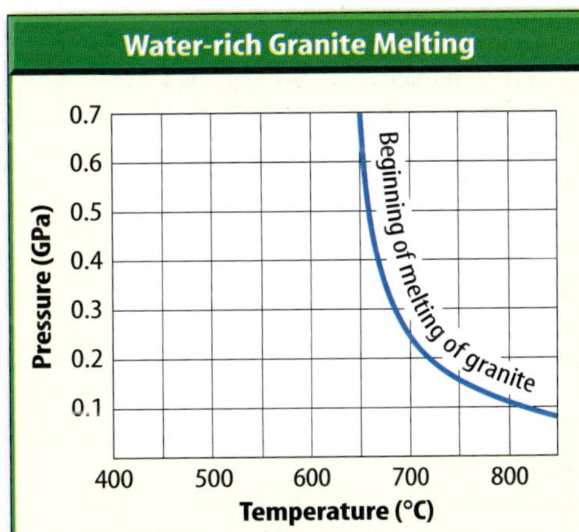

29. Melting Granite Determine the melting temperature of a water-rich granite at a pressure of 0.2 GPa.

Pressure conversions:
1 GPa, or gigapascal, = 10,000 bars
1 bar = 0.9869 atmospheres

30. Melting Pressure At about what pressure will a water-rich granite melt at 680°C?

Science Online bookf.msscience.com/chapter_review

Assessment Resources

Reproducible Masters

Chapter *Fast File* Resources
Chapter Review, pp. 35–36
Chapter Tests, pp. 37–40
Assessment Transparency Activity, p. 49

Glencoe Science Web site
Chapter Review Test
Standardized Test Practice

Glencoe Technology
Assessment Transparency
Exam*View*® Pro Testmaker
MindJogger Videoquiz
Interactive Chalkboard

chapter 2 Review

Thinking Critically

20. Because granite is an intrusive rock. Pumice and scoria form above ground where escaping gases can form bubbles in lava.

21. Marble is metamorphic. The processes that form it often destroy organic remains.

22. Sandstone; quartz grains in sandstone are held together by cement, whereas quartz grains in quartzite have grown in size and have become interlocking.

23. Both are molten materials that form deep within Earth. Basaltic magma is fluid and is rich in iron and magnesium. Granitic magma is more viscous and is rich in silica compared with basaltic magma.

24. Sample hypothesis: sedimentary layer forms in ocean basin; more sediments accumulate; mountain building processes uplift sedimentary layers originally formed in ocean.

25. See student page.

Performance Activities

26. Posters should classify rocks as igneous, sedimentary, or metamorphic, and list their names. Use **Performance Assessment in the Science Classroom,** p. 145.

Applying Math

National Math Standards

1, 2, 9

27. approximately 9 mm. Accept reasonable answers.

28. 75 percent

29. approximately 720 °C

30. approximately 0.3 GPa

FAST FILE

Answer Sheet A practice answer sheet can be found at bookf.msscience.com/answer_sheet.

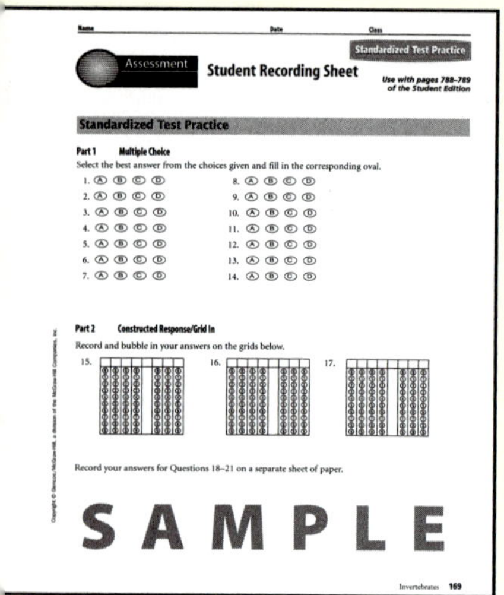

Part 1 Multiple Choice

1. C
2. A
3. C
4. B
5. C
6. B
7. A
8. B
9. D

Part 2 Short Response

10. A rock is a mixture of minerals, volcanic glass, organic matter, or other materials. A rock is mixture; minerals are not.
11. Students should state that igneous rocks that cool quickly are fine grained, and igneous rocks that cool slowly are coarse grained. Some other factors, such as magma viscosity, also affect grain size.
12. Foliation is planar layering caused by the orientation and distribution of mineral grains in a metamorphic rock. Foliation is caused by pressure-induced deformation within Earth.

Part 1 Multiple Choice

Record your answers on the answer sheet provided by your teacher or on a sheet of paper.

Use the illustration below to answer question 1.

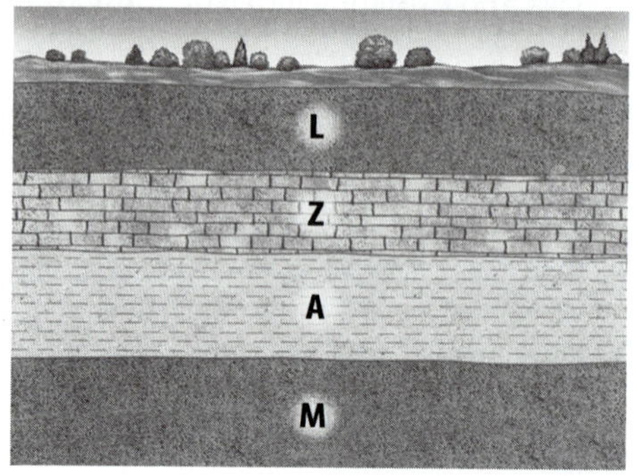

1. These layers of sedimentary rock were not disturbed after they were deposited. Which layer was deposited first?
A. layer L **C.** layer M
B. layer Z **D.** layer A

2. Who realized that rocks undergo changes through long periods of time after observing rocks at Siccar Point, Scotland?
A. James Hutton **C.** Galileo Galilei
B. Neil Armstrong **D.** Albert Einstein

3. During which process do minerals precipitate in the spaces between sediment grains?
A. compaction **C.** cementation
B. weathering **D.** conglomerate

4. Which rock often is sculpted to create statues?
A. shale **C.** coquina
B. marble **D.** conglomerate

Test-Taking Tip

Careful Reading Read each question carefully for full understanding.

5. Which of the following rocks is a metamorphic rock?
A. shale **C.** slate
B. granite **D.** pumice

6. Which rock consists mostly of pieces of seashell?
A. sandstone **C.** pumice
B. coquina **D.** granite

Use the diagram below to answer questions 7–9.

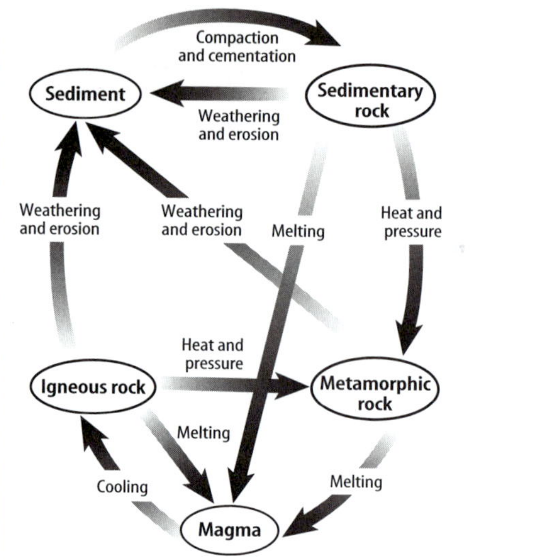

7. Which process in the rock cycle causes magma to form?
A. melting **C.** weathering
B. erosion **D.** cooling

8. What forms when rocks are weathered and eroded?
A. igneous rock **C.** sedimentary rock
B. sediment **D.** metamorphic rock

9. Which type of rock forms because of high heat and pressure without melting?
A. igneous rock **C.** sedimentary rock
B. intrusive rock **D.** metamorphic rock

Part 2 Short Response/Grid In

13. Chemical sedimentary rocks form when minerals precipitate from solution. Rock salt forms when halite precipitates from evaporating seawater.
14. Some sedimentary rocks contain fossils because the fossils were deposited, along with any other sediment in the rock, and lithified by compaction and cementation.
15. Both chemical sedimentary rocks and cement in detrital rocks can form as dissolved mineral material precipitates out of solution.
16. about 110 km; accept reasonable answers
17. Temperature increases with depth.

Part 2 Short Response/Grid In

Record your answers on the answer sheet provided by your teacher or on a sheet of paper.

10. What is a rock? How is a rock different from a mineral?

11. Explain why some igneous rocks are coarse and others are fine.

12. What is foliation? How does it form?

13. How do chemical sedimentary rocks, such as rock salt, form?

14. Why do some rocks contain fossils?

15. How is the formation of chemical sedimentary rocks similar to the formation of cement in detrital sedimentary rocks?

Use the graph below to answer questions 16–17.

Temperature Beneath the Continents

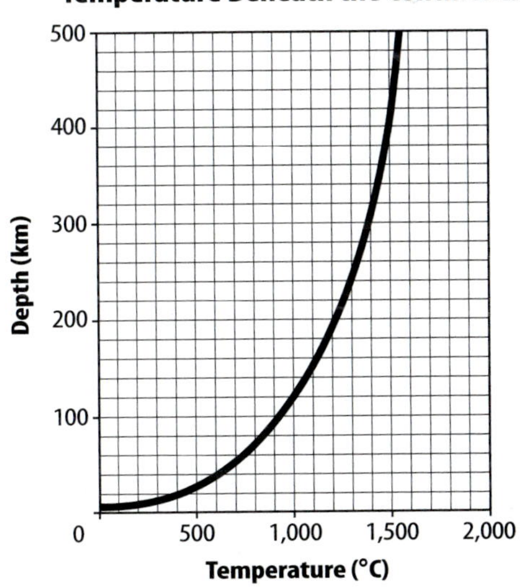

16. According to the graph, about how deep below a continent does the temperature reach 1,000°C?

17. In general, what happens to temperature as depth below Earth's surface increases?

Part 3 Open Ended

Record your answers on a sheet of paper.

Use the table below to answer questions 18 and 19.

Magma Type	Basaltic	Andesitic	Granitic
Intrusive	Gabbro	Diorite	Granite
Extrusive	Basalt or scoria	Andesite	Rhyolite, pumice, or obsidian

18. Copy the table on your paper. Then, fill in the empty squares with a correct rock name.

19. Explain how igneous rocks are classified.

20. Explain how loose sediment can become sedimentary rock.

21. Why does pressure increase with depth in Earth? How does higher pressure affect rocks?

22. Why is slate sometimes used as shingles for roofs? What other rocks are used for important purposes in society?

23. How are organic sedimentary rocks different from other rocks? List an example of an organic sedimentary rock.

24. Why is the rock cycle called a cycle?

25. A geologist found a sequence of rocks in which 200-million-year-old shales were on top of 100-million-year-old sandstones. Hypothesize how this could happen.

26. Explain why coquina could be classified in more than one way.

Part 3 Open Ended

18. See student page.

19. Igneous rocks are classified according to composition and crystal size.

20. The sediment can be buried, compacted, and/or cemented to become sedimentary rock.

21. Pressure increases with depth because rocks deeper in Earth have more overburden above them. Higher pressure can cause metamorphism to occur. It also affects melting. Higher pressure inhibits melting. This is why much of Earth's mantle is solid in spite of the high temperature.

22. Slate is flat like a shingle. It also is very impermeable to water. Other rocks include marble, limestone, sandstone, granite, and many others.

23. Organic sedimentary rocks could not form if life did not exist on Earth. Possible examples include coal, chalk, coquina, fossiliferous limestone.

24. It has no beginning or end. Matter is continually cycled and recycled by the processes included in the rock cycle.

25. Sample hypothesis: Tectonic forces caused the older layer to be thrust over the younger layer.

26. Coquina is classified as an organic rock, but it also resembles a detrital rock. Its granular texture results from weathered pieces of shells cemented together.

Rubrics

For more help evaluating open-ended assessment questions, see the rubric on p. 10T.

chapter 3 Organizer

Section/Objectives	Standards		Labs/Features
	National	State/Local	
Chapter Opener	See pp. 9T–10T for a Key to Standards.		**Launch Lab:** Finding Energy Reserves, p. 65 **Foldables,** p. 65
Section 1 Nonrenewable Energy Resources 2 sessions 1 block 1. **Identify** examples of nonrenewable energy resources 2. **Describe** the advantages and disadvantages of using fossil fuels. 3. **Explain** the advantages and disadvantages of using nuclear energy.	National Content Standards: UCP.1–3, UCP.5, A.1, A.2, D.1, D.2, F.2, F.5		**Integrate Life Science,** p. 67 **Science Online,** p. 71 **Visualizing Methane Hydrates,** p. 72 **MiniLAB:** Practicing Energy Conservation, p. 73
Section 2 Renewable Energy Resources 1 session 0.5 block 4. **Compare and contrast** inexhaustible and renewable energy resources. 5. **Explain** why inexhaustible and renewable resources are used less than nonrenewable resources.	National Content Standards: UCP.1–3, UCP.5, A.1, A.2, D.1, F.1, F.2		**Integrate Career,** p. 77 **Science Online,** p. 79 **Lab:** Soaking up Solar Energy, p. 82
Section 3 Mineral Resources 3 sessions 1.5 blocks 6. **Explain** the conditions needed for a mineral to be classified as an ore. 7. **Describe** how market conditions can cause a mineral to lose its value as an ore. 8. **Compare and contrast** metallic and nonmetallic mineral resources.	National Content Standards: UCP.1–3, UCP.5, A.1, A.2, D.1, D.3, F.2		**MiniLAB:** Observing the Effects of Insulation, p. 85 **Applying Science:** Why should you recycle?, p. 86 **Lab:** Home Sweet Home, p. 88 **Oops! Accidents in Science:** Black Gold!, p. 90

Glencoe Exclusive! TeacherWorks™ All-In-One Planner and Resource Center

Lab Materials	Reproducible Resources	Section Assessment	Technology
Launch Lab: sandstone, shale, shallow baking pan, dropper, cooking oil, timer	**Chapter *FAST FILE* Resources** Foldables Worksheet, p. 17 Directed Reading Overview, p. 19 Note-taking Worksheets, pp. 33–35	Glencoe's Assessment Advantage	TeacherWorks includes: • Interactive Teacher Edition • Lesson Planner with calendar • Access to all program blacklines • Correlations to standards • Web links
MiniLAB: electric meter at home *Need materials?* Contact Science Kit at 1-800-828-7777 or www.sciencekit.com on the Internet.	**Chapter *FAST FILE* Resources** Transparency Activity, p. 44 MiniLAB, p. 3 Enrichment, p. 30 Reinforcement, p. 27 Transparency Activity, pp. 47–48 Lab Activity, pp. 9–12 Directed Reading, p. 20 **Cultural Diversity,** pp. 45, 49	**Portfolio** Cultural Diversity, p. 68 Curriculum Connection, p. 71 **Performance** MiniLAB, p. 73 Applying Math, p. 75 **Content** Section Review, p. 75	Section Focus Transparency Teaching Transparency Virtual Labs CD-ROM Guided Reading Audio Program Interactive Chalkboard CD-ROM
Lab: dry black, brown, and sandy white soils, thermometers (3), ring stand, graph paper, colored pencils (3), metric ruler, clear-glass or plastic dishes (3), 200-watt gooseneck lamp, watch or clock with second hand	**Chapter *FAST FILE* Resources** Transparency Activity, p. 45 Enrichment, p. 31 Reinforcement, p. 28 Directed Reading, p. 20 Lab Activity, pp. 13–16 Lab Worksheet, pp. 5–6 **Earth Science Critical Thinking/ Problem Solving,** pp. 10, 17	**Portfolio** Activity, p. 79 **Performance** Applying Skills, p. 81 **Content** Section Review, p. 81	Section Focus Transparency Virtual Labs CD-ROM Guided Reading Audio Program Interactive Chalkboard CD-ROM Video Lab
MiniLAB: warm and cold water, thermos bottle, ice, glasses (2), thermal cup holder, cup, thermometers (or temperature probes) (4) **Lab:** paper, ruler, pencils, cardboard, glue, aluminum foil	**Chapter *FAST FILE* Resources** Transparency Activity, p. 46 MiniLAB, p. 4 Enrichment, p. 32 Reinforcement, p. 29 Directed Reading, pp. 21, 22 Lab Worksheet, pp. 7–8 **Lab Management and Safety,** p. 71	**Portfolio** Differentiated Instruction, p. 86 **Performance** MiniLAB, p. 85 Applying Science, p. 86 Applying Skills, p. 87 **Content** Section Review, p. 87	Section Focus Transparency Virtual Labs CD-ROM Guided Reading Audio Program Interactive Chalkboard CD-ROM

End of Chapter Assessment

Glencoe's Assessment Advantage

Blackline Masters	Technology	Professional Series
Chapter *FAST FILE* Resources Chapter Review, pp. 37–38 Chapter Tests, pp. 39–42 **Standardized Test Practice,** pp. 15–18	MindJogger Videoquiz Virtual Labs CD-ROM Exam*View*® Pro Testmaker TeacherWorks CD-ROM Interactive Chalkboard CD-ROM	**Performance Assessment in the Science Classroom (PASC)**

Transparencies

Section Focus

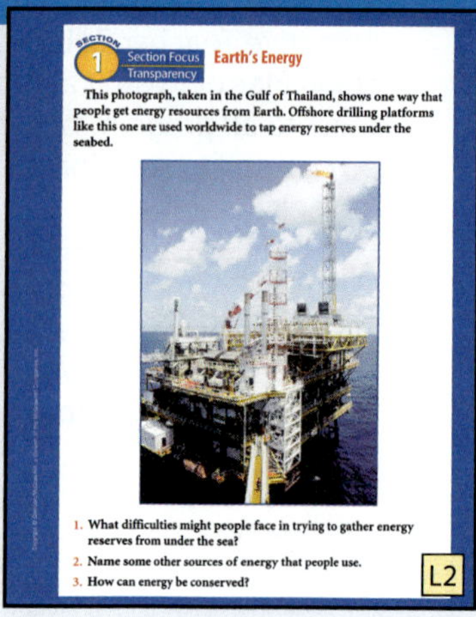
Section 1 Section Focus Transparency — Earth's Energy

This photograph, taken in the Gulf of Thailand, shows one way that people get energy resources from Earth. Offshore drilling platforms like this one are used worldwide to tap energy reserves under the seabed.

1. What difficulties might people face in trying to gather energy reserves from under the sea?
2. Name some other sources of energy that people use.
3. How can energy be conserved?

L2

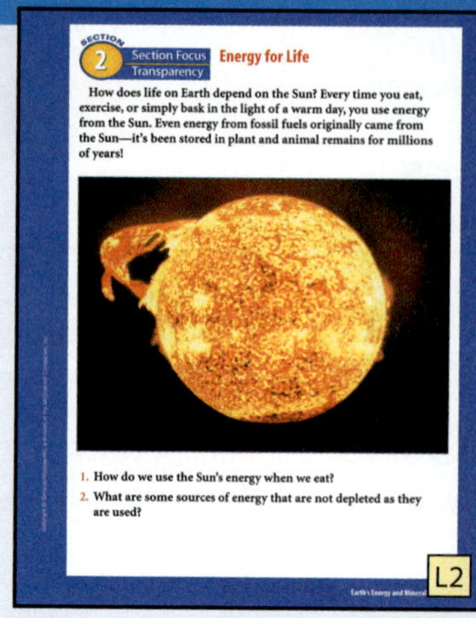
Section 2 Section Focus Transparency — Energy for Life

How does life on Earth depend on the Sun? Every time you eat, exercise, or simply bask in the light of a warm day, you use energy from the Sun. Even energy from fossil fuels originally came from the Sun—it's been stored in plant and animal remains for millions of years!

1. How do we use the Sun's energy when we eat?
2. What are some sources of energy that are not depleted as they are used?

L2

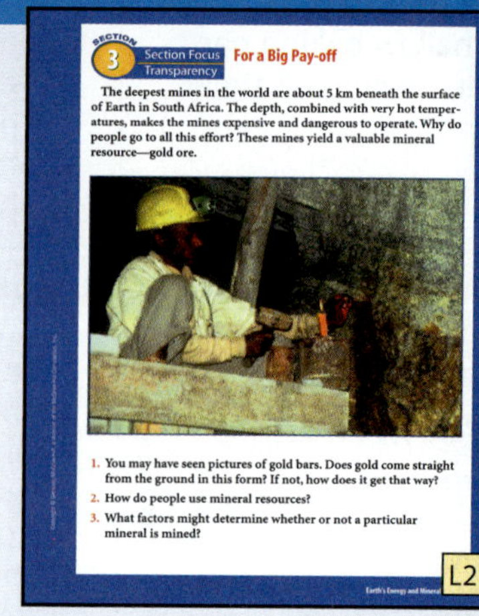
Section 3 Section Focus Transparency — For a Big Pay-off

The deepest mines in the world are about 5 km beneath the surface of Earth in South Africa. The depth, combined with very hot temperatures, makes the mines expensive and dangerous to operate. Why do people go to all this effort? These mines yield a valuable mineral resource—gold ore.

1. You may have seen pictures of gold bars. Does gold come straight from the ground in this form? If not, how does it get that way?
2. How do people use mineral resources?
3. What factors might determine whether or not a particular mineral is mined?

L2

This is a representation of key blackline masters available in the Teacher Classroom Resources. See Resource Manager boxes within the chapter for additional information.

Assessment

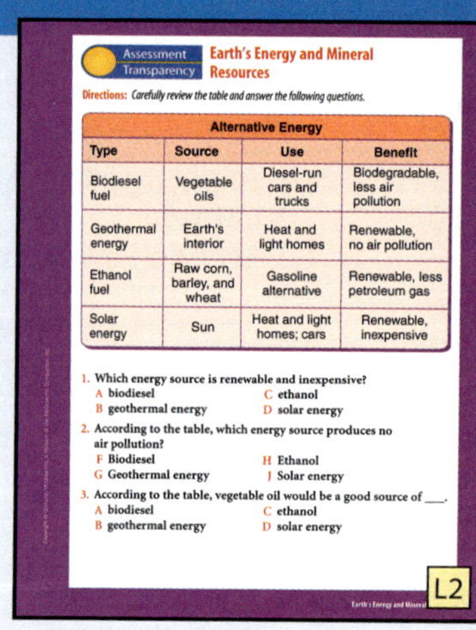
Assessment Transparency — Earth's Energy and Mineral Resources

Directions: Carefully review the table and answer the following questions.

Alternative Energy			
Type	Source	Use	Benefit
Biodiesel fuel	Vegetable oils	Diesel-run cars and trucks	Biodegradable, less air pollution
Geothermal energy	Earth's interior	Heat and light homes	Renewable, no air pollution
Ethanol fuel	Raw corn, barley, and wheat	Gasoline alternative	Renewable, less petroleum gas
Solar energy	Sun	Heat and light homes; cars	Renewable, inexpensive

1. Which energy source is renewable and inexpensive?
A biodiesel C ethanol
B geothermal energy D solar energy
2. According to the table, which energy source produces no air pollution?
F Biodiesel H Ethanol
G Geothermal energy J Solar energy
3. According to the table, vegetable oil would be a good source of ___.
A biodiesel C ethanol
B geothermal energy D solar energy

L2

Teaching

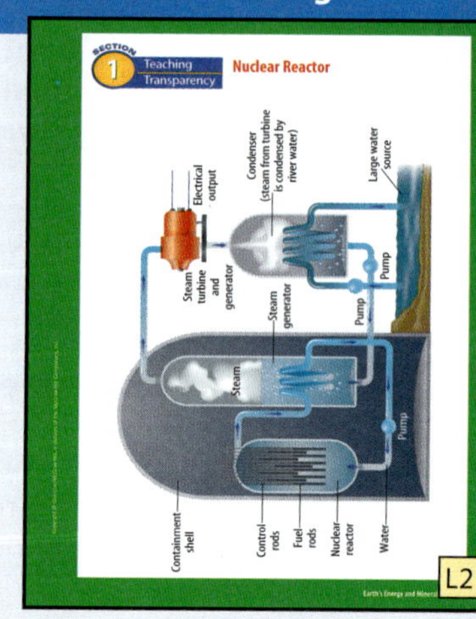

Key to Teaching Strategies

The following designations will help you decide which activities are appropriate for your students.

- L1 Level 1 activities should be appropriate for students with learning difficulties.
- L2 Level 2 activities should be within the ability range of all students.
- L3 Level 3 activities are designed for above-average students.
- ELL ELL activities should be within the ability range of English Language Learners.
- COOP LEARN Cooperative Learning activities are designed for small group work.
- LS Multiple Learning Styles logos, as described on page 6T, are used throughout to indicate strategies that address different learning styles.
- P These strategies represent student products that can be placed into a best-work portfolio.
- PBL Problem-Based Learning activities apply real-world situations to learning.

Hands-on Activities

Student Text Lab Worksheet

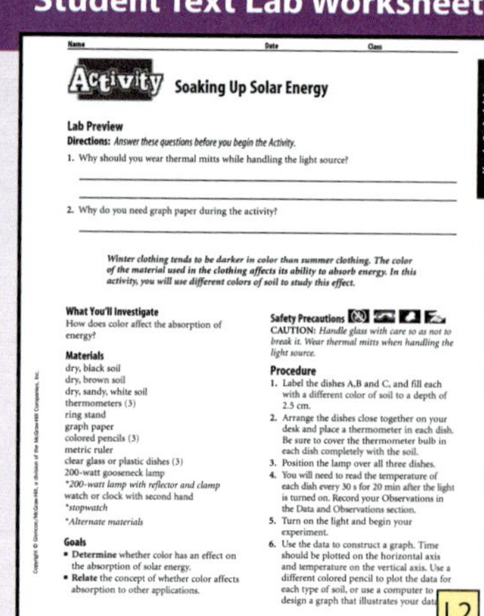
Activity — Soaking Up Solar Energy

Lab Preview

Directions: Answer these questions before you begin the Activity.

1. Why should you wear thermal mitts while handling the light source?
2. Why do you need graph paper during the activity?

L2

Laboratory Activities

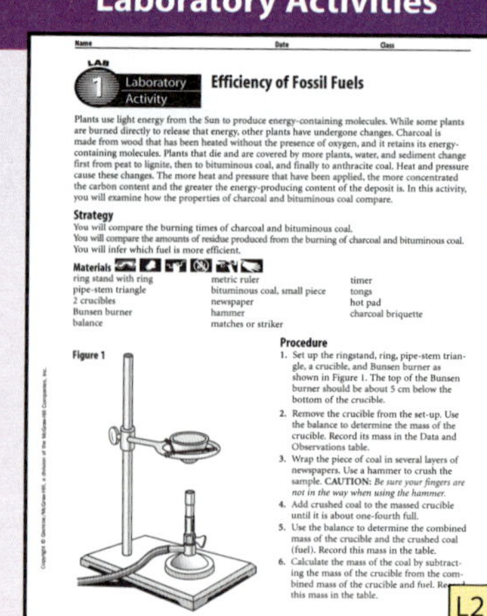
Lab 1 Laboratory Activity — Efficiency of Fossil Fuels

L2

Meeting Different Ability Levels

Content Outline

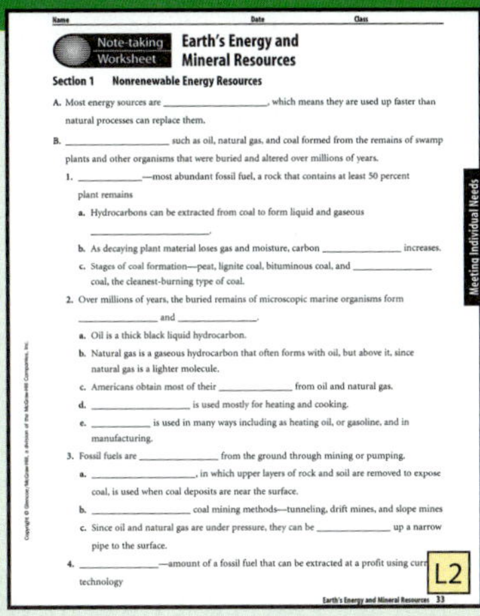

Reinforcement

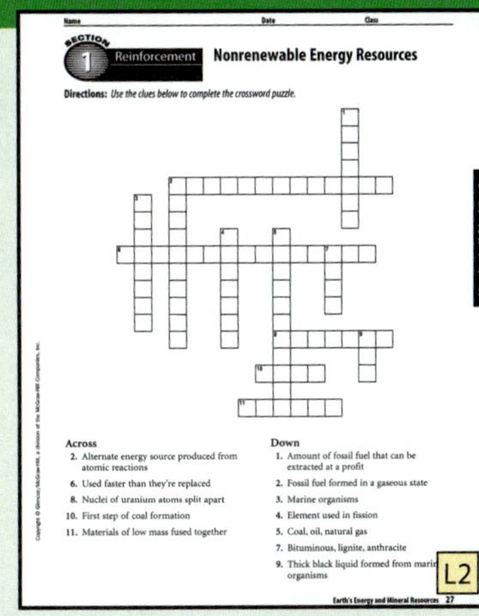

Enrichment

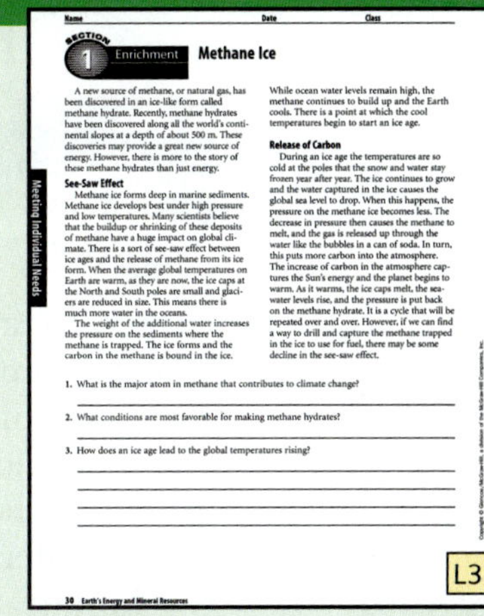

Directed Reading (English/Spanish)

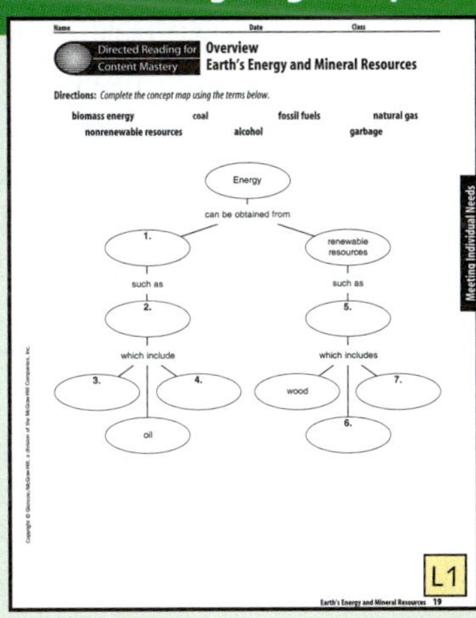

Study Guide

Reading Essentials

Assessment

Test Practice Workbook

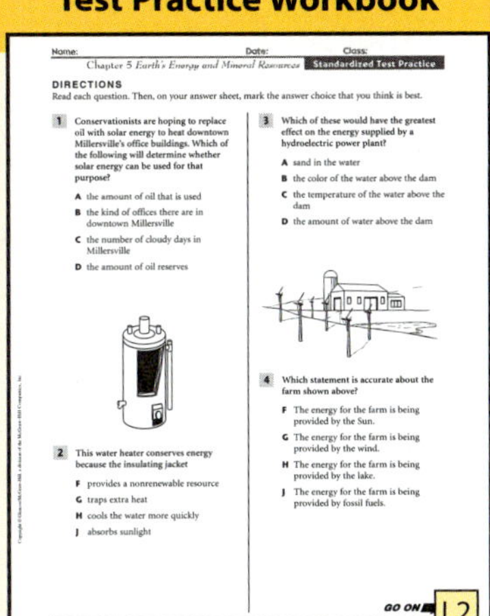

Chapter Review

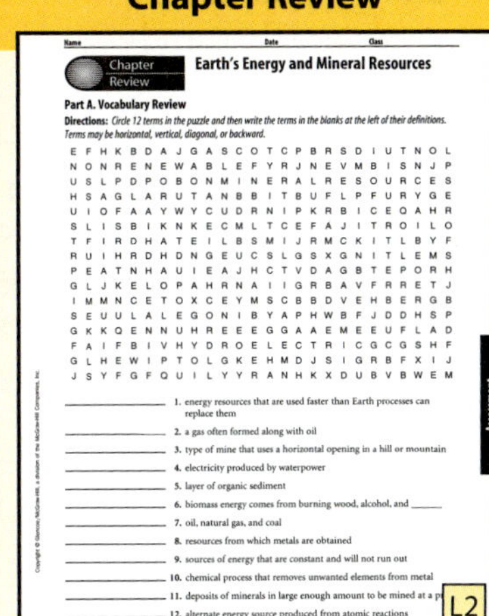

Chapter Tests

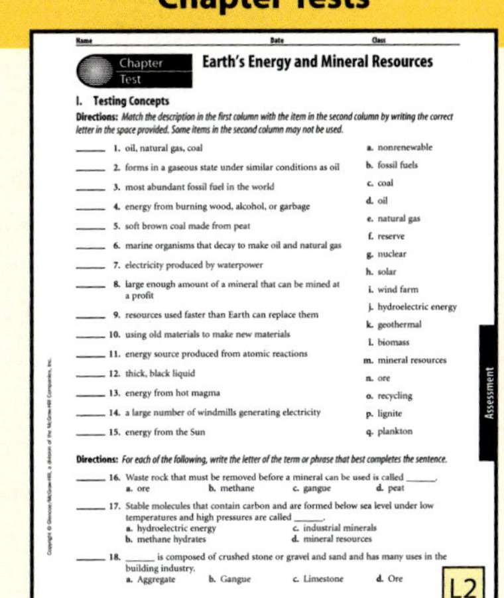

Science Content Background

section 1 Nonrenewable Energy Resources

Synthetic Fuels

Coal is gasified when steam and oxygen are introduced under heat and pressure. The end product is a gas that can be burned as fuel consisting of carbon monoxide, hydrogen, and methane. This process can be done below ground without extracting the coal. There are several different methods that can be utilized in the liquification of coal. One method begins with the gasification of coal. The carbon monoxide and hydrogen produced from this step are then recovered from the gaseous mixture. The union of the carbon monoxide and hydrogen forms a liquid fuel. Another method takes place under high pressure as hydrogen gas is added to solid coal. Next, a chemical reaction between the hydrogen and the coal takes place under high heat. As the product cools, it forms a liquid fuel.

Methane Hydrates

The name of the group of compounds in which methane hydrates are classified, the *clathrates*, gives an apt description of its structure in nature. The word clathrate is derived from the Latin word *clathratus* which means "furnished with a lattice." The structure of a methane hydrate is just that—methane trapped within a lattice of water molecules, frozen into a formation closely resembling ice. Methane hydrates were first studied by French scientists in the late 1800s. In the 1930s they were found forming in and blocking gas pipelines in colder areas. Methane hydrates were not discovered in nature until the 1960s. At that time, Russian scientists found methane hydrates in the permafrost of Siberia while drilling for natural gas. A decade later, methane hydrates were discovered in marine sediments. Today, methane hydrates have been found worldwide on continental margins.

section 2 Renewable Energy Resources

Energy from Earth

The use of geothermal energy can be divided into three categories depending on the amount of technology involved. The use of geothermal energy without first changing it to electricity is called direct use. Another way to use geothermal energy is through heat pumps. Heat pumps are used to heat and cool buildings with the help of a heat exchanger. Power plants use the heat from Earth's interior as an energy source to generate electricity.

Direct use is the simplest application. Evidence suggests that humans have been using hot springs for thousands of years. In North America, hot springs were used for cooking, bathing, and relaxing. In some areas of the world, such as New Guinea, hot springs still are used to boil potatoes and eggs. The Japanese macaque, a monkey that lives in the mountainous areas of Japan and China, sits for hours in the hot springs of those regions, taking refuge from the cold during winter. Aside from recreation, other direct-use applications include heating homes, greenhouses, and fish farms and using the water as a source of heat to dehydrate fruits and vegetables and to pasteurize milk.

Alcohol—A Form of Energy from the Sun

Alcohol was used as a source of fuel to run internal combustion engines in the late 1800s. Although alcohol does not contain as much energy as gasoline, gasoline-alcohol blends have been used as fuel for cars in Europe during times of oil shortages. Brazil produces large amounts of ethanol by fermenting sugar cane. The ethanol is used either directly as fuel for automobiles, or it is mixed with gasoline in a 20:80 ratio and the resulting gasohol is used as fuel.

section 3

Mineral Resources

Ores

Although aluminum is one of the most abundant elements in Earth's crust, it is not retrievable in a pure state. Bauxite is the main source of commercial aluminum. When rocks containing large stores of aluminum are weathered in tropical regions, bauxite is produced. Large bauxite deposits are found in France, West Africa, Jamaica, and Brazil. In the United States bauxite deposits are found in Arkansas, Alabama, and Georgia, areas that had a tropical climate millions of years ago.

Iron is another element that is abundant in Earth's crust. However, iron is not found in a pure state. Hematite, which is a combination of iron and oxygen, is a commonly mined iron ore. In the United States, hematite is found in Minnesota, Wisconsin, and Michigan, as well as in the Appalachian Mountains.

Building Materials

Limestone is a sedimentary rock made of calcium carbonate. Limestone can be formed from large deposits of dead marine organisms whose shells or *tests*, were composed of calcium carbonate. In some areas of the world, such as Bermuda, a type of limestone called *coquina*, which is composed of broken shells held together by calcium carbonate cement, is used to build homes.

Teacher to Teacher

Ralph M. Feather, Jr., Teacher
Derry Area High School
Derry, PA

"To help students think about how important Earth resources are to them, I ask them to compile a list of eight items from the classroom, their lockers, or their homes that are made from Earth resources. Next to each item, I have them write the Earth resource from which the product is produced."

Ralph M. Feather

chapter content resources

Internet Resources

For additional content background, visit bookf.msscience.com to:

- access your book online
- find references to related articles in popular science magazines
- access Web links with related content background
- access current events with science journal topics

Print Resources

Physical Geology, by Charles C. Pummer and David McGeary, William C. Brown Publishers, 1993

Rocks and Minerals, by Steve Parker, DK Publishing, Inc., February 1997

Essentials of Geology, by Frederick K. Lutgens and Edward J. Tarbuck, Pearson Education, April 2002

chapter 3

Chapter Vocabulary

fossil fuel, p. 66
coal, p. 67
oil, p. 69
natural gas, p. 69
reserve, p. 71
nuclear energy, p. 73
solar energy, p. 76
wind farm, p. 77
hydroelectric energy, p. 78
geothermal energy, p. 78
biomass energy, p. 79
mineral resource, p. 83
ore, p. 83
recycling, p. 87

Science Journal The photo shows a drill bit that is used on the end of the string of pipe that is lowered down to drill for oil. Student responses will vary, but may include responses such as coal, oil, water (hydropower), or wind.

INTERACTIVE CHALKBOARD with Image Bank

PowerPoint® Presentations

This CD-ROM is an editable Microsoft® PowerPoint® presentation that includes:

- a pre-made presentation for every chapter
- interactive graphics
- animations
- audio clips
- image bank
- all new section and chapter questions
- Standardized Test Practice
- transparencies
- pre-lab questions for all labs
- Foldables directions
- links to bookf.msscience.com

chapter 3

Earth's Energy and Mineral Resources

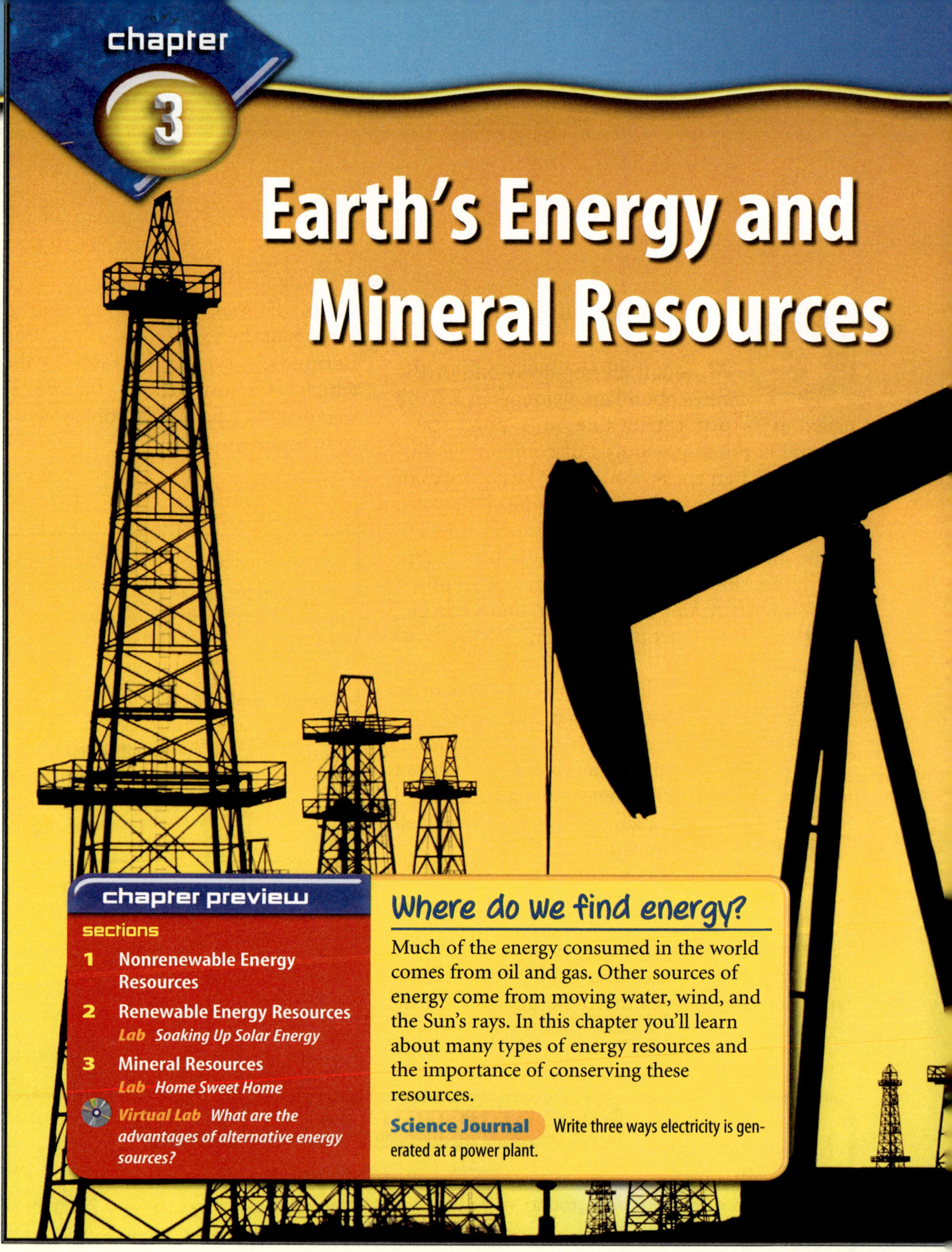

chapter preview

sections

Where do we find energy?

Much of the energy consumed in the world comes from oil and gas. Other sources of energy come from moving water, wind, and the Sun's rays. In this chapter you'll learn about many types of energy resources and the importance of conserving these resources.

Science Journal Write three ways electricity is generated at a power plant.

Theme Connection

Energy The theme is developed throughout the chapter in an exploration of Earth's renewable (biomass, geothermal), inexhaustible (solar, wind), and nonrenewable (fossil fuels, nuclear) energy resources.

About the Photo

Drilling For Oil Drill rig derricks like those in the background and grasshopper pumps (so called grasshoppers because they look like grasshoppers when they move) are used to extract oil from the ground. The derricks are used to drill long strings of pipe into the ground to a desired depth. Once the desired depth is reached the derrick is removed and the pump brings oil to the surface.

Start-Up Activities

Launch LAB

Finding Energy Reserves

The physical properties of Earth materials determine how easily liquids and gases move through them. Geologists use these properties, in part, to predict where reserves of energy resources like petroleum or natural gas can be found.

1. Obtain a sample of sandstone and a sample of shale from your teacher.
2. Make sure that your samples can be placed on a tabletop so that the sides facing up are reasonably flat and horizontal.
3. Place the two samples side by side in a shallow baking pan.
4. Using a dropper, place three drops of cooking oil on each sample.
5. For ten minutes, observe what happens to the oil on the samples.
6. **Think Critically** Write your observations in your Science Journal. Infer which rock type might be a good reservoir for petroleum.

Science Online — Preview this chapter's content and activities at bookf.msscience.com

FOLDABLES Study Organizer

Energy Resources Make the following Foldable to help you identify energy resources.

STEP 1 **Fold** a sheet of paper in half lengthwise. Make the back edge about 1.25 cm longer than the front edge.

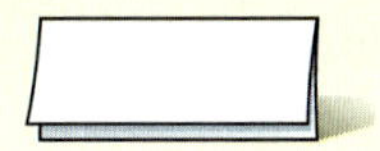

STEP 2 **Turn** lengthwise and **fold** into thirds.

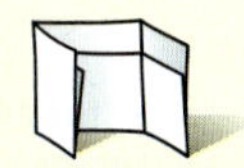

STEP 3 **Unfold and cut** only the top layer along both folds to make three tabs.

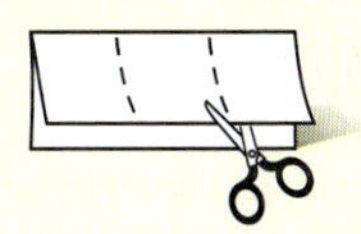

STEP 4 **Label** each tab as shown.

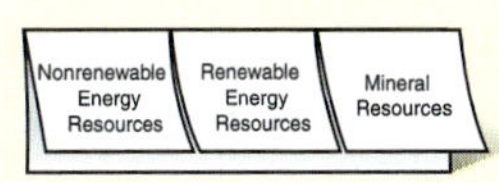

Find Main Ideas As you read the chapter, list examples on the front of the tabs and write about each type of resource under the tabs.

Launch LAB

Purpose Students compare the permeability of sandstone and shale. L2 LS **Kinesthetic**

Preparation Be sure the rock samples can be placed so that the upper side is reasonably flat.

Materials small samples of sandstone and shale, small amount of cooking oil, dropper, baking pan, timer

Teaching Strategy Students may need to place the rocks on a small amount of modeling clay to make sure that a flat side is horizontal.

Think Critically

Sandstone is permeable, so oil seeps through it easily. Oil pools on top of shale. Sandstone is a good reservoir for petroleum because it absorbs petroleum like a sponge.

Assessment

Portfolio Have students make posters showing how petroleum might be trapped in sandstone sandwiched between layers of shale. Use **Performance Assessment in the Science Classroom,** p. 145.

FOLDABLES Study Organizer — Dinah Zike Study Fold

Student preparation materials for this Foldable are available in the Chapter *FAST FILE* Resources.

section 1

Nonrenewable Energy Resources

1 Motivate

INTERACTIVE CHALKBOARD
PowerPoint® Presentations

Bellringer

Section Focus Transparencies also are available on the Interactive Chalkboard CD-ROM. L2 ELL

Section 1 Section Focus Transparency — Earth's Energy

This photograph, taken in the Gulf of Thailand, shows one way that people get energy resources from Earth. Offshore drilling platforms like this one are used worldwide to tap energy reserves under the seabed.

1. What difficulties might people face in trying to gather energy reserves from under the sea?
2. Name some other sources of energy that people use.
3. How can energy be conserved?

Earth's Energy and Mineral Resources

Tie to Prior Knowledge

Fossil Fuels Ask students to think of ways they use fossil fuels (coal, oil, natural gas) in their everyday lives. Possible answers: heating with oil or natural gas, gasoline fuel for cars made from petroleum, coal or oil burned in power plants to make electricity, petroleum made into plastic Tell students they will learn more about nonrenewable energy resources in this section.

as you read

***What* You'll Learn**

- **Identify** examples of nonrenewable energy resources.
- **Describe** the advantages and disadvantages of using fossil fuels.
- **Explain** the advantages and disadvantages of using nuclear energy.

***Why* It's Important**

Nonrenewable resources should be conserved to ensure their presence for future generations.

Review Vocabulary
fuel: a material that provides useful energy

New Vocabulary
- fossil fuel
- coal
- oil
- natural gas
- reserve
- nuclear energy

Energy

The world's population relies on energy of all kinds. Energy is the ability to cause change. Some energy resources on Earth are being used faster than natural Earth processes can replace them. These resources are referred to as nonrenewable energy resources. Most of the energy resources used to generate electricity are nonrenewable.

Fossil Fuels

Nonrenewable energy resources include fossil fuels. **Fossil fuels** are fuels such as coal, oil, and natural gas that form from the remains of plants and other organisms that were buried and altered over millions of years. Coal is a sedimentary rock formed from the compacted and transformed remains of ancient plant matter. Oil is a liquid hydrocarbon that often is referred to as petroleum. Hydrocarbons are compounds that contain hydrogen and carbon atoms. Other naturally occurring hydrocarbons occur in the gas or semisolid states. Fossil fuels are processed to make gasoline for cars, to heat homes, and for many other uses, as shown in **Table 1.**

Table 1 Uses of Fossil Fuels

Coal	■ To generate electricity
Oil	■ To produce gasoline and other fuels ■ As lubricants ■ To make plastics, home shingles, and other products
Natural Gas	■ To heat buildings ■ As a source of sulfur

Section 1 Resource Manager

Chapter *FAST FILE* Resources
Transparency Activity, pp. 44, 47–48
Directed Reading for Content Mastery, pp. 19, 20
Note-taking Worksheets, pp. 33–35
Reinforcement, p. 27
MiniLAB, p. 3
Lab Activity, pp. 9–12
Enrichment, p. 30

Earth Science Critical Thinking/Problem Solving, p. 2
Science Inquiry Labs, p. 0
Physical Science Critical Thinking/Problem Solving, pp. 14, 20

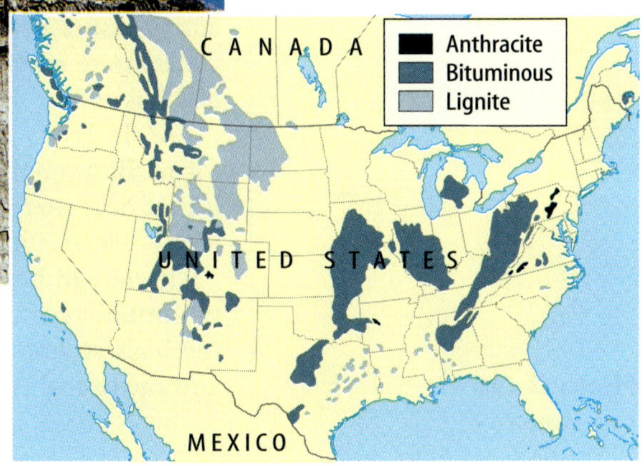

Figure 1 This coal layer is located in Castle Gate, Utah.
Analyze and Conclude *Using the map and legend below, can you determine what type of coal it is?*

Coal The most abundant fossil fuel in the world is coal, shown in **Figure 1.** If the consumption of coal continues at the current rate, it is estimated that the coal supply will last for about another 250 years.

Coal is a rock that contains at least 50 percent plant remains. Coal begins to form when plants die in a swampy area. The dead plants are covered by more plants, water, and sediment, preventing atmospheric oxygen from coming into contact with the plant matter. The lack of atmospheric oxygen prevents the plant matter from decaying rapidly. Bacterial growth within the plant material causes a gradual breakdown of molecules in the plant tissue, leaving carbon and some impurities behind. This is the material that eventually will become coal after millions of years. Bacteria also cause the release of methane gas, carbon dioxide, ammonia, and water as the original plant matter breaks down.

Reading Check *What happens to begin the formation of coal in a swampy area?*

Synthetic Fuels Unlike gasoline, which is refined from petroleum, other fuels called synthetic fuels are extracted from solid organic material. Synthetic fuels can be created from coal—a sedimentary rock containing hydrocarbons. The hydrocarbons are extracted from coal to form liquid and gaseous synthetic fuels. Liquid synthetic fuels can be processed to produce gasoline for automobiles and fuel oil for home heating. Gaseous synthetic fuels are used to generate electricity and heat buildings.

INTEGRATE Life Science

Coal Formation The coal found in the eastern and midwestern United States formed from plants that lived in great swamps about 300 million years ago during the Pennsylvanian Period of geologic time. Research the Pennsylvanian Period to find out what types of plants lived in these swamps. Describe the plants in your Science Journal.

2 Teach

Use Science Words

Word Usage Have students use the term *nonrenewable energy resources* in a sentence describing their characteristics. Possible answer: Nonrenewable energy resources are those that require great amounts of time for Earth's natural processes to make.

Teacher FYI

Today's Coal Much of the coal burned today formed from plants that lived during the Mississippian and Pennsylvanian Periods of the Paleozoic Era.

INTEGRATE Life Science

Coal Formation There were ancient forms of cone-bearing plants, ferns, and mosses that grew in the "coal swamps".

Reading Check

Answer Plants die in a swampy area and are quickly covered.

Caption Answer

Figure 1 Bituminous

Differentiated Instruction

English-Language Learner Have students list the states where major coal reserves are found. Ask them if they live close to one of these reserves. Answers could include: Ohio, West Virginia, Illinois, and Pennsylvania. ELL **Visual-Spatial**

Challenge Have students find the heat value of each type of coal in BTU/pound. Ask them to decide which type of coal provides the most heat per pound. BTU stands for British Thermal Unit, a unit of heat. Peat—3,000 to 5,000; lignite coal—7,000; bituminous coal—12,000; anthracite coal—14,000; anthracite provides the most heat per pound. L3 LS **Logical-Mathematical**

Stages of Coal Formation As decaying plant material loses gas and moisture, the concentration of carbon increases. The first step in this process, shown in Figure 2, results in the formation of peat. Peat is a layer of organic sediment. When peat burns, it releases large amounts of smoke because it has a high concentration of water and impurities.

As peat is buried under more sediment, it changes into lignite, which is a soft, brown coal with much less moisture. Heat and pressure produced by burial force water out of peat and concentrate carbon in the lignite. Lignite releases more energy and less smoke than peat when it is burned.

As the layers are buried deeper, bituminous coal, or soft coal, forms. Bituminous coal is compact, black, and brittle. It provides lots of heat energy when burned. Bituminous coal contains various levels of sulfur, which can pollute the environment.

If enough heat and pressure are applied to buried layers of bituminous coal, anthracite coal forms. Anthracite coal contains the highest amount of carbon of all forms of coal. Therefore, anthracite coal is the cleanest burning of all coals.

Figure 2 Coal is formed in four basic stages.

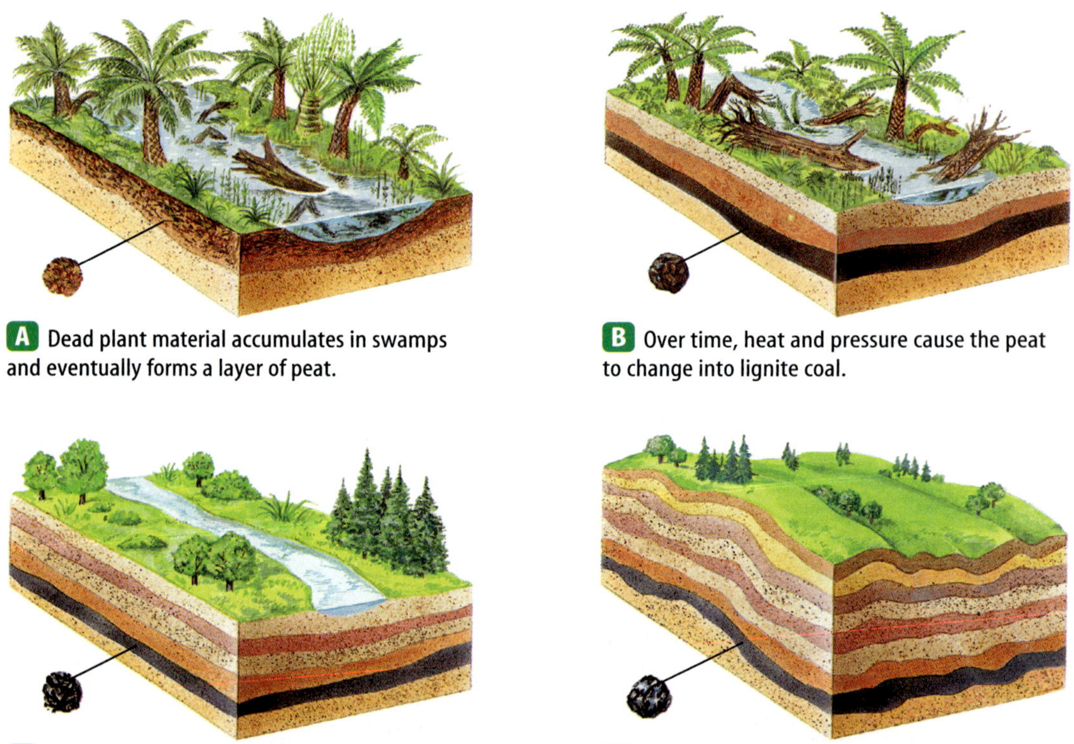

A Dead plant material accumulates in swamps and eventually forms a layer of peat.

B Over time, heat and pressure cause the peat to change into lignite coal.

C As the lignite coal becomes buried by more sediments, heat and pressure change it into bituminous coal.

D When bituminous coal is heated and squeezed during metamorphism, anthracite coal forms.

Visual Learning

Figure 2 Ask students to compare the appearance of peat and anthracite in the illustrations. How has the appearance of coal changed between its first stage of development, peat, and its last stage of development, anthracite? Peat looks soft and brownish, and the plant material from which it forms is visible. Anthracite is black and dense like a rock.

Activity

Classifying Coal Have student groups examine samples of peat, lignite coal, bituminous coal, and anthracite coal, but do not tell them what the samples are. Have them contrast the color, texture, and hardness of each sample. Explain that the specimens represent the four different stages in the formation of coal. Have students arrange the samples from least to most changed. L2 COOP LEARN LS **Visual-Spatial**

Discussion

Bacteria and Coal Why are bacteria needed to produce coal? Bacteria breaks down plant matter in the first stage of coal formation.

Cultural Diversity

Limiting Technology The Amish are a religious group with communities in the United States and Canada. Amish principles involve separation from worldly things, including most technology. As a result, the energy crisis that affects most other Americans who use electricity and drive cars is not a concern for them. Most Amish farm the land, relying extensively on human and animal power. Fossil fuel use is limited. Kerosene is used for lamps, and gasoline-driven motors may be used for some farm equipment. But machinery must move via horse-drawn wagons. Have students research to find out more about how the Amish use (or avoid) modern energy resources. Have them write a report on their findings. P

Oil and Natural Gas Coal isn't the only fossil fuel used to obtain energy. Two other fossil fuels that provide large quantities of the energy used today are oil and natural gas. **Oil** is a thick, black liquid formed from the buried remains of microscopic marine organisms. **Natural gas** forms under similar conditions and often with oil, but it forms in a gaseous state. Oil and natural gas are hydrocarbons. However, natural gas is composed of hydrocarbon molecules that are lighter than those in oil.

Residents of the United States burn vast quantities of oil and natural gas for daily energy requirements. As shown in **Figure 3,** Americans obtain most of their energy from these sources. Natural gas is used mostly for heating and cooking. Oil is used in many ways, including as heating oil, gasoline, lubricants, and in the manufacture of plastics and other important compounds.

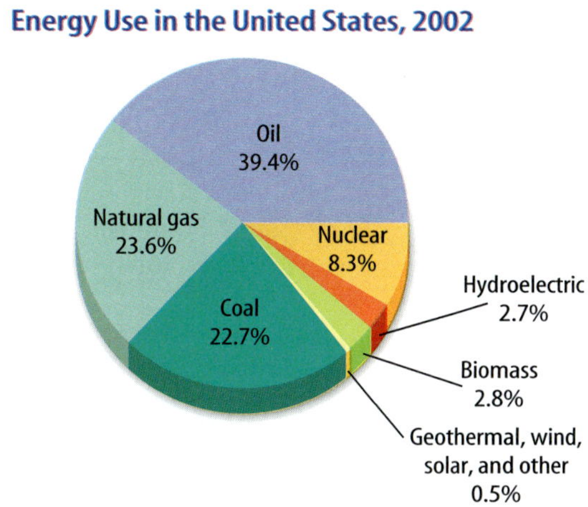

Figure 3 This circle graph shows the percentages of energy that the United States derives from various energy resources.
Calculate *What percentage is from nonrenewable energy resources?*

Formation of Oil and Natural Gas Most geologists agree that petroleum forms over millions of years from the remains of tiny marine organisms in ocean sediment. The process begins when marine organisms called plankton die and fall to the seafloor. Similar to the way that coal is buried, sediment is deposited over them. The temperature rises with depth in Earth, and increased heat eventually causes the dead plankton to change to oil and gas after they have been buried deeply by sediment.

Oil and natural gas often are found in layers of rock that have become tilted or folded. Because they are less dense than water, oil and natural gas are forced upward. Rock layers that are impermeable, such as shale, stop this upward movement. When this happens, a folded shale layer can trap the oil and natural gas below it. Such a trap for oil and gas is shown in **Figure 4.** The rock layer beneath the shale in which the petroleum and natural gas accumulate is called a reservoir rock.

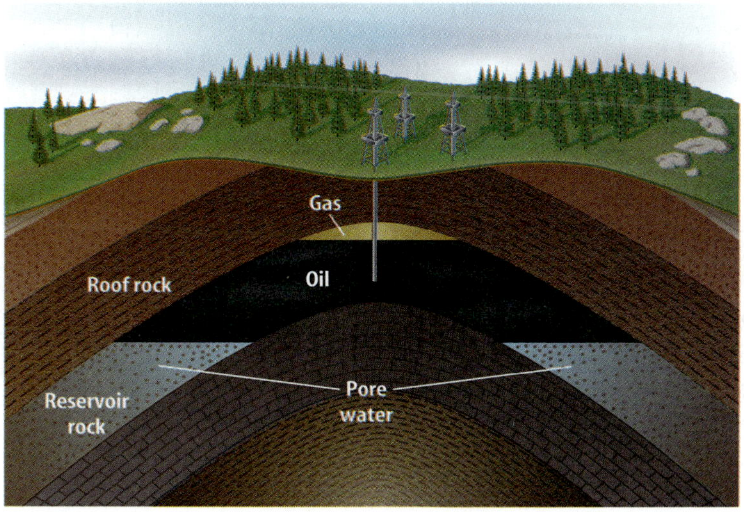

Figure 4 Oil and natural gas are fossil fuels formed by the burial of marine organisms. These fuels can be trapped and accumulate beneath Earth's surface.

Caption Answer

Figure 3 85.7 percent

Make a Model

Oil-Bearing Rock Have students use clay, small gravel, and pieces of wet, balled-up paper towels (representing oil deposits) to model layers of oil-bearing rock, such as in **Figure 4.** Models should show the layers and trapped oil in cross section. Instruct students to use this model to show where oil and natural gas would collect in the porous layer between two nonporous clay layers. The models should show that oil and natural gas have moved up through the porous rock and have collected at the highest location under the nonporous clay. L2

LS **Kinesthetic**

Differentiated Instruction

Learning Disabled Ask students if they have ever seen a parent or guardian pump gas into their car. Explain that the gasoline used by their cars comes from oil pumped from the ground. Explain that oil can be used for making things like jet fuel, diesel fuel, and plastics.

Science Journal

Fossil Fuels Have students make a table in their Science Journals that lists the advantages and disadvantages of using fossil fuels. Possible advantages: produce energy inexpensively compared to some other sources, presently available; possible disadvantages: pollutes the environment, nonrenewable

Visual Learning

Figure 5 Why might open-pit mining be of concern to those who are trying to protect endangered species? Open-pit mining disturbs Earth's surface in a way that could destroy animal habitats.

Use an Analogy

Coal Mines Use the following analogy to help students understand the two basic ways to remove coal from the ground. Ask students how they would go about looking for an object that their dog has buried behind a stone wall. Some students might dig a hole behind the wall and remove the object from the hole. Other students might remove stones from the wall and dig a tunnel into the ground behind the wall, hoping to find the object. Small pieces of wood could be used to support the top and sides of the tunnel. The first example is analogous to an open-pit mine, and the second example is analogous to an underground mine.

Caption Answer

Figure 5 Possible answer: with rail cars along the tracks

Removing Fossil Fuels from the Ground

Coal is removed from the ground using one of several methods of excavation. The two most common methods are strip mining, also called open-pit mining, and underground mining, shown in **Figure 5.** Oil and natural gas are removed by pumping them out of the ground.

Coal Mining During strip mining, as shown in **Figure 5,** layers of soil and rock above coal are removed and piled to one side. The exposed coal then is removed and loaded into trucks or trains and transported elsewhere. After the coal has been removed, mining companies often return the soil and rock to the open pit and cover it with topsoil. Trees and grass are planted in a process called land reclamation. If possible, animals native to the area are reintroduced. Strip mining is used only when the coal deposits are close to the surface.

In one method of underground coal mining, tunnels are dug and pillars of rock are left to support the rocks surrounding the tunnels. Two types of underground coal mines are drift mines and slope mines. Drift mining, shown in the **Figure 5** inset photo, is the removal of coal that is not close to Earth's surface through a horizontal opening in the side of a hill or mountain. In slope mining, an angled opening and air shaft are made in the side of a mountain to remove coal.

Figure 5 Coal is a fossil fuel that can be removed from Earth in many different ways.

During strip mining, coal is accessed by removing the soil and rock above it.

During drift mining, tunnels are made into Earth.
Explain *how you think the coal is removed from these tunnels.*

LAB DEMONSTRATION

Purpose to contrast auto emissions from new and old cars that burn gasoline for fuel

Materials three white socks

Preparation Select a fairly new car and an ... running car. Place a sock over ...e of each car. Keep the third ... Run each car for two minutes.

Procedure Show students the three socks, and explain how the material on each sock was obtained.

Expected Outcome Students should see that the old car deposits more particulates on the sock than the new car.

Assessment

Ask students to predict whether an old car that needs maintenance or a new car that is well maintained will have a more negative impact on air quality. The older car that needs maintenance will have a more negative impact on air quality.

Drilling for Oil and Gas Oil and natural gas are fossil fuels that can be pumped from underground deposits. Geologists and engineers drill wells through rocks where these resources might be trapped, as shown in **Figure 6.** As the well is being drilled, it is lined with pipe to prevent it from caving in. When the drill bit reaches the rock layer containing oil, drilling is stopped. Equipment is installed to control the flow of oil. The surrounding rock then is fractured to allow oil and gas to flow into the well. The oil and gas are pumped to the surface.

Reading Check *How are oil and natural gas brought to Earth's surface?*

Figure 6 Oil and natural gas are recovered from Earth by drilling deep wells.

Fossil Fuel Reserves

The amount of a fossil fuel that can be extracted at a profit using current technology is known as a **reserve.** This is not the same as a fossil fuel resource. A fossil fuel resource has fossil fuels that are concentrated enough that they can be extracted from Earth in useful amounts. However, a resource is not classified as a reserve unless the fuel can be extracted economically. What might cause a known fossil fuel resource to become classified as a reserve?

Methane Hydrates You have learned that current reserves of coal will last about 250 years. Enough natural gas is located in the United States to last about 60 more years. However, recent studies indicate that a new source of methane, which is the main component of natural gas, might be located beneath the seafloor. Icelike substances known as methane hydrates could provide tremendous reserves of methane.

Methane hydrates are stable molecules found hundreds of meters below sea level in ocean floor sediment. They form under conditions of relatively low temperatures and high pressures. The hydrocarbons are trapped within the cagelike structure of ice, as described in **Figure 7.** Scientists estimate that more carbon is contained in methane hydrates than in all current fossil fuel deposits combined. Large accumulations of methane hydrates are estimated to exist off the eastern coast of the United States. Can you imagine what it would mean to the world's energy supply if relatively clean-burning methane could be extracted economically from methane hydrates?

Science Online

Topic: Methane Hydrates
Visit bookf.msscience.com for Web links to information about methane hydrates.

Activity Identify which oceans might contain significant amounts of methane hydrates.

Reading Check

Answer They are pumped to Earth's surface through wells.

Text Question Answer

an increase in the price of the resource that would make it profitable to extract; a change in technology that would make a resource previously too difficult or expensive to extract easier to extract and therefore profitable

Identifying Misconceptions

Natural Gas and Coal Producers Because Saudi Arabia is the leading producer of oil, students may also think Saudi Arabia is the leading producer of natural gas and coal. Inform students that Russia and the United States are the top two producers of natural gas and that China and the United States are the leading producers of coal.

Quick Demo

Drilling Operation

Materials water, vegetable oil, 250-mL beakers (2), transparent straw

Estimated Time 20 minutes

Procedure Have students fill a beaker with approximately 50 mL of tap water. Pour an additional 150 mL of vegetable oil on the water. Explain to students that vegetable oil represents a large oil reserve and the water is underlying pore water. Have students extract vegetable oil with the straw by inserting the straw into oil, placing a finger over top of the straw, and withdrawing. Have students extract as much of the oil as they can while leaving the water. Explain that the straw represents the oil well and removing the vegetable oil represents pumping.

Curriculum Connection

Geography Have students research the world's five leading oil producers in terms of barrels per day. Tell them to make a bar graph that displays the data. Saudi Arabia—8,885,000; United States—8,290,000; Russia—6,200,000; Iran—3,705,000; Mexico—3,065,000 P LS **Logical-Mathematical**

SECTION 1 Nonrenewable Energy Resources F ◆ 71

NATIONAL GEOGRAPHIC

Visualizing Methane Hydrates

Have students examine the pictures and read the captions. Then ask the following questions.

Why is it important that methane hydrates are flammable? They represent an enormous source of potential energy.

What environmental problems could be associated with harvesting methane hydrates? Possible answers: Both permafrost and oceans are environmentally sensitive areas. Mining ocean sediments and digging in permafrost implies the need for equipment and manpower—both bringing wastes, as well as the disruption of habitat for resident organisms.

Activity

Methane Hydrates Have students make posters depicting an imaginary mining site for methane hydrates. Encourage them to include information on necessary equipment, types of professionals, and the environments in which methane hydrates are likely to be found. LS **Visual-Spatial**

NATIONAL GEOGRAPHIC **VISUALIZING METHANE HYDRATES**

Figure 7

Reserves of fossil fuels—such as oil, coal, and natural gas—are limited and will one day be used up. Methane hydrates could be an alternative energy source. This icelike substance, background, has been discovered in ocean floor sediments and in permafrost regions worldwide. If scientists can harness this energy, the world's gas supply could be met for years to come.

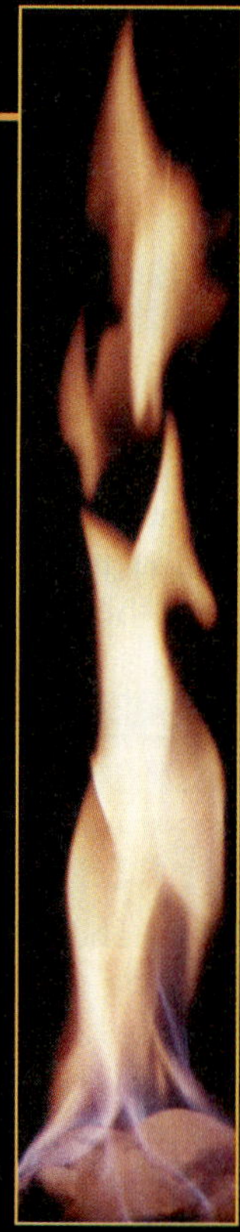

Methane hydrates are highly flammable compounds made up of methane—the main component of natural gas—trapped in a cage of frozen water. Methane hydrates represent an enormous source of potential energy. However, they contain a greenhouse gas that might intensify global warming. More research is needed to determine how to safely extract them from the seafloor.

In the photo above, a Russian submersible explores a site in the North Atlantic that contains methane hydrate deposits.

Conserving Fossil Fuels Do you sometimes forget to turn off the lights when you walk out of a room? Wasteful habits might mean that electricity to run homes and industries will not always be as plentiful and cheap as it is today. Fossil fuels take millions of years to form and are used much faster than Earth processes can replenish them.

Today, coal provides about 25 percent of the energy that is used worldwide and 22 percent of the energy used in the United States. Oil and natural gas provide almost 61 percent of the world's energy and about 65 percent of the U.S. energy supply. At the rate these fuels are being used, they could run out someday. How can this be avoided?

By remembering to turn off lights and appliances, you can avoid wasting fossil fuels. Another way to conserve fossil fuels is to make sure doors and windows are shut tightly during cold weather so heat doesn't leak out of your home. If you have air-conditioning, run it as little as possible. Ask the adults you live with if more insulation could be added to your home or if an insulated jacket could be put on the water heater.

Energy from Atoms

Most electricity in the United States is generated in power plants that use fossil fuels. However, alternate sources of energy exist. **Nuclear energy** is an alternate energy source produced from atomic reactions. When the nucleus of a heavy element is split, lighter elements form and energy is released. This energy can be used to light a home or power the submarines shown in **Figure 8.**

The splitting of heavy elements to produce energy is called nuclear fission. Nuclear fission is carried out in nuclear power plants using a type of uranium as fuel.

Mini LAB

Practicing Energy Conservation

Procedure

1. Have an adult help you find the **electric meter** for your home and record the reading in your **Science Journal.**
2. Do this for several days, taking your meter readings at about the same time each day.
3. List things you and your family can do to reduce your electricity use.
4. Encourage your family to try some of the listed ideas for several days.

Analysis

1. Keep taking meter readings and infer whether the changes make any difference.
2. Have you and your family helped conserve energy?

Try at Home

Figure 8 Atoms can be a source of energy.

These submarines are powered by nuclear fission.

During nuclear fission, energy is given off when a heavy atom, like uranium, splits into lighter atoms.

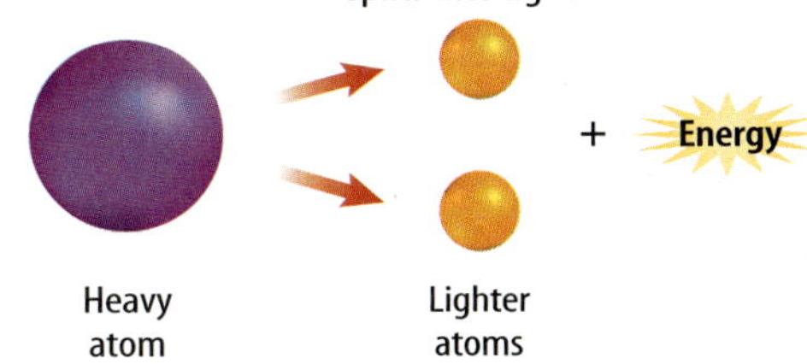

Text Question Answer

Conservation could make the resources last longer.

Mini LAB

Purpose Students will observe the results of conserving energy. L2 LS **Logical-Mathematical**

Teaching Strategy Students should practice taking meter readings before doing the activity. You might contact the local electric utility for a publication that shows how to read meters.

Analysis

1. Students should note a reduction in electricity use after the household tries conservation measures.
2. If students are successful, the family should have conserved energy and saved money on the electric bill.

Assessment

Performance Have students come up with a list of conservation measures that could be used at school. Encourage them to get permission to put their measures into effect. Use **Performance Assessment in the Science Classroom,** p. 89.

Try at Home

Fun Fact

The most commonly used fuel in fission power plants is uranium-235.

Active Reading

Speculation About Effects/Prediction Journal This strategy allows students to examine events and speculate about their possible long-term effects. Have students divide sheets of paper in half. On the left side they should record "what happened." On the right, have them write "What might/should happen as a result of this." Have students write a Speculation About Effects/Prediction Journal as they study the use of energy sources and the consequences, both good and bad, of using the different types of energy.

Use Science Words

Word Meaning Have students find out why the current process of producing energy from nuclear reactions is called *fission.* The term fission means "splitting apart," which is what happens to the atoms.

Discussion

Living Near a Nuclear Plant Would you want to live next to a nuclear power plant? Why or why not? Accept all opinions as a basis for discussion. Some students would be concerned because of possible danger from radiation leakage. Others might feel that safety procedures would keep the area around the plant safe.

Caption Answer

Figure 9 Water is used as a coolant and for making steam.

Reading Check

Answer Heat is released as atoms split. The heat is used to boil water, which gives off steam. The steam drives a turbine, which turns a generator, producing electricity.

Teacher FYI

Control Systems Nuclear reactors have several different systems that control their reactions. One system consists of control rods made of materials that absorb neutrons. If the reactions speed up too much, the control rods are lowered between the fuel rods, slowing or stopping the reactions. A reactor's temperature also is controlled by a water-cooling system.

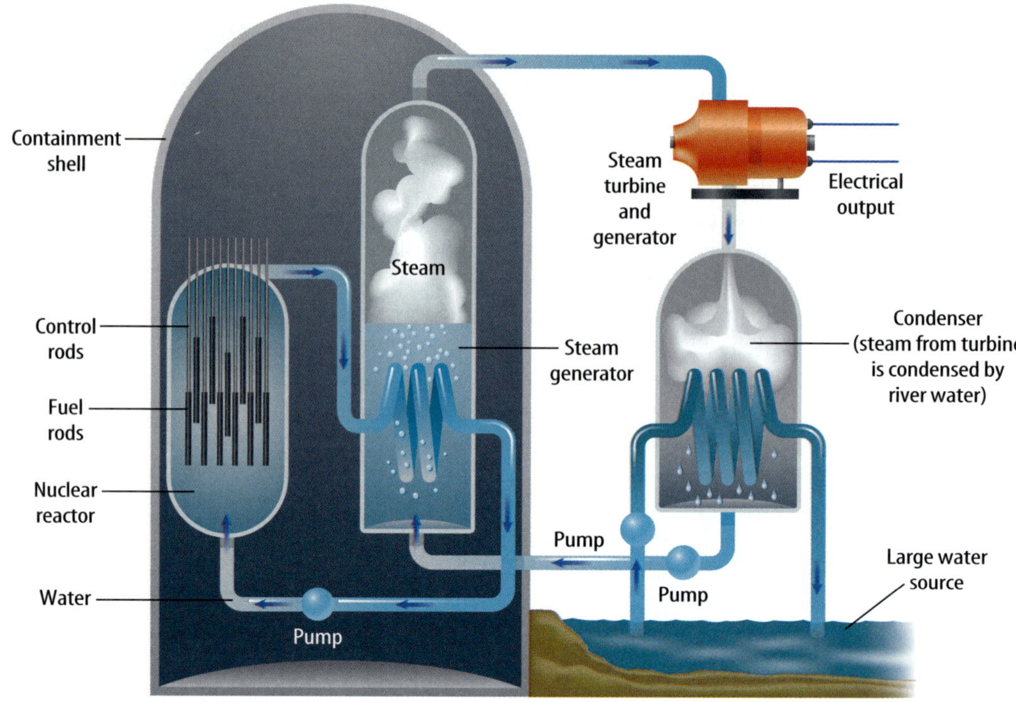

Figure 9 Heat released in nuclear reactors produces steam, which in turn is used to produce electricity. This is an example of transforming nuclear energy into electrical energy.
Infer *Why do you think nuclear power plants are located near rivers and lakes?*

Electricity from Nuclear Energy A nuclear power plant, shown in **Figure 9,** has a large chamber called a nuclear reactor. Within the nuclear reactor, uranium fuel rods sit in a pool of cooling water. Neutrons are fired into the fuel rods. When the uranium-235 atoms are hit, they break apart and fire out neutrons that hit other atoms, beginning a chain reaction. As each atom splits, it not only fires neutrons but also releases heat that is used to boil water to make steam. The steam drives a turbine, which turns a generator that produces electricity.

Reading Check *How is nuclear energy used to produce electricity?*

Nuclear energy from fission is considered to be a nonrenewable energy resource because it uses uranium-235 as fuel. A limited amount of uranium-235 is available for use. Another problem with nuclear energy is the waste material that it produces. Nuclear waste from power plants consists of highly radioactive elements formed by the fission process. Some of this waste will remain radioactive for thousands of years. The Environmental Protection Agency (EPA) has determined that nuclear waste must be stored safely and contained for at least 10,000 years before reentering the environment.

74 ◆ F CHAPTER 3 Earth's Energy and Mineral Resources

Fusion Environmental problems related to nuclear power could be eliminated if usable energy could be obtained from fusion. The Sun is a natural fusion power plant that provides energy for Earth and the solar system. Someday fusion also might provide energy for your home.

During fusion, materials of low mass are fused together to form a substance of higher mass. No fuel problem exists if the low-mass material is a commonly occurring substance. Also, if the end product is not radioactive, storing nuclear waste is not a problem. In fact, fusion of hydrogen into helium would satisfy both of these conditions. However, technologies do not currently exist to enable humans to fuse hydrogen into helium at reasonably low temperatures in a controlled manner. But research is being conducted, as shown in **Figure 10.** If this is accomplished, nuclear energy could be considered an inexhaustible fuel resource. You will learn the importance of inexhaustible and renewable energy resources in the next section.

Figure 10 Lasers are used in research facilities to help people understand and control fusion.

section 1 review

Summary

Fossil Fuels

- Coal, natural gas, and oil are all nonrenewable energy sources.
- Synthetic fuels are human-made fuels that can be derived from coal.
- The four stages of coal formation are peat, lignite, bituminous coal, and anthracite coal.
- Oil and gas are made from the decay of ancient marine organisms.
- Strip mining and underground mining are two common methods that are used to extract coal reserves.

Energy from Atoms

- Energy is released during a fission reaction when a heavy atom is split into lighter atoms.
- Fusion occurs when two atoms come together to form a single atom.

Self Check

1. **Explain** why coal, oil, and natural gas are fossil fuels.
2. **Explain** why fossil fuels are considered to be nonrenewable energy resources.
3. **Describe** two disadvantages of nuclear energy.
4. **Think Critically** Why are you likely to find natural gas and oil deposits in the same location, but less likely to find coal and petroleum deposits at the same location?

Applying Math

5. **Design a Graph** Current energy consumption by source in the U.S. is as follows: oil, 39%; natural gas, 24%; coal, 23%; nuclear energy, 8%; renewable resources, 6%. Design a bar graph to show the energy consumption by source in the U.S. Display the sources from greatest to least.

3 Assess

DAILY INTERVENTION

Check for Understanding

Visual-Spatial Have students draw maps of the United States that show the location of coal, oil, and natural gas reserves. Discuss with students where each type of resource is generally located. coal: Appalachian and Rocky Mountains as well as parts of the Midwest and Great Plains; petroleum: mid-Appalachians, Texas, Oklahoma and parts of the Great Plains, southwestern California; natural gas: northern Appalachians, Gulf Coast, Great Plains, California's Central Valley L3 LS

Reteach

Supply of Fuel Ask students to hypothesize how their lives might change if the supply of fossil fuels ended or were drastically reduced. Possible answers: Travel might be restricted because of shortages of gasoline; home heating would be dependent on other energy sources, such as wood; materials made of plastic would be less plentiful. L2 LS **Logical-Mathematical**

Assessment

Performance Have students use the graph they make in Question 6 to compare the energy consumption of nonrenewable fuels with that of inexhaustible and renewable resources. Energy consumption from nonrenewable fuels is 94 percent. Inexhaustible and renewable fuel sources total about 6 percent. Use **Performance Assessment in the Science Classroom,** p. 109.

section 1 review

1. Coal, oil, and natural gas form from the remains of plants and animals that lived millions of years ago and were buried in soil and rock.
2. It takes fossil fuels millions of years to form. Once we use up the present supply, they are gone.
3. Nuclear fission releases energy when a larger or heavier atom is split into lighter atoms. When this reaction occurs it is used to produce electricity. Possible answer: Nuclear energy is nonrenewable and can produce radioactive waste.
4. Petroleum and natural gas form from the decay of buried marine organisms; coal forms from the decay of buried plants that grew on land.
5. Energy sources should be on the *x*-axis and consumption percentages should be on the *y*-axis.

1 Motivate

INTERACTIVE CHALKBOARD PowerPoint® Presentations

Bellringer

Section Focus Transparencies also are available on the Interactive Chalkboard CD-ROM. L2 ELL

Section 2 Section Focus Transparency **Energy for Life**

How does life on Earth depend on the Sun? Every time you eat, exercise, or simply bask in the light of a warm day, you use energy from the Sun. Even energy from fossil fuels originally came from the Sun—it's been stored in plant and animal remains for millions of years!

1. How do we use the Sun's energy when we eat?
2. What are some sources of energy that are not depleted as they are used?

Earth's Energy and Mineral Resources

Tie to Prior Knowledge

Windmills Ask students whether they've ever seen a windmill. Ask them what they think it was used for. Tell students that some windmills were used long ago to grind grain, but now they are sometimes used for making electricity. Tell students they will find out in this section how wind, the Sun, and heat from Earth can produce electric power.

section 2

Renewable Energy Resources

as you read

What You'll Learn

- **Compare and contrast** inexhaustible and renewable energy resources.
- **Explain** why inexhaustible and renewable resources are used less than nonrenewable resources.

Why It's Important

As fossil fuel reserves continue to diminish, alternate energy resources will be needed.

Review Vocabulary
energy: the ability to cause change

New Vocabulary
- solar energy
- wind farm
- hydroelectric energy
- geothermal energy
- biomass energy

Inexhaustible Energy Resources

How soon the world runs out of fossil fuels depends on how they are used and conserved. Fortunately, there are inexhaustible energy resources. These sources of renewable energy are constant and will not run out in the future. Inexhaustible energy resources include the Sun, wind, water, and geothermal energy.

Energy from the Sun When you sit in the Sun, walk into the wind, or sail against an ocean current, you are experiencing the power of solar energy. **Solar energy** is energy from the Sun. You already know that the Sun's energy heats Earth, and it causes circulation in Earth's atmosphere and oceans. Global winds and ocean currents are examples of nature's use of solar energy. Thus, solar energy is used indirectly when the wind and some types of moving water are used to do work.

People can use solar energy in a passive way or in an active way. South-facing windows on buildings act as passive solar collectors, warming exposed rooms. Solar cells actively collect energy from the Sun and transform it into electricity. Solar cells were invented to generate electricity for satellites. Now they also are used to power calculators, streetlights, and experimental cars. Some people have installed solar energy cells on their roofs, as shown in **Figure 11.**

Figure 11 Solar panels, such as on this home in Laguna Niguel, California, can be used to collect inexhaustible solar energy to power appliances and heat water.

Section 2 Resource Manager

Chapter FAST FILE Resources
- Transparency Activity, p. 45
- Directed Reading for Content Mastery, p. 20
- Enrichment, p. 31
- Reinforcement, p. 28
- Lab Worksheet, pp. 5–6
- Lab Activity, pp. 13–16

Cultural Diversity, p. 41

Physical Science Critical Thinking/Problem Solving, p. 17

Figure 12 Wind farms are used to produce electricity.
Evaluate *Some people might argue that windmills produce visual pollution. Why do you think this is?*

Disadvantages of Solar Energy Solar energy is clean and inexhaustible, but it does have some disadvantages. Solar cells work less efficiently on cloudy days and cannot work at all at night. Some systems use batteries to store solar energy for use at night or on cloudy days, but it is difficult to store large amounts of energy in batteries. Worn out batteries also must be discarded. This can pollute the environment if not done properly.

Energy from Wind What is better to do on a warm, windy day than fly a kite? A strong wind can lift a kite high in the sky and whip it around. The pull of the wind is so great that you wonder if it will whip the kite right out of your hands. Wind is a source of energy. It was and still is used to power sailing ships. Windmills have used wind energy to grind corn and pump water. Today, windmills can be used to generate electricity. When a large number of windmills are placed in one area for the purpose of generating electricity, the area is called a **wind farm,** as shown in **Figure 12.**

Wind energy has advantages and disadvantages. Wind is nonpolluting and free. It does little harm to the environment and produces no waste. However, only a few regions of the world have winds strong enough to generate electricity. Also, wind isn't steady. Sometimes it blows too hard and at other times it is too weak or stops entirely. For an area to use wind energy consistently, the area must have a persistent wind that blows at an appropriate speed.

Reading Check *Why are some regions better suited for wind farms than others?*

INTEGRATE Career

Physicists The optimal speed of wind needed to rotate blades on a windmill is something a physicist would study. They can calculate the energy produced based on the speed at which the blades turn. Some areas in the country are better suited for wind farms than others. Find out which areas utilize wind farms and report in your Science Journal how much electric-ity is produced and what it is used for. What kinds of organizations would a physicist work for in these locations?

2 Teach

Caption Answer

Figure 12 If a windmill is exposed to the wind, it will also be very visible.

Quick Demo

Solar Power

Materials magnifying lens, foil pie pan, ball of tissue paper

Estimated Time 20 minutes

Procedure Place the ball of tissue paper in the pie pan. Hold the hand lens in the sunlight to focus the rays on the paper. Within a few minutes, the paper should begin to burn. Discuss the reason the paper burns. Have students consider how solar panels focus and collect solar energy for use in heating buildings. **WARNING:** *Warn students NOT to repeat this demo on their own.*

INTEGRATE Career

Physicists Student reports should include the wind farms in California, but also may mention those in the American Midwest. Physicists may work for power companies, planners, or universities.

Reading Check

Answer These areas have fairly strong and constant wind.

Teacher FYI

Wind Power Some energy experts project that wind power could provide more than 20 percent of the world's electricity, and 10 to 25 percent of the electricity generated in the United States, by the middle of the twenty-first century.

SECTION 2 Renewable Energy Resources F ◆ 77

Make a Model

Wind and Water Energy Have student groups make models that demonstrate how energy from wind or water can be used. There are several possibilities, including: a pinwheel toy, a model sailboat, a model water wheel, or a model dam with water flowing over a wheel that models a turbine. Have groups explain their models in short classroom presentations. LS **Kinesthetic**

Visual Learning

Figure 13 How was the area behind the dam probably different before the dam was built? The huge lake behind the dam was not there, and the land that is under the lake was dry, exposed ground. A river (the one that still flows downstream from the dam) flowed through a now-drowned river valley behind the dam.

Fun Fact

In some coastal areas where the difference between high tide and low tide is significant, power plants have been built near shore in order to produce electricity using the energy of raising and ebbing tides.

Energy from Water For a long time, waterwheels steadily spun next to streams and rivers. The energy in the flowing water powered the wheels that ground grain or cut lumber. More than a pretty picture, using a waterwheel in this way is an example of microhydropower. Microhydropower has been used throughout the world to do work.

Running water also can be used to generate electricity. Electricity produced by waterpower is called **hydroelectric energy.** To generate electricity from water running in a river, a large concrete dam is built to retain water, as illustrated in **Figure 13.** A lake forms behind the dam. As water is released, its force turns turbines at the base of the dam. The turbines then turn generators that make electricity.

At first it might appear that hydroelectric energy doesn't create any environmental problems and that the water is used with little additional cost. However, when dams are built, upstream lakes fill with sediment and downstream erosion increases. Land above the dam is flooded, and wildlife habitats are damaged.

Energy from Earth Erupting volcanoes and geysers like Old Faithful are examples of geothermal energy in action. The energy that causes volcanoes to erupt or water to shoot up as a geyser also can be used to generate electricity. Energy obtained by using hot magma or hot, dry rocks inside Earth is called **geothermal energy.**

Bodies of magma can heat large reservoirs of groundwater. Geothermal power plants use steam from the reservoirs to produce electricity, as shown in **Figure 14.** In a developing method, water becomes steam when it is pumped through broken, hot, dry rocks. The steam then is used to turn turbines that run generators to make electricity. The advantage of using hot, dry rocks is that they are found just about everywhere. Geothermal energy presently is being used in Hawaii and in parts of the western United States.

Figure 13 Hydroelectric power is important in many regions of the United States. Hoover Dam was built on the Colorado River to supply electricity for a large area.

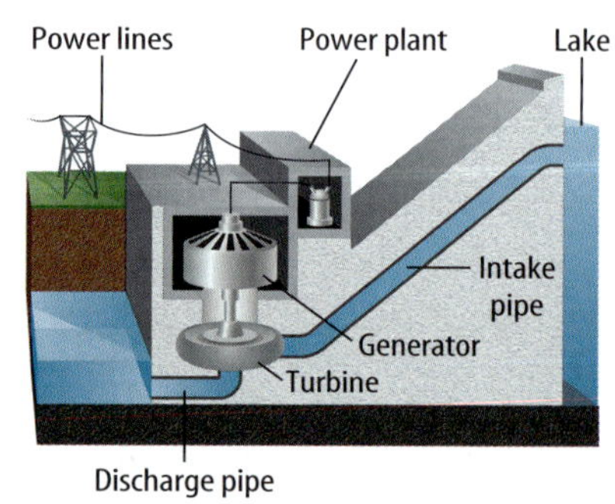

The power of running water is converted to usable energy in a hydroelectric power plant.

LAB DEMONSTRATION

Purpose to show students that sunlight is a source of energy

Materials 1.5-volt battery, 1.5-volt solar cell, small electric motor, two wires

Preparation Explain to the class how a solar cell works.

Procedure Connect the motor to the battery with the two wires. Demonstrate that the battery powers the motor. Disconnect the battery and connect the solar cell. Point the solar cell toward the Sun. Have students observe the motor. Block the light with a piece of cardboard. Again, observe the motor.

Expected Outcome Sunlight will provide energy to run the motor.

Assessment

Ask students which disadvantage of solar cells is shown in this demonstration. Solar cells only work when light shines on them.

Figure 14 Geothermal energy is used to supply electricity to industries and homes.

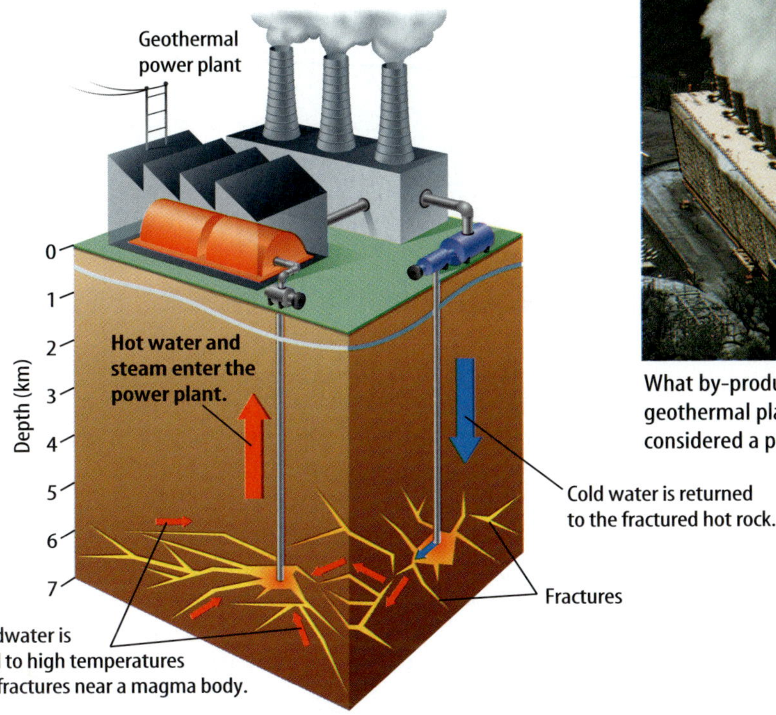

What by-product is produced in this geothermal plant in California? Is it considered a pollutant?

Renewable Energy Resources

Energy resources that can be replaced in nature or by humans within a relatively short period of time are referred to as renewable energy resources. This short period of time is defined generally as being within a human life span. For example, trees can be considered a renewable energy resource. As one tree is cut down, another can be planted in its place. The new tree might be left untouched or harvested.

You have learned that most energy used in the United States—about 90 percent—comes from fossil fuels, which are nonrenewable energy resources. Next, you'll look at some renewable energy resources and how they might fit into the world's total energy needs now and in the future.

Biomass Energy

A major renewable energy resource is biomass materials. **Biomass energy** is energy derived from burning organic material such as wood, alcohol, and garbage. The term *biomass* is derived from the words *biological* and *mass*.

Science Online

Topic: Biomass Energy

Visit bookf.msscience.com for Web links to information about biomass energy.

Activity List three new technologies that turn biomass into useable energy. Give two examples of each type of biomass and its energy technology.

Activity

Meeting Energy Needs Have students draw a plan for a town in which all of the energy needs are met by inexhaustible and renewable energy sources. Have them show the main energy utility (if there is one) and how it works, as well as individual initiatives such as solar panels on roofs, passive solar heating, waterwheels, or private windmills. Have students share their plans with the class. Discuss whether any of these ideas could be used in your community. L3 LS **Visual-Spatial** P

Inquiry Lab

Wind to Energy

Purpose To illustrate wind power as a natural resource for electricity generation.

Possible Materials pinwheel, electric fan, push pin, stopwatch, protractor, pen and paper

Estimated Time 30 minutes

Teaching Strategies

- Students can replicate simple wind powered electricity generators using pinwheels as windmills.
- Students should place a push pin through a straw rod, such that as the pinwheel tines pass the pin, a "click" is heard.
- Students can record the number of "clicks" that are heard by placing the pinwheels closer, farther, and at different angles to the electric fan that represents the wind.
- Students should infer the amount of electricity produced when the wind is stronger or when the wind approaches at different angles.

For additional inquiry activities, see *Science Inquiry Labs*.

Differentiated Instruction

Learning Disabled Ask students how many of them have ever sat in front of a wood-burning stove or fireplace. Emphasize that the wood is biomass. Ask students to explain why wood is considered a biomass fuel. It comes from trees, which are organic.

Caption Answer

Figure 15 Its availability and high energy content makes it very useful to many people.

Visual Learning

Figure 15 What else, in addition to energy for cooking and light, is the wood providing for the campers? It is providing heat energy to keep the campers warm.

Discussion

Energy Dependence Would it make sense for Americans to switch from a dependence on fossil fuels to a dependence on biomass? Have students explain their answers. Some students will think it is a good idea because biomass is renewable, especially if trees or crops are planted for use as fuel. Other students will think that biomass is not a good solution because it could require cutting forests (wood) or using a larger amount of fossil fuel energy to distill the fuel (ethanol) than would be produced burning it.

Teacher FYI

Biomass Use About 14 percent of the energy used worldwide to cook food and provide heat comes from burning wood and manure. In the United States and Canada, only 4 to 5 percent of energy used for those purposes comes from wood and manure.

Reading Check

Answer Air pollution; producing liquid biomass fuels takes more energy than is produced.

Figure 15 These campers are using wood, a renewable energy resource, to produce heat and light.
Discuss *Why do you think wood is the most commonly used biomass fuel?*

Energy from Wood If you've ever sat around a campfire, like the campers shown in **Figure 15,** or close to a wood-burning fireplace to keep warm, you have used energy from wood. The burning wood is releasing stored solar energy as heat energy. Humans have long used wood as an energy resource. Much of the world still cooks with wood. In fact, firewood is used more widely today than any other type of biomass fuel.

Using wood as a biomass fuel has its problems. Gases and small particles are released when wood is burned. These materials can pollute the air. When trees are cut down for firewood, natural habitats are destroyed. However, if proper conservation methods are employed or if tree farms are maintained specifically for use as fuel, energy from wood can be a part of future energy resources.

Energy from Alcohol Biomass fuel can be burned directly, such as when wood or peat is used for heating and cooking. However, it also can be transformed into other materials that might provide cleaner, more efficient fuels.

For example, during distillation, biomass fuel, such as corn, is changed to an alcohol such as ethanol. Ethanol then can be mixed with another fuel. When the other fuel is gasoline, the mixture is called gasohol. Gasohol can be used in the same way as gasoline, as shown in **Figure 16,** but it cuts down on the amount of fossil fuel needed to produce gasoline. Fluid biomass fuels are more efficient and have more uses than solid biomass fuels do.

The problem with this process is that presently, growing the corn and distilling the ethanol often uses more energy from burning fossil fuels than the amount of energy that is derived from burning ethanol. At present, biomass fuel is best used locally.

Figure 16 Gasohol sometimes is used to reduce dependence on fossil fuels.

Reading Check *What are the drawbacks of biomass fuels?*

Differentiated Instruction

Challenge Have students research the production and use of methanol gas, another biomass fuel. Methanol is an alcohol fuel produced mostly from natural gas, but which can also be produced from wood, garbage, coal, and farm waste such as corncobs. When used as a vehicle fuel, methanol produces less total air pollution than many other fuels. Making it can be expensive, however, and there are increased emissions of toxic formaldehyde. L3

80 ◆ F CHAPTER 3 Earth's Energy and Mineral Resources

Energy from Garbage Every day humans throw away a tremendous amount of burnable garbage. As much as two thirds of what is thrown away could be burned. If more garbage were used for fuel, as shown in **Figure 17,** human dependence on fossil fuels would decrease. Burning garbage is a cheap source of energy and also helps reduce the amount of material that must be dumped into landfills.

Compared to other nations, the United States lags in the use of municipal waste as a renewable energy resource. For example, in some countries in Western Europe, as much as half of the waste generated is used for biomass fuel. When the garbage is burned, heat is produced, which turns water to steam. The steam turns turbines that run generators to produce electricity.

Unfortunately, some problems can be associated with using energy from garbage. Burning municipal waste can produce toxic ash residue and air pollution. Substances such as heavy metals could find their way into the smoke from garbage and thus into the atmosphere.

Figure 17 Garbage can be burned to produce electricity at trash-burning power plants such as this one in Virginia.

section 2 review

Summary

Inexhaustible Energy Resources

- Solar cells are used to collect the Sun's energy.
- Wind energy produces no waste or pollution, however only a few areas are conducive for creating significant energy supplies.
- Dams are used to help provide running water, which is used to produce electricity.
- Energy obtained by using heat from inside Earth is called geothermal energy.

Renewable Energy Resources

- Biomass energy is produced when organic material such as wood, alcohol, or garbage is burned.
- Trash-burning power plants convert waste into electricty by burning garbage.

Self Check

1. **List** three advantages and disadvantages of using solar energy, wind energy, and hydroelectric energy.
2. **Explain** the difference between inexhaustible and renewable energy resources. Give two examples of each.
3. **Describe** how geothermal energy is used to create electricity.
4. **Infer** why nonrenewable resources are used more than inexhaustible and renewable resources.
5. **Think Critically** How could nuclear energy, which normally is classified as a nonrenewable energy resource, be reclassified as an inexhaustible energy resource?

Applying Skills

6. **Use a Spreadsheet** Make a table of energy resources. Include an example of how each resource is used. Then describe how you could reduce the use of energy resources at home.

Science Online bookf.msscience.com/self_check_quiz

3 Assess

DAILY INTERVENTION

Check for Understanding

Linguistic Have students draw a map of the United States that locates the areas where geothermal energy is predominant. Ask them to research how that energy is used and write a brief report on their findings. L2 LS

Reteach

Other Resources Ask students why garbage is considered a renewable energy resource and wind is considered inexhaustible. Garbage is renewable because people will always generate more. Wind is inexhaustible because the winds will never stop blowing. The only problem is finding a place where wind is strong and steady. L2

Assessment

Content Ask students to classify the following fuels as nonrenewable, renewable, or inexhaustible: wind, natural gas, ethanol, biomass, solar, nuclear, coal. nonrenewable: natural gas, nuclear, coal; renewable: ethanol, biomass; inexhaustible: wind, solar Use **Performance Assessment in the Science Classroom,** p. 89.

section 2 review

1. solar: advantages—nonpolluting, disadvantages—sunshine needed; wind: advantages—nonpolluting, disadvantages—need steady, strong wind; hydroelectric: advantages—no pollution, disadvantages—upstream lakes fill with sediment, animal habitat is destroyed
2. Inexhaustible energy resources (solar, wind, water, geothermal) will never run out; renewable energy resources (wood, alcohol, garbage) can be replenished in a short period of time.
3. Heat from magma or hot dry rocks changes water to steam that runs turbines that generate electricity.
4. Possible answer: The technology for using nonrenewable resources is more common and widespread.
5. Fusion would be virtually inexhaustible.
6. Table should include energy sources discussed in text and give an example of how each resource could be used.

BENCH TESTED

Real-World Question

Purpose Students observe how color affects energy absorption. L2 LS **Kinesthetic**

Process Skills use variables, constants, and controls; collect and organize data; observe; make and use tables; make and use graphs; analyze data; infer

Time Required 40–45 minutes

Alternate Materials 200-watt lamp with reflector and clamp, stopwatch

Safety Precautions Tell students to use thermal mitts when handling the light source.

Teaching Strategy Have students predict what will happen and then check their predictions after the activity.

Conclude and Apply

1. black soil; white sandy soil
2. At a certain point, the amount of energy radiated from the soils equaled the amount absorbed.
3. The black plates absorb the most solar energy.
4. The darker the color, the greater its ability to absorb energy.
5. Darker clothing absorbs more energy, keeping people warmer in cold weather.

LAB

Soaking Up Solar Energy

Winter clothing tends to be darker in color than summer clothing. The color of the material used in the clothing affects its ability to absorb energy. In this lab, you will use different colors of soil to study this effect.

Real-World Question

How does color affect the absorption of energy?

Goals

- **Determine** whether color has an effect on the absorption of solar energy.
- **Relate** the concept of whether color affects absorption to other applications.

Materials

dry, black soil
dry, brown soil
dry, sandy, white soil
thermometers (3)
ring stand
graph paper
colored pencils (3)
metric ruler
clear-glass or plastic dishes (3)
200-watt gooseneck lamp
**200-watt lamp with reflector and clamp*
watch or clock with second hand
**stopwatch*

**Alternate materials*

Safety Precautions

WARNING: *Handle glass with care so as not to break it. Wear thermal mitts when handling the light source.*

Procedure

1. Fill each dish with a different color of soil to a depth of 2.5 cm.
2. Arrange the dishes close together on your desk and place a thermometer in each dish. Be sure to cover the thermometer bulb in each dish completely with the soil.
3. Position the lamp over all three dishes.
4. **Design** a data table for your observations similar to the sample table above. You will need to read the temperature of each dish every 30 s for 20 min after the light is turned on.
5. Turn on the light and begin your experiment.
6. Use the data to construct a graph. Time should be plotted on the horizontal axis and temperature on the vertical axis. Use a different colored pencil to plot the data for each type of soil, or use a computer to design a graph that illustrates your data.

Time and Temperature

Time (min)	Temperature Dish A (°C)	Temperature Dish B (°C)	Temperature Dish C (°C)
0.0		Answers will vary.	
0.5			
1.0			
1.5			

Conclude and Apply

1. **Observe** which soil had the greatest temperature change. The least?
2. **Explain** why the curves on the graph flatten.
3. **Infer** Why do flat-plate solar collectors have black plates behind the water pipes?
4. **Explain** how the color of a material affects its ability to absorb energy.
5. **Infer** Why is most winter clothing darker in color than summer clothing?

Assessment

Performance To further assess how color affects the ability of substances to absorb energy, have students place thermometers under three different-colored pieces of cloth. Have students compare the results. Use **Performance Assessment in the Science Classroom,** p. 97.

Communicating Your Data

Encourage students to use an electronic spreadsheet program to help them design a data table.

section 3

Mineral Resources

Metallic Mineral Resources

If your room at home is anything like the one shown in **Figure 18,** you will find many metal items. Metals are obtained from Earth materials called metallic mineral resources. A **mineral resource** is a deposit of useful minerals. See how many metals you can find. Is there anything in your room that contains iron? What about the metal in the frame of your bed? Is it made of iron? If so, the iron might have come from the mineral hematite. What about the framing around the windows in your room? Is it aluminum? Aluminum, like that in a soft-drink can, comes from a mixture of minerals known as bauxite. Many minerals contain these and other useful elements. Which minerals are mined as sources for the materials you use every day?

Ores Deposits in which a mineral or minerals exist in large enough amounts to be mined at a profit are called **ores.** Generally, the term ore is used for metallic deposits, but this is not always the case. The hematite that was mentioned earlier as an iron ore and the bauxite that was mentioned earlier as an aluminum ore are metallic ores.

Reading Check *What is an ore?*

as you read

What You'll Learn

- **Explain** the conditions needed for a mineral to be classified as an ore.
- **Describe** how market conditions can cause a mineral to lose its value as an ore.
- **Compare and contrast** metallic and nonmetallic mineral resources.

Why It's Important

Many products you use are made from mineral resources.

Review Vocabulary

metal: a solid material that is generally hard, shiny, pliable and a good electrical conductor

New Vocabulary

- mineral resource
- ore
- recycling

Figure 18 Many items in your home are made from metals obtained from metallic mineral resources.

Section 3 Resource Manager

Chapter *Fast File* Resources

Transparency Activity, p. 46
Directed Reading for Content Mastery, pp. 21, 22
MiniLAB, pp. 21, 22
Enrichment, p. 32
Reinforcement, p. 29
Lab Worksheet, pp. 7–8

Lab Management and Safety, p. 71

1 Motivate

INTERACTIVE CHALKBOARD PowerPoint® Presentations

Bellringer

Section Focus Transparencies also are available on the Interactive Chalkboard CD-ROM. L2 ELL

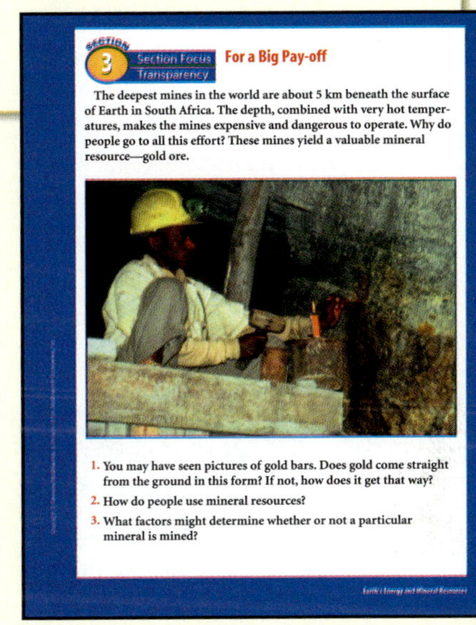
Section 3 Section Focus Transparency **For a Big Pay-off**

The deepest mines in the world are about 5 km beneath the surface of Earth in South Africa. The depth, combined with very hot temperatures, makes the mines expensive and dangerous to operate. Why do people go to all this effort? These mines yield a valuable mineral resource—gold ore.

1. You may have seen pictures of gold bars. Does gold come straight from the ground in this form? If not, how does it get that way?
2. How do people use mineral resources?
3. What factors might determine whether or not a particular mineral is mined?

Earth's Energy and Mineral Resources

Tie to Prior Knowledge

Mineral Use Ask students to brainstorm materials or products they use that are made from minerals. possible answers: gold or silver jewelry, metal coins, iron or steel cars or appliances, salt [halite] for food Tell students they will learn about the many minerals they use each day in this section.

Text Question Answer

Possible answer: Hematite (iron) and bauxite (aluminum) are among the materials mined.

Reading Check

Answer a deposit in which a mineral or minerals exist in amounts large enough to be mined at a profit

SECTION 3 Mineral Resources F ◆ 83

2 Teach

Caption Answer

Figure 19 Examples could be part of the frame of your bed, part of your school desk, or any product made from iron or steel

Activity

Metals List the following metals on the chalkboard: copper, bronze, gold, tin, steel, and silver. Have students find out which metals are minerals, and which mineral ore each metal comes from. Copper, gold, tin, and silver are minerals. Copper comes from chalcopyrite and native copper; gold comes from native gold; tin comes from cassiterite and native tin; silver comes from argentite and native silver; iron in steel comes from hematite; bronze is an alloy of copper and tin.

Fun Fact

The world's main source of phosphate is the mineral apatite. The main source of potassium is the mineral sylvite.

Figure 19 Iron ores are smelted to produce nearly pure iron.
List *three examples of what this iron could be used for.*

Economic Effects When is a mineral deposit considered an ore? The mineral in question must be in demand. Enough of it must be present in the deposit to make it worth removing. Some mining operations are profitable only if a large amount of the mineral is needed. It also must be fairly easy to separate the mineral from the material in which it is found. If any one of these conditions isn't met, the deposit might not be considered an ore.

Supply and demand is an important part of life. You might have noticed that when the supply of fresh fruit is down, the price you pay for it at the store goes up. Economic factors largely determine what an ore is.

Refining Ore The process of extracting a useful substance from an ore involves two operations—concentrating and refining. After a metallic ore is mined from Earth's crust, it is crushed and the waste rock is removed. The waste rock that must be removed before a mineral can be used is called gangue (GANG).

INTEGRATE Chemistry Refining produces a pure or nearly pure substance from ore. For example, iron can be concentrated from the ore hematite, which is composed of iron oxide. The concentrated ore then is refined to be as close to pure iron as possible. One method of refining is smelting, illustrated in **Figure 19.** Smelting is a chemical process that removes unwanted elements from the metal that is being processed. During one smelting process, a concentrated ore of iron is heated with a specific chemical. The chemical combines with oxygen in the iron oxide, resulting in pure iron. Note that one resource, fossil fuel, is burned to produce the heat that is needed to obtain the finished product of another resource, in this case iron.

Nonmetallic Mineral Resources

Any mineral resources not used as fuels or as sources of metals are nonmetallic mineral resources. These resources are mined for the nonmetallic elements contained in them and for the specific physical and chemical properties they have. Generally, nonmetallic mineral resources can be divided into two different groups—industrial minerals and building materials. Some materials, such as limestone, belong to both groups of nonmetallic mineral resources, and others are specific to one group or the other.

Curriculum Connection

Geography Have students list the top five countries for the production of copper, gold, and silver. Tell them to place symbols for each metal on a world map. copper: Chile, U.S., Canada, Russia, Indonesia; gold: South Africa, U.S., Australia, China, Russia; silver: Mexico, Peru, U.S., Canada, Chile

Teacher FYI

Mineral Consumption The consumption of minerals in the United States has been calculated to be about 8,100 kg per person annually. This includes metals such as steel, iron, aluminum, copper, and zinc, as well as nonmetals such as salt, cement, sand, and stone.

Industrial Minerals Many useful chemicals are obtained from industrial minerals. Sandstone is a source of silica (SiO_2), which is a compound that is used to make glass. Some industrial minerals are processed to make fertilizers for farms and gardens. For example, sylvite, a mineral that forms when seawater evaporates, is used to make potassium fertilizer.

Many people enjoy a little sprinkle of salt on french fries and pretzels. Table salt is a product derived from halite, a nonmetallic mineral resource. Halite also is used to help melt ice on roads and sidewalks during winter and to help soften water.

Other industrial minerals are useful because of their characteristic physical properties. For example, abrasives are made from deposits of corundum and garnet. Both of these minerals are hard and able to scratch most other materials they come into contact with. Small particles of garnet can be glued onto a sheet of heavy paper to make abrasive sandpaper. **Figure 20** illustrates just a few ways in which nonmetallic mineral resources help make your life more convenient.

Figure 20 You benefit from the use of industrial minerals every day.

Road salt melts ice on streets.

Many important chemicals are made from industrial minerals.

An industrial mineral called trona is important for making glass.

Mini LAB

Observing the Effects of Insulation

Procedure

1. Pour **warm water** into a **thermos bottle.** Cap it and set it aside.
2. Pour **cold water** with **ice** into a **glass** surrounded by a **thermal cup holder.**
3. Pour warm water—the same temperature as in step 1—into an **uncovered cup.** Pour cold water with ice into a glass container that is not surrounded by a thermal cup holder.
4. After 2 h, measure the temperature of each of the liquids.

Analysis

1. Infer how the insulation affected the temperature of each liquid.
2. Relate the usefulness of insulation in a thermos bottle to the usefulness of fiberglass insulation in a home.

Mini LAB

Purpose Students will observe the benefits of insulation. L2

LS **Kinesthetic**

Materials thermos bottle, thermal cup holder, two glasses, cup, 4 thermometers, water

Teaching Strategy Review the reading of thermometers before beginning the activity.

Safety Precautions Students should use care when handling glass equipment.

Analysis

1. The thermos bottle kept the water in it warmer than the water in the open cup. The water in the glass with the thermal cup holder stayed colder than the water in the glass without one.
2. Home insulation helps keep heat from a furnace or heater inside a home, the way insulation in a thermos holds in the heat of warm liquids.

Assessment

Performance Have students repeat the experiment with the warm water. Tell them to measure the temperature of the water in the thermos bottle and the water in the open cup periodically until both have reached a temperature you determine. Have them determine how much longer the thermos bottle kept the warm water warm. Use **Performance Assessment in the Science Classroom,** p. 97.

Discussion

Nonrenewable Minerals Would you consider mineral resources such as copper, quartz, gold, and tin to be nonrenewable resources? Some students would say that these are nonrenewable resources because once they are extracted from the ground and turned into something then the resource is consumed, especially if the product is thrown away. Other students will argue that mineral resources are renewable through recycling techniques.

Differentiated Instruction

Visually Impaired Have these students dip one hand into the cold water from the glass in the thermal cup holder and the other hand into the cold water from the glass without the holder. Have students determine which water was kept colder.

Reading Check

Answer sandstone, limestone, sylvite, halite, corundum, garnet, aggregate, gypsum

Applying Science

Teaching Strategy

Discuss science and technology, the price for each, the "good" and "bad" of each, and the decisions that are made, what they are based on, and the effects of the decisions made.

Answers

1. It has increased. Steel and aluminum are recycled the most. Plastics are the least recycled material. Plastics are more difficult to recycle because of the many different types of plastics.
2. Recycling can save electricity, petroleum, natural gas, and coal. This cuts down on pollution and conserves natural resources.

Quick Demo

Minerals for Health

Materials bottle of multivitamin supplements

Estimated Time 10 minutes

Procedure Have students read the nutritional facts on the back label of a multivitamin supplement bottle. Items such as zinc, magnesium, chromium, potassium, copper, and selenium should show up on the label. Have the students also read the ingredients to see the form or chemical name of the minerals. Explain that these minerals need to be extracted from Earth to be put into these multivitamins. Also explain that humans need minerals and vitamins to maintain good health.

Building Materials One of the most important nonmetallic mineral resources is aggregate. Aggregate is composed of crushed stone or a mixture of gravel and sand and has many uses in the building industry. For example, aggregates can be mixed with cement and water to form concrete. Quality concrete is vital to the building industry. Limestone also has industrial uses. It is used as paving stone and as part of concrete mixtures. Have you ever seen the crushed rock in a walking path or driveway? The individual pieces might be crushed limestone. Gypsum, a mineral that forms when seawater evaporates, is soft and lightweight and is used in the production of plaster and wallboard. If you handle a piece of broken plaster or wallboard, note its appearance, which is similar to the mineral gypsum.

Rock also is used as building stone. You might know of buildings in your region that are made from granite, limestone, or sandstone. These rocks and others are quarried and cut into blocks and sheets. The pieces then can be used to construct buildings. Some rock also is used to sculpt statues and other pieces of art.

Reading Check *What are some important nonmetallic mineral resources?*

Applying Science

Why should you recycle?

Recycling in the United States has become a way of life. In 2000, 88 percent of Americans participated in recycling. Recycling is important because it saves precious raw materials and energy. Recycling aluminum saves 95 percent of the energy required to obtain it from its ore. Recycling steel saves up to 74 percent in energy costs, and recycling glass saves up to 22 percent.

Identifying the Problem

The following table includes materials that currently are being recycled and rates of recycling for the years 1995, 1997, and 2001. Examine the table to determine materials for which recycling increased or decreased between 1995 and 2001.

Recycling Rates in the United States

Material	1995 (%)	1997 (%)	2001 (%)
Glass	24.5	24.3	27.2
Steel	36.5	38.4	43.5 (est.)
Aluminum	34.6	31.2	33.0
Plastics	5.3	5.2	7.0

Solving the Problem

1. Has the recycling of materials increased or decreased over time? Which materials are recycled most? Which materials are recycled least? Discuss why some materials might be recycled more than others.
2. How can recycling benefit society? Explain your answer.

Differentiated Instruction

Challenge Challenge students to do research using print resources and the Internet to make a mineral resource map of the United States, showing the locations of major minerals with symbols they design. Students can work in groups of two or three. Display the maps in class. L3 COOP LEARN P

Teacher FYI

Aggregate Aggregate is used in the building industry. When concrete is mixed, aggregate is added to provide it with strength, and cement is added as a binder to make it hard as a rock. More than 85 metric tons of aggregate are used to build every two kilometers of a four-lane highway.

Recycling Mineral Resources

Mineral resources are nonrenewable. You've learned that nonrenewable resources are those that Earth processes cannot replace within an average human's lifetime. Most mineral resources take millions of years to form. Have you ever thrown away an empty soft-drink can? Many people do. These cans become solid waste. Wouldn't it be better if these cans and other items made from mineral resources were recycled into new items?

Recycling is using old materials to make new ones. Recycling has many advantages. It reduces the demand for new mineral resources. The recycling process often uses less energy than it takes to obtain new material. Because supplies of some minerals might become limited in the future, recycling could be required to meet needs for certain materials, as shown in **Figure 21.**

Recycling also can be a profitable experience. Some companies purchase scrap metal and empty soft-drink cans for the aluminum and tin content. The seller receives a small amount of money for turning in the material. Schools and other groups earn money by recycling soft-drink cans.

Percentage Comparison of Mineral Resources

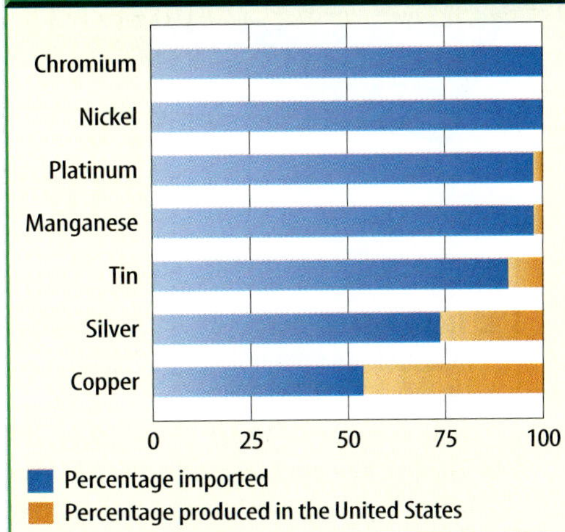

Figure 21 The United States produces only a small percentage of the metallic resources it consumes.

section 3 review

Summary

Metallic Mineral Resources

- Minerals found in rocks that can be mined for an economic profit are called ores.

Nonmetallic Mineral Resources

- Nonmetallic mineral resources can be classified into groups: industrial minerals and building materials.
- Sedimentary rocks such as limestone and sandstone can be used as building materials to make things like buildings and statues.

Recycling Mineral Resources

- Recycling materials helps to preserve Earth's resources by reusing old or used materials without extracting new resources from Earth.
- The recycling process may use fewer resources than it takes to obtain new material.

Self Check

1. **Explain** how metals obtained from metallic mineral resources are used in your home and school. Which of these products could be recycled easily?
2. **List** two industrial uses for nonmetallic mineral resources.
3. **Explain** how supply and demand of a material can cause a mineral to become an ore.
4. **Think Critically** Gangue is waste rock remaining after a mineral ore is removed. Why is gangue sometimes reprocessed?

Applying Skills

5. **Classify** the following mineral resources as metallic or nonmetallic: *hematite, limestone, bauxite, sandstone, garnet,* and *chalcopyrite.* Explain why you classified each one as you did.

Science Online bookf.msscience.com/self_check_quiz

3 Assess

DAILY INTERVENTION

Check for Understanding

Logical-Mathematical Have students investigate world production of copper and graph their findings.

Reteach

Mineral Walk Take a brief mineral identification walk through the neighborhood near the school with students. Point out buildings or objects as you walk, and ask students to identify the minerals that could have gone into building or producing the object in question. Have students take notes and write a paragraph summing up what they have discovered.

Assessment

Content Have students work in groups of three to write a poem using the terms *mineral resource* and *ore* in a way that shows their meaning. Use **Performance Assessment in the Science Classroom,** p. 151.

Visual Learning

Figure 21 If the United States only produces a small percentage of the metal resources it consumes, then where does the remainder of metal resources come from? The U.S. imports the materials from other countries.

section 3 review

1. Answers will vary; desks, bedframes, kitchen utensils, soft-drink cans, window or chalkboard casings. Soft drink cans can be recycled easily.
2. manufacture of chemicals, abrasives
3. A material that can be mined at a profit is an ore. If there is no demand, there is no profit. If there is a high demand, there is a profit.
4. If the price of the metal goes up or technology improves, the amount of metal still in the gangue might be removed at a profit.
5. Metallic: hematite, bauxite, chalcopyrite; nonmetallic: limestone, sandstone, garnet; metals are derived from hematite, bauxite, and chalcopyrite but not from the others.

BENCH TESTED

Real-World Question

Purpose Students will design and build a model of an energy-efficient home. L2 LS **Logical-Mathematical**

Process Skills observe and infer, recognize cause and effect, make models, communicate

Time Required two weeks

Discussion Ask students to list the desirable characteristics of an energy-efficient home.

Possible Materials Be sure the cardboard used in the construction of the model homes is sturdy. If using poster board, students may need to reinforce it or use a double layer.

Alternate Materials Instead of cardboard, students may use corrugated cardboard from packing boxes.

Make a Model

Teaching Strategies

- Allow students to work in small groups when designing and building their models.
- Encourage communication between groups. Ask selected students to research different aspects of the topic of energy-efficient homes and present their findings to the class.

LAB **Model and Invent**

Home Sweet Home

Goals

- **Research** various inexhaustible and other energy resources available to use in the home.
- **Design** blueprints for an energy-efficient home and/or design and build a model of an energy-efficient home.

Possible Materials

paper
ruler
pencils
cardboard
glue
aluminum foil

Real-World Question

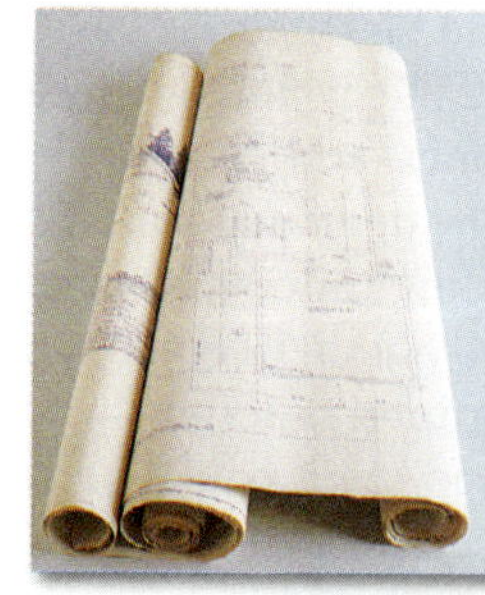

As fossil fuel supplies continue to be depleted, an increasing U. S. population has recognized the need for alternative energy sources. United States residents might be forced to consider using inexhaustible and other renewable energy resources to meet some of their energy needs. The need for energy-efficient housing is more relevant now than ever before. A designer of energy-efficient homes considers proper design and structure, a well chosen building site with wise material selection, and selection of efficient energy generation systems to power the home. Energy-efficient housing uses less energy and produces fewer pollutants. What does the floor plan, building plan, or a model of an energy efficient home look like? How and where should your house be designed and built to efficiently use the alternative energy resources you've chosen?

Using Scientific Methods

Make a Model

Plan

1. **Research** current information about energy-efficient homes.
2. **Research** inexhaustible energy resources such as wind, hydroelectric power, or solar power, as well as energy conservation. Decide which energy resources are most efficient for your home design.
3. Decide where your house should be built to use energy efficiently.
4. Decide how your house will be laid out and draw mock blueprints for your home. Highlight energy issues such as where solar panels can be placed.
5. Build a model of your energy-efficient home.

Do

1. Ask your peers for input on your home. As you research, become an expert in one area of alternative energy generation and share your information with your classmates.
2. **Compare** your home's design to energy-efficient homes you learn about through your research.

Test Your Model

1. Think about how most of the energy in a home is used. Remember as you plan your home that energy-efficient homes not only generate energy—they also use it more efficiently.
2. Carefully consider where your home should be built. For instance, if you plan to use wind power, will your house be built in an area that receives adequate wind?
3. Be sure to plan for backup energy generation. For instance, if you plan to use mostly solar energy, what will you do if it's a cloudy day?

Analyze Your Data

Devise a budget for building your home. Could your energy-efficient home be built at a reasonable price? Could anyone afford to build it?

Conclude and Apply

Create a list of pro and con statements about the use of energy-efficient homes. Why aren't inexhaustible and other renewable energy sources widely used in homes today?

Communicating Your Data

Present your model to the class. Explain which energy resources you chose to use in your home and why. Have an open house. Take prospective home owners/classmates on a tour of your home and sell it.

Tie to Prior Knowledge Have students name and describe the benefits of energy-efficient designs in their homes, such as double-paned windows and storm doors.

Troubleshooting Encourage students to think about energy losses through the roof, ceiling, doors, and windows, as well as the home's location and source of energy.

Test Your Model

Expected Outcome Students will learn about the many ways a home can be made energy efficient.

Analyze Your Data

Budgets should list major supplies needed as well as their costs. Students should realize that some energy-efficient designs are prohibitively expensive, but many methods for improving energy efficiency are affordable.

Conclude and Apply

Possible answer: Energy-efficient homes conserve Earth's natural resources and reduce environmental damage, but they are often expensive. Readily available and affordable renewable energy sources are yet to be developed.

Differentiated Instruction

Learning Disabled Working in small groups will allow students to work off each other's ideas as they design and build the model homes. Allow class time for them to discuss the project. Provide students with suggestions as needed to help them develop their ideas.

Communicating Your Data

Have students use computer graphics software to prepare a half-page advertisement for their home. The advertisement should describe details of the energy-efficient design of the home. These pages can be available to prospective buyers during the open-house tour of their home.

Assessment

Oral After completing the project, have students discuss the different energy-efficient technologies they investigated, describing the benefits and disadvantages of each technology. Use **Performance Assessment in the Science Classroom,** p. 89.

Content Background

Petroleum from ground seepages, often in the form of asphalt or tar, has been used since ancient times for boat sealing, as a medicine, and for lighting purposes. Before the 1850s, lamp oil was derived from whales. However, whales were overhunted and lamp oil became more expensive, leading to the search for substitutes. Oil began to be dug from surface seeps and refined by heating into kerosene, which was used as a lamp oil.

After water and food, petroleum may be the most important substance used in modern society. It provides the raw materials for products ranging from plastics and many other synthetics to those used in medicine and agriculture, as well as fuel for heating, industry, and transportation.

Discussion

Oil Importance What are some ways that oil has played a part in your preparation for school today? Possible answers: some homes are heated by fuel oil; some clothing is made from raw materials that come from oil; vehicles powered by gasoline or other petroleum-based fuel transported some of them to school.

Analyze the Event

Discuss Ask students to brainstorm how their lives might be different today if people hadn't realized the importance of the discovery of oil. Possible answers: Most forms of transportation use fuels derived from petroleum—gasoline, diesel fuel, and jet fuel—so transportation would be different and probably difficult. Many power plants use oil to generate electricity, so electricity would not be as readily available and would probably be more expensive. Homes and businesses would not be able to use fuel oil for heating. Many products, such as plastics, synthetic fibers, and Plexiglas, which come from petrochemicals, wouldn't exist.

Oops! Accidents in SCIENCE

SOMETIMES GREAT DISCOVERIES HAPPEN BY ACCIDENT!

BLACK GOLD!

What if you went out to your backyard, started digging a hole, and all of the sudden oil spurted out of the ground? Dollar signs might flash before your eyes.

It wasn't quite that exciting for Charles Tripp. Tripp, a Canadian, is credited with being the first person to strike oil. And he wasn't even looking for what has become known as "black gold."

In 1851, Tripp built a factory in Ontario, Canada, not far from Lake Erie. He used a natural, black, thick, sticky substance that could be found nearby to make asphalt for paving roads and to construct buildings.

In 1855, Tripp dug a well looking for freshwater for his factory. After digging just 2 m or so, he unexpectedly came upon liquid. It wasn't clear, clean, and delicious; it was smelly thick, and black. You guessed it—oil! Tripp didn't understand the importance of his find. Two years after his accidental discovery, Tripp sold his company to James Williams. In 1858,

The Titusville, Pennsylvania, oil well drilled by Edwin Drake. This photo was taken in 1864.

Some people used TNT to search for oil. This photo was taken in 1943.

Williams continued to search for water for the factory, but, as luck would have it, diggers kept finding oil.

Some people argue that the first oil well in North America was in Titusville, Pennsylvania, when Edwin Drake hit oil in 1859. However, most historians agree that Williams was first in 1858. But they also agree that it was Edwin Drake's discovery that led to the growth of the oil industry. So, Drake and Williams can share the credit!

Today, many oil companies are drilling beneath the sea for oil.

Make a Graph Research the leading oil-producing nations and make a bar graph of the top five producers. Research how prices of crude oil affect the U.S. and world economies. Share your findings with your class.

Science Online For more information, visit bookf.msscience.com/oops

Make a Graph Students should find that the world's major producers of oil are Saudi Arabia, United States, Russia, Iran, and China. During the Arab oil embargo of the 1970s, the United States gross national product declined and unemployment doubled. In 1990, in response to a threat to Middle East oil supplies, the Gulf War was fought.

Resources for Teachers and Students

Business Builders in Oil, by Nathan Aaseng, Oliver Press, 2000

Where Does Oil Come From? By C. Vance Cast, Barron's, 1993

Reviewing Main Ideas

Section 1 Nonrenewable Energy Resources

1. Fossil fuels are considered to be nonrenewable energy resources.
2. The higher the concentration of carbon in coal is, the cleaner it burns.
3. Oil and natural gas form from altered and buried marine organisms and often are found near one another.
4. Nuclear energy is obtained from the fission of heavy isotopes.

Section 2 Renewable Energy Resources

1. Inexhaustible energy resources—solar energy, wind energy, hydroelectric energy, and geothermal energy—are constant and will not run out.
2. Renewable energy resources are replaced within a relatively short period of time.
3. Biomass energy is derived from organic material such as wood and corn.

Section 3 Mineral Resources

1. Metallic mineral resources provide metals.
2. Ores are mineral resources that can be mined at a profit.
3. Smelting is a chemical process that removes unwanted elements from a metal that is being processed.
4. Nonmetallic mineral resources are classified as industrial minerals or building materials.

Visualizing Main Ideas

Copy and complete the following table that lists advantages and disadvantages of energy resources.

Energy Resources

Resource	Advantages	Disadvantages
Fossil fuels	economical, works with current technology	nonrenewable
Nuclear energy	alternate source as fossil fuels are depleted	Fission produces radioactive waste.
Solar energy	inexhaustible	not always sunny
Wind energy	inexhaustible	not always windy
Geothermal energy	inexhaustible	not available everywhere
Biomass fuel	readily available in different forms	can produce air pollution

chapter 3 Study Guide

Reviewing Main Ideas

Summary statements can be used by students to review the major concepts of the chapter.

Visualizing Main Ideas

See student page.

Science Online

Visit bookf.msscience.com
/self_check_quiz
/interactive_tutor
/vocabulary_puzzlemaker
/chapter_review
/standardized_test

Assessment Transparency

For additional assessment questions, use the *Assessment Transparency* located in the transparency book.

Assessment

Assessment Transparency **Earthquakes**

Directions: *Carefully review the table and answer the following questions.*

Worldwide Earthquakes, 1990–1994

Magnitude	1990	1991	1992	1993	1994
8.0–9.9	0	0	0	1	2
7.0–7.9	12	11	23	15	13
6.0–6.9	115	105	104	141	161
5.0–5.9	1,635	1,469	1,541	1,449	1,542
4.0–4.9	4,493	4,372	5,196	5,034	4,544
3.0–3.9	3,457	2,952	4,643	4,263	5,000
2.0–2.9	2,364	2,927	3,068	5,390	5,369
1.0–1.9	474	801	887	1,177	779
0.1–0.9	0	1	2	9	17
No Magnitude	5,062	3,878	4,084	3,997	1944
Total	16,612	16,516	19,548	21,476	19,371
Estimated Deaths	51,916	2,326	3,814	10,036	1,038

1. According to this information, which magnitude earthquakes were the most common between 1990 and 1994?
   A No magnitude C 4.0–4.9
   B 3.0–3.9 D 5.0–5.9
2. According to the information in the table, the year that experienced the most deadly earthquake activity was____.
   F 1992 H 1993
   G 1994 J 1990
3. According to the table, the year that had the greatest number of earthquakes was____.
   A 1992 C 1993
   B 1991 D 1990

Earthquakes

chapter 3 Review

Using Vocabulary

1. ore
2. coal
3. recycling
4. solar energy
5. nuclear energy
6. oil

Checking Concepts

7. D	12. B
8. A	13. A
9. C	14. A
10. B	15. D
11. A	16. D

chapter 3 Review

Using Vocabulary

biomass energy p. 79	nuclear energy p. 73
coal p. 67	oil p. 69
fossil fuel p. 66	ore p. 83
geothermal energy p. 78	recycling p. 87
hydroelectric energy p. 78	reserve p. 71
mineral resource p. 83	solar energy p. 76
natural gas p. 69	wind farm p. 77

Each phrase below describes a vocabulary word from the list. Write the word that matches the phrase describing it.

1. mineral resource mined at a profit
2. fuel that is composed mainly of the remains of dead plants
3. method of conservation in which items are processed to be used again
4. inexhaustible energy resource that is used to power the *Hubble Space Telescope*
5. energy resource that is based on fission
6. liquid from remains of marine organisms

Checking Concepts

Choose the word or phrase that best answers the question.

7. Which has the highest content of carbon?
   A) peat C) bituminous coal
   B) lignite D) anthracite coal

8. Which is the first step in coal formation?
   A) peat C) bituminous coal
   B) lignite D) anthracite

9. Which of the following is an example of a fossil fuel?
   A) wind C) natural gas
   B) water D) uranium-235

10. What is the waste material that must be separated from an ore?
    A) smelter C) mineral resource
    B) gangue D) petroleum

11. What common rock structure can trap oil and natural gas under it?
    A) folded rock C) porous rock
    B) sandstone rock D) permeable rock

Use the figure below to answer question 12.

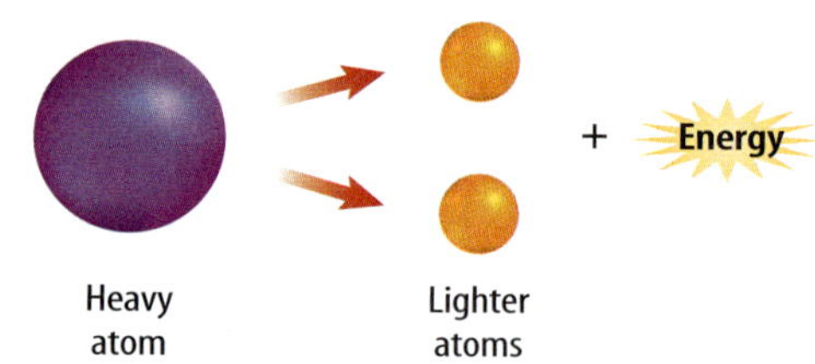

12. What other particles are released in the reaction above?
    A) protons C) uranium atoms
    B) neutrons D) heavy atoms

13. What is a region where many windmills are located in order to generate electricity from wind called?
    A) wind farm
    B) hydroelectric dam
    C) oil well
    D) steam-driven turbine

14. Which of the following is a deposit of hematite that can be mined at a profit?
    A) ore C) gangue
    B) anthracite D) energy resource

15. What is an important use of petroleum?
    A) making plaster C) as abrasives
    B) making glass D) making gasoline

16. Which of the following is a nonrenewable energy resource?
    A) water C) geothermal
    B) wind D) petroleum

Science Online bookf.msscience.com/vocabulary_puzzlemaker

Use the ExamView® Pro Testmaker CD-ROM to:

- create multiple versions of tests
- create modified tests with one mouse click for inclusion students
- edit existing questions and add your own questions
- build tests aligned with state standards using built-in State Curriculum Tags
- change English tests to Spanish with one mouse click and vice versa

Thinking Critically

17. Describe the major problems associated with generating electricity using nuclear power plants.

18. Explain why wind is considered to be an inexhaustible energy resource.

19. Determine which type of energy resources are considered to be biomass fuels. List three biomass fuels.

20. Discuss two conditions which could occur to cause gangue to be reclassified as an ore.

21. Predict If a well were drilled into a rock layer containing petroleum, natural gas, and water, which substance would be encountered first? Illustrate your answer with a labeled diagram.

22. Compare and contrast solar energy and wind energy by creating a table.

23. Concept Map Copy and complete the following concept map about mineral resources.

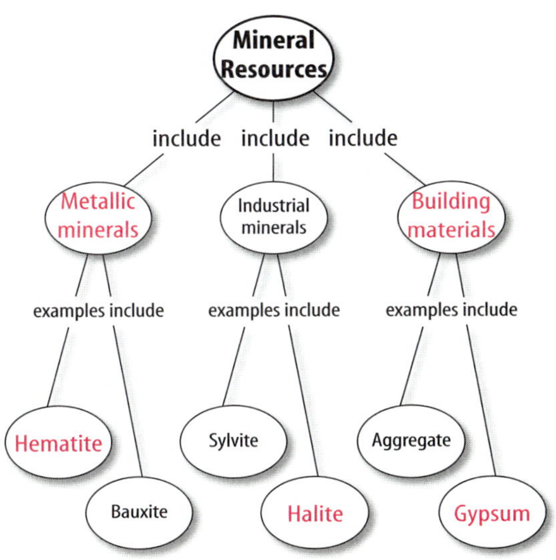

Performance Activities

24. Make Models Make a blueprint of a house that has been built to use passive solar energy. Which side of the house will face the sun?

25. Letter Write a letter to the Department of Energy asking how usable energy might be obtained from methane hydrates in the future. Also inquire about methods to extract methane hydrates.

Applying Math

Use the table below to answer questions 26–28.

Big Canyon Mine			
Ore Mineral	Metal	Percent Composition	Value (dollars/kg)
Bauxite	Aluminum	5	1.00
Hematite	Iron	2	4.00
Chalcopyrite	Copper	1	6.00
Galena	Lead	7	1.00

26. Ore Composition If 100 kg of rock are extracted from this mine, what percentage of the rock is gangue?

A. 15% **C.** 74%

B. 26% **D.** 85%

27. Total Composition Graph the total composition of the extracted rock using a circle graph. Label each component clearly. Provide a title for your graph.

28. Economic Geology Of that 100 kg of extracted rock, determine how many kilograms of each ore mineral is extracted. List the total dollar value for each metal after the gangue has been eliminated and the ore mineral extracted.

Assessment Resources

Reproducible Masters

Chapter *Fast File* Resources
Chapter Review, pp. 37–38
Chapter Tests, pp. 39–42
Assessment Transparency Activity, p. 49

Glencoe Science Web site
Chapter Review Test
Standardized Test Practice

Glencoe Technology

Assessment Transparency
Exam*View*® Pro Testmaker
MindJogger Videoquiz
Interactive Chalkboard

Thinking Critically

17. the possibility of uncontrolled reactions, thermal pollution, storage and disposal of radioactive wastes

18. Winds will always blow, so wind energy will never run out.

19. Biomass fuels are made from organic material. They include wood, alcohols made from plants, and certain types of garbage.

20. If the price of a metal being mined increases, or new technology makes it easier and cheaper to extract, it might become profitable to remove the metal in the gangue.

21. from top to bottom: natural gas, petroleum, water.

22. Tables should show that both are low-cost fuels, nonpolluting, inexhaustible, and limited by weather and geographic location. Unlike wind, which can be used at any time of day or night, solar energy is limited to daylight hours when the Sun shines.

23. See student page.

Performance Activities

24. The house plans should include large windows facing the Sun.

25. Letters should refer to gas hydrate deposits. Students can search the Internet for an address at the Department of Energy web site. Use **Performance Assessment in the Science Classroom,** p. 139.

Applying Math

National Math Standards

5

26. D

27. Graph will include percent compositions for each of the minerals in the table, gangue will be 85 percent.

28. bauxite-5 kg-aluminum-$5; hematite-2 kg-iron-$8; chalcopyrite-1 kg-copper-$6; galena-7 kg-lead-$7

FAST FILE

Answer Sheet A practice answer sheet can be found at bookf.msscience.com/answer_sheet.

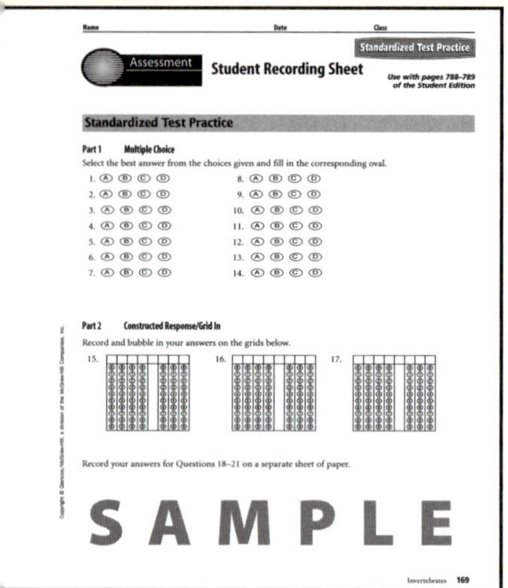

Part 1 Multiple Choice

1. B	6. B
2. D	7. C
3. C	8. A
4. B	9. A
5. D	10. C

Part 2 Short Response

11. Burning garbage is a cheap source of energy and also helps reduce the amount of material that must be dumped into landfills.
12. persistent wind that blows at an appropriate speed
13. Coal is removed from the ground by either strip mining or underground mining. Oil and natural gas are removed by pumping them out of the ground.
14. Bacterial growth within the decaying plant material causes the gradual breakdown of molecules in the plant tissue, leaving carbon and some impurities behind. This is the material that will eventually become coal.

chapter 3 Standardized Test Practice

Part 1 Multiple Choice

Record your answers on the answer sheet provided by your teacher or on a sheet of paper.

1. Which is a sedimentary rock formed from decayed plant matter?
   A. biomass **C.** natural gas
   B. coal **D.** oil

Use the graph below to answer questions 2 and 3.

Energy Use in the United States, 2002

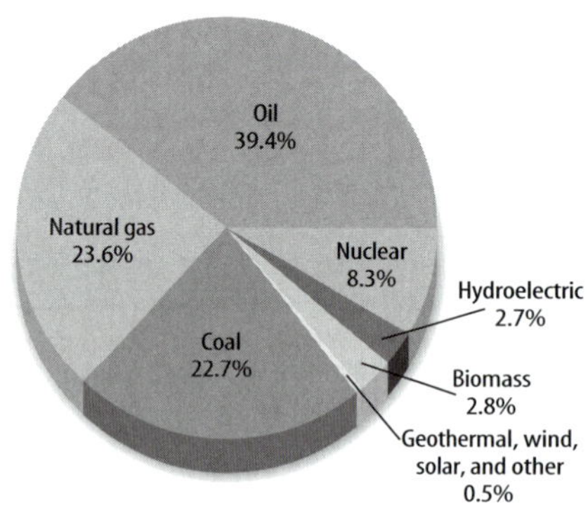

2. What percentage of the energy used in the United States comes from fossils fuels?
   A. 6% **C.** 47%
   B. 24% **D.** 86%

3. What percentage of our energy sources would be lost if coal runs out?
   A. 6% **C.** 23%
   B. 8% **D.** 39%

4. Which is a new potential source of methane?
   A. coal **C.** hydrocarbons
   B. hydrates **D.** petroleum

5. Which is an inexhaustible resource?
   A. coal
   B. nuclear
   C. oil
   D. solar

Use the illustration below to answer questions 6 and 7.

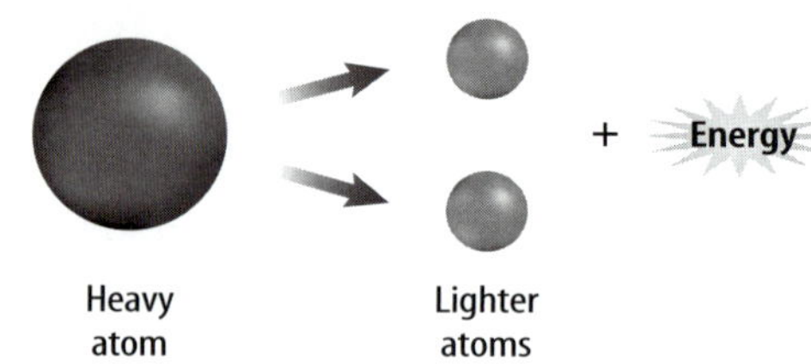

6. Which type of energy source is shown in this diagram?
   A. coal **C.** fusion
   B. fission **D.** natural gas

7. What is produced that drives a turbine, which turns a generator?
   A. atoms **C.** steam
   B. neutrons **D.** waste

8. Which type of energy uses magma or hot dry rocks to generate electricity?
   A. geothermal
   B. hydroelectric
   C. nuclear
   D. solar

9. What is combined to make gasohol?
   A. ethanol and gasoline
   B. oil and gasoline
   C. oil and petroleum
   D. wood and gasoline

10. Which helps to reduce the demand for new mineral resources?
   A. generating **C.** recycling
   B. mining **D.** refining

Test-Taking Tip

Circle Graphs If the question asks about the sum of multiple segments of a circle graph, do your addition on scratch paper and double-check your math before selecting an answer.

15. Methane hydrates are highly flammable and contain greenhouse gases, which could be harmful to the environment.
16. No fuel problem exists if the low-mass material is a commonly occurring substance. If the end product is not radioactive, storing nuclear waste is not a problem.
17. A mineral resource is a deposit of useful minerals. Deposits in which a mineral or minerals exist in large enough amounts to be mined at a profit are called ores. Yes, if the demand for a mineral increases, the mineral resource could then be mined at a profit. If a product made from an ore was replaced with another product, the ore might not be able to be mined at a profit.
18. halite; Halite can also be used as table salt and to help soften water.

Part 2 Short Response/Grid In

Record your answers on the answer sheet provided by your teacher or on a sheet of paper.

11. What are two advantages of burning garbage for fuel?

12. What conditions are necessary for a wind farm?

13. Contrast methods used to remove coal with methods used to remove oil and natural gas.

14. How are bacteria involved in the formation of coal?

15. Why are methane hydrates so difficult to extract from the seafloor?

16. List two advantages of using fusion as an energy source.

17. Compare and contrast a mineral resource and an ore. How could a mineral resource become an ore? Is it possible for an ore to become just a mineral resource? Explain your answers.

Use the photo below to answer question 18.

18. What type of nonmetallic mineral resource is being used in this picture? List two other uses for this nonmetallic mineral.

Part 3 Open Ended

Record your answers on a sheet of paper.

19. Contrast the amount of heat released and smoke produced when burning peat, lignite, bituminous coal, and anthracite coal.

20. What are some household ways to help conserve fossil fuels?

Use the illustration below to answer question 21.

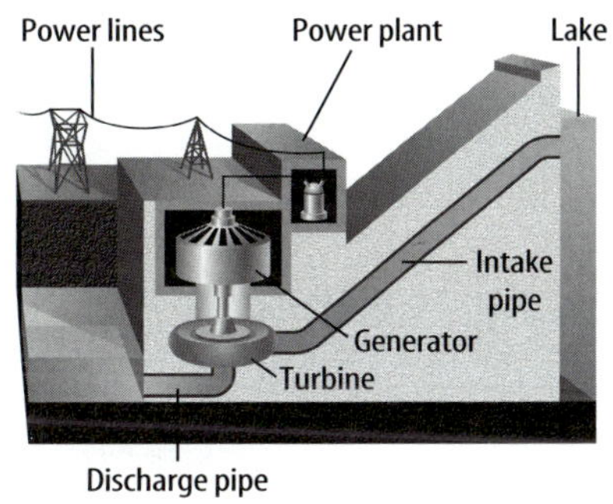

21. How is energy to run the turbine being produced? Discuss environmental issues associated with this energy source.

22. Design a 4-part, time-lapse illustration to show the path of iron from the hematite mine to pure iron.

23. How do population growth and technology affect the use of nonrenewable resources?

24. Some sources describe the Sun, wind, water, and geothermal energy as inexhaustible energy resources. What might be some limitations to these resources?

25. Are mineral resources considered to be renewable or nonrenewable? Explain your answer.

Part 3 Open Ended

19. As you burn peat, lignite, bituminous coal, and anthracite coal, each fuel produces more heat and less smoke than the fuel before it.

20. turning off lights and appliances that are not in use, make sure doors and windows are shut tightly during cold weather, run the air conditioning as little as possible, and add additional insulation to your home and water heater

21. To generate electricity from running water, a large concrete dam is built to retain water. A lake forms behind the dam. As water is released, its force turns the turbines at the base of the dam. The turbines then turn generators to make electricity. However, when dams are built, upstream lakes fill with sediment and downstream erosion increases. Land above the dam is flooded and wildlife habitats are destroyed.

22. Student illustrations should show each of the steps involved in concentrating and refining hematite. Hematite is mined from Earth's crust and crushed to remove waste rock. Refining produces a pure or nearly pure substance. Iron can be concentrated from the ore hematite by smelting. During one smelting process, a concentrated iron ore is heated with a specific chemical. The chemical combines with oxygen from the iron oxide, resulting in pure iron.

23. Some energy resources on Earth are being used faster than natural Earth processes can replace them. Changes in population and technology affect the rate in which nonrenewable energy resources become depleted.

24. solar—geographic location, time of day, cloud cover, battery limitations; wind—geographic location, fluctuations in wind speed, amount of space; water—geographic location, sedimentation, erosion, destruction of habitats; geothermal—differences in economics and environmental sensitivities of different geographic locations

25. nonrenewable; Nonrenewable resources are those that Earth processes cannot replace within an average human's lifetime. Most mineral resources take millions of years to form.

Rubrics

For more help evaluating open-ended assessment questions, see the rubric on p. 10T.

chapter 4 Organizer

Section/Objectives	Standards		Labs/Features
	National	State/Local	
Chapter Opener	See p. 9T–10T for a Key to Standards.		**Launch Lab:** Reassemble an Image, p. 97 **Foldables,** p. 97
Section 1 Continental Drift 2 sessions 1 block 1. **Describe** the hypothesis of continental drift. 2. **Identify** evidence supporting continental drift.	National Content Standards: UCP.1–3, UCP.5, A.1, A.2, B.2, D.1, D.2, F.3		**Science Online,** p. 99 **MiniLAB:** Interpreting Fossil Data, p. 100
Section 2 Seafloor Spreading 2 sessions 1 block 3. **Explain** seafloor spreading. 4. **Recognize** how age and magnetic clues support seafloor spreading.	National Content Standards: UCP.1–3, UCP.5, A.1, A.2, B.2, D.1, F.3		**Integrate Chemistry,** p. 103 **Lab:** Seafloor Spreading Rates, p. 105
Section 3 Theory of Plate Tectonics 4 sessions 2 blocks 5. **Compare** and contrast different types of plate boundaries. 6. **Explain** how heat inside Earth causes plate tectonics. 7. **Recognize** features caused by plate tectonics.	National Content Standards: UCP.1–3, UCP.5, A.1, A.2, B.2, D.1, F.3, G.2		**Science Online,** p. 108 **Applying Science:** How well do the continents fit together?, p. 108 **Visualizing Plate Boundaries,** p. 109 **MiniLAB:** Modeling Convection Currents, p. 111 **Integrate Career,** p. 113 **Integrate Physics,** p. 114 **Lab:** Predicting Tectonic Activity, p. 116 **Science and Language Arts:** Listening In, p. 118

Glencoe Exclusive! TeacherWorks™ All-In-One Planner and Resource Center

Lab Materials	Reproducible Resources	Section Assessment	Technology
Launch Lab: photographs from discarded magazines, scissors	**Chapter *FAST FILE* Resources** Foldables Worksheet, p. 17 Directed Reading Overview, p. 19 Note-taking Worksheets, pp. 33–35	Glencoe's Assessment Advantage	TeacherWorks includes: • Interactive Teacher Edition • Lesson Planner with calendar • Access to all program blacklines • Correlations to standards • Web links
MiniLAB: clay or modeling dough; objects for "fossils," such as macaroni, small buttons or peanuts; spatula	**Chapter *FAST FILE* Resources** Transparency Activity, p. 44 MiniLAB, p. 3 Enrichment, p. 30 Reinforcement, p. 27 Directed Reading, p. 20 Lab Activity, pp. 9–11	**Portfolio** Differentiated Instruction, p. 100 **Performance** MiniLAB, p. 100 Applying Skills, p. 104 **Content** Section Review, p. 104	Section Focus Transparency Virtual Labs CD-ROM Guided Reading Audio Program Interactive Chalkboard CD-ROM
Lab: metric ruler, pencil *Need materials?* Contact Science Kit at 1-800-828-7777 or www.sciencekit.com on the Internet.	**Chapter *FAST FILE* Resources** Transparency Activity, p. 45 Enrichment, p. 31 Reinforcement, p. 28 Directed Reading, p. 20 Lab Worksheets, pp. 5–6 **Mathematics Skill Activities,** p. 5 **Science Inquiry Labs,** pp. 23–24	**Portfolio** Integrate Chemistry, p. 103 **Performance** Applying Math, p. 104 **Content** Section Review, p. 104	Section Focus Transparency Virtual Labs CD-ROM Guided Reading Audio Program Interactive Chalkboard CD-ROM
MiniLAB: water; clear, colorless casserole dish; hot plate; thermal mitts; food coloring **Lab:** Internet sites and other resources on earthquake and volcanic activity, world map	**Chapter *FAST FILE* Resources** Transparency Activity, p. 46 MiniLAB, p. 4 Enrichment, p. 32 Reinforcement, p. 29 Directed Reading, pp. 21, 22 Lab Activity, pp. 13–15 Lab Worksheets, pp. 7–8 Transparency Activity, pp. 47–48 **Lab Management and Safety,** p. 65 **Reading and Writing Skill Activities,** p. 27	**Portfolio** Science Journal, p. 107 **Performance** Applying Science, p. 108 MiniLAB, p. 111 Applying Skills, p. 115 **Content** Section Review, p. 115	Section Focus Transparency Teaching Transparency Virtual Labs CD-ROM Guided Reading Audio Program Interactive Chalkboard CD-ROM Video Lab

End of Chapter Assessment

Blackline Masters	Technology	Professional Series
Chapter *FAST FILE* Resources Chapter Review, pp. 37–38 Chapter Tests, pp. 39–42 **Standardized Test Practice,** pp. 19–22	MindJogger Videoquiz Virtual Labs CD-ROM Exam*View*® Pro Testmaker TeacherWorks CD-ROM Interactive Chalkboard CD-ROM	**Performance Assessment in the Science Classroom (PASC)**

Transparencies

Section Focus

Section Focus Transparency 1 — A Cold Dig

If you were interested in the fossils of an animal that liked warm weather, would you think of digging in Antarctica? Archaeologists have found many interesting fossils there, including parts of a hadrosaur, a dinosaur previously found only in the Americas.

1. Antarctica has a very inhospitable climate. Why might fossils of warm-weather animals be found there?
2. What are some reasons that the climate of Antarctica might change in the future?

L2

Section Focus Transparency 2 — The Main Event

Until recently, the bottom of the sea was impossible to see. New technology has improved the view, and today we have a better idea of what is going on there. This photo shows one feature of the ocean floor—a deep-sea vent.

1. What is occurring in the photograph?
2. What features on land are similar to this deep-sea vent?
3. Judging from the photo, what do you think conditions around this vent are like?

L2

Section Focus Transparency 3 — Valley of Ten Thousand Smokes

One of the most massive volcanic eruptions ever investigated occurred in a valley in southern Alaska in 1912. The eruption covered over forty square miles with ash as deep as 210 meters and left thousands of vents (called fumaroles) in the valley spewing steam and gas.

1. How did this valley get its name, the Valley of Ten Thousand Smokes?
2. Why don't you see any smoke in the photograph?
3. Name some other places where there are volcanoes.

L2

This is a representation of key blackline masters available in the Teacher Classroom Resources. See Resource Manager boxes within the chapter for additional information.

Assessment

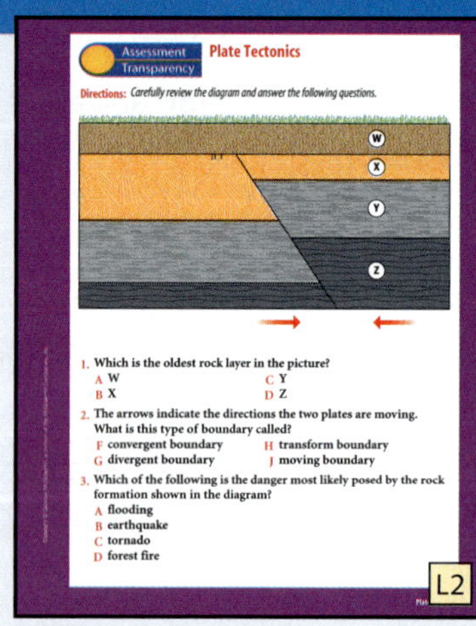
Assessment Transparency — Plate Tectonics

Directions: Carefully review the diagram and answer the following questions.

1. Which is the oldest rock layer in the picture?
 A W C Y
 B X D Z
2. The arrows indicate the directions the two plates are moving. What is this type of boundary called?
 F convergent boundary H transform boundary
 G divergent boundary J moving boundary
3. Which of the following is the danger most likely posed by the rock formation shown in the diagram?
 A flooding
 B earthquake
 C tornado
 D forest fire

L2

Teaching

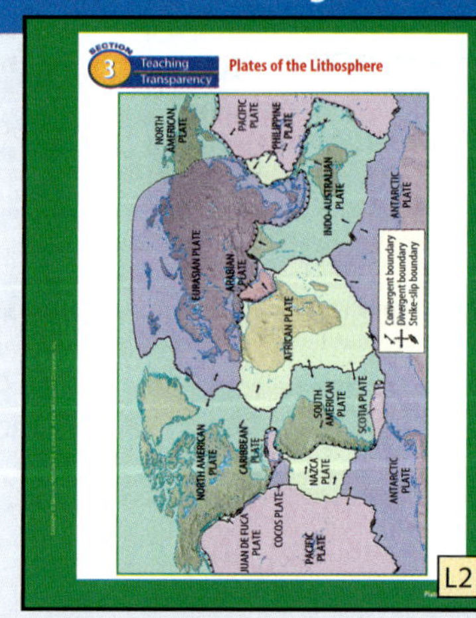
Teaching Transparency 3 — Plates of the Lithosphere

L2

Key to Teaching Strategies

The following designations will help you decide which activities are appropriate for your students.

- **L1** Level 1 activities should be appropriate for students with learning difficulties.
- **L2** Level 2 activities should be within the ability range of all students.
- **L3** Level 3 activities are designed for above-average students.
- **ELL** ELL activities should be within the ability range of English Language Learners.
- **COOP LEARN** Cooperative Learning activities are designed for small group work.
- **LS** Multiple Learning Styles logos, as described on page 6T, are used throughout to indicate strategies that address different learning styles.
- **P** These strategies represent student products that can be placed into a best-work portfolio.
- **PBL** Problem-Based Learning activities apply real-world situations to learning.

Hands-on Activities

Student Text Lab Worksheet

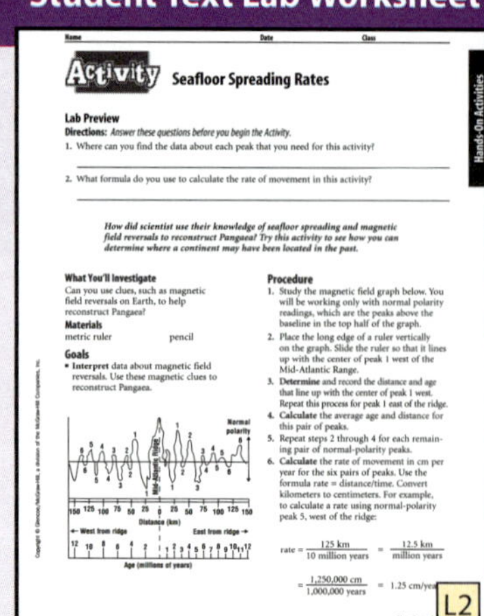
Activity — Seafloor Spreading Rates

L2

Laboratory Activities

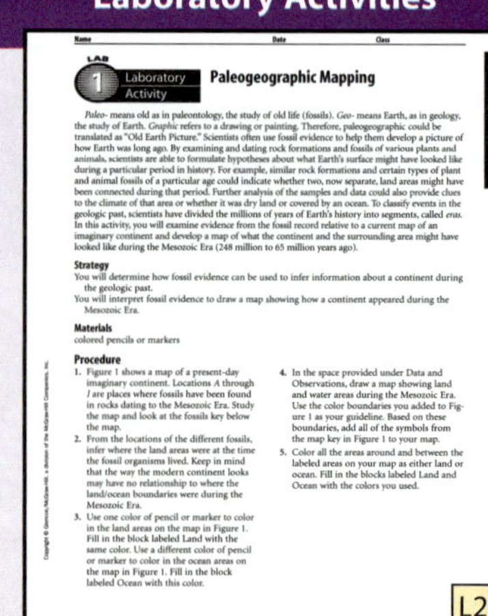
Laboratory Activity 1 — Paleogeographic Mapping

L2

Meeting Different Ability Levels

Content Outline

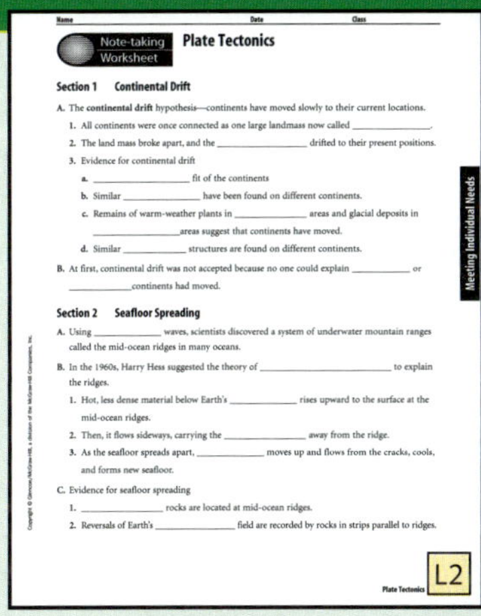

Reinforcement

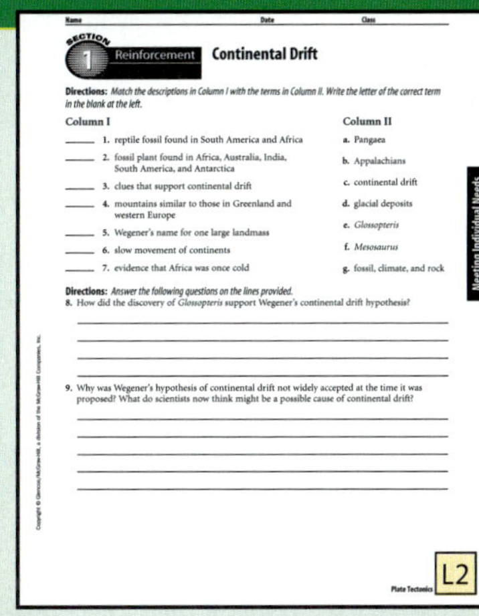

Enrichment

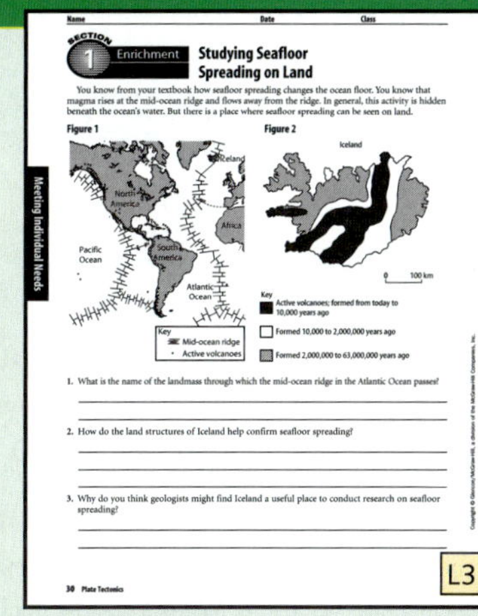

Directed Reading (English/Spanish)

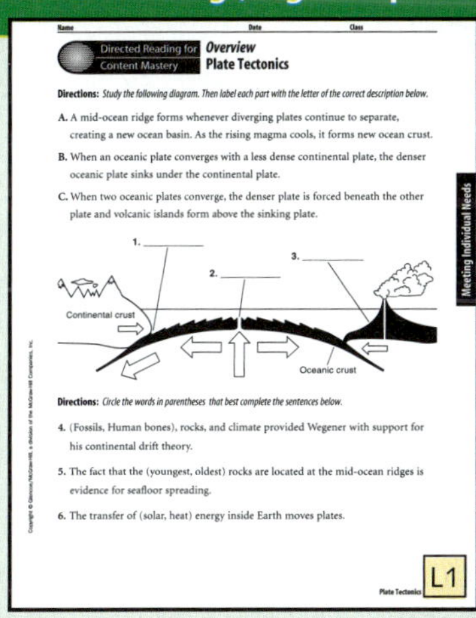

Study Guide

Reading Essentials

Assessment

Test Practice Workbook

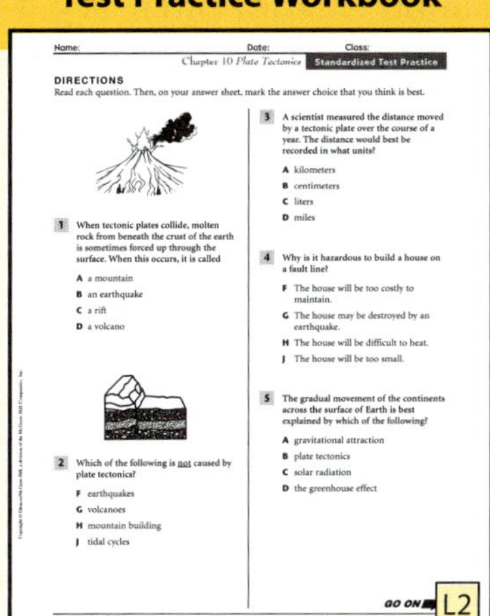

Chapter Review

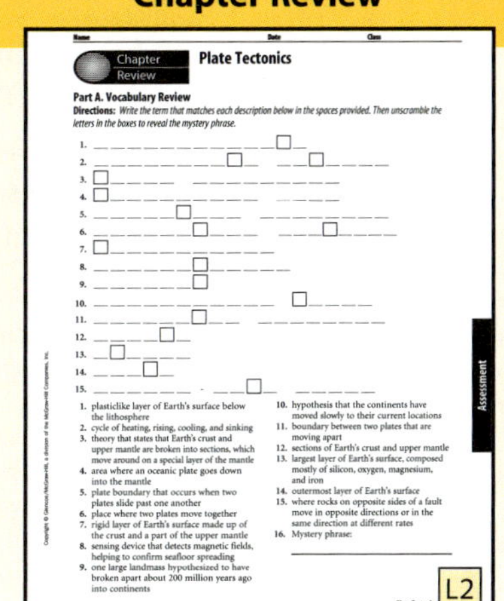

Chapter Tests

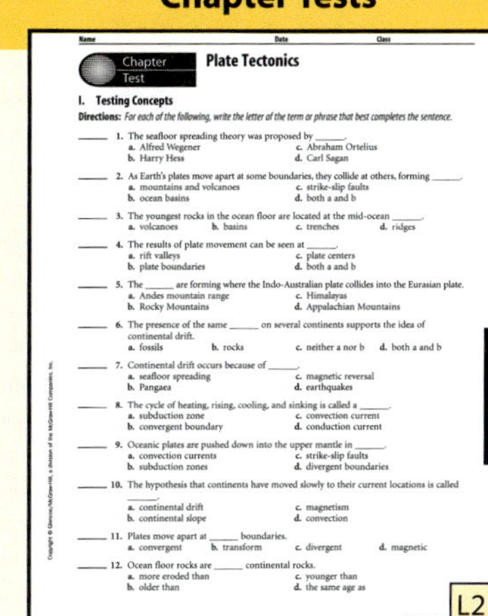

Science Content Background

section 1 Continental Drift

Pangaea

About 200 million years ago, a continent called Pangaea covered around 30% of Earth's surface, while a large ocean, Panthalassa, covered the rest of the planet. Pangaea broke up to form Laurasia and Gondwana, which were separated by the Tethys Sea.

Climate Clues

The study of ancient climates is called paleoclimatology. Alfred Wegener, a meteorologist, collected data about ancient climates in hopes of finding supporting evidence for his hypothesis. During his study of paleoclimates, he noted that glaciers had covered much of the southern hemisphere. This occurred between 320 million and 260 million years ago.

section 2 Seafloor Spreading

The Seafloor Moves

As the ocean floor slowly separates, new rocks form at a mid-ocean ridge. It is estimated that the Atlantic Ocean grows about 2.5 cm wider every year as a result of new rock forming at the Mid-Atlantic Ridge. Around 200 million years ago, a rift began to form between Greenland and Scotland. This rift led to the formation of the North Atlantic, which appeared only 65 million years ago, and continued spreading to form the modern Atlantic Ocean.

Teacher to Teacher
Kevin Finnegan, Teacher
McCord Middle School
Worthington, OH

"The geographic locations of volcanoes are an important part of understanding plate tectonics. Have students plot the locations of volcanoes on a class map. This activity helps to either introduce or reinforce the concept of plate tectonics."

Kevin Finnegan

Magnetic Time Scale

Paleomagnetism is the study of the magnetic properties of ancient rocks. When rock material is heated above the Curie point, magnetic minerals lose their magnetic properties. As they cool they align with the current magnetic field.

Tom Van Sant/Photo Researchers, Inc.

Thus the record of the reversals of Earth's magnetic field is recorded in the magnetic characteristics of rocks formed at that time.

section 3 Theory of Plate Tectonics

Plate Boundaries

Motion of Earth's plates is accommodated at the plate edges. Boundaries between lithospheric plates may be divergent, convergent, or transform. The intersection of any three plate boundaries is called a triple junction.

Divergent plate boundaries often begin as continental rifts such as the modern-day African Rift Valley. Tensional forces may stretch and thin the continental crust, forming a depression that fills with ocean water. Tensional forces continue to act on the area until a mid-ocean ridge develops. At this point, full-fledged seafloor spreading begins and the continental fragments begin to separate.

In order for Earth to maintain a constant size, lithosphere produced at mid-ocean ridges must be consumed elsewhere. This occurs at convergent boundaries. At these boundaries, older, cooler, and denser oceanic plates sink beneath less-dense plates in subduction zones. The denser plate descends into the mantle along a plane that may dip at angles ranging from 30° to 90°. This plane is defined by the foci of earthquakes associated with the subducting slab.

Transform boundaries may occur on land or on the seafloor. These faults most commonly connect offset segments of mid-ocean ridges. These ridge-to-ridge transform faults are a conspicuous feature of any mid-ocean ridge system. Transform faults also may connect a ridge to a trench or a trench to another trench.

Mountains and Volcanoes

Most of the world's spectacular mountain ranges were formed at collision-type convergent plate boundaries. When two plates carrying continents collide, rocks are folded and faulted, which results in a thickened and uplifted continental crust. The Himalaya are a classic example of a collision-type mountain range. This range started to form about 50 million years ago when India began colliding with Tibet and the Eurasian Plate. Even today, India is pushing northward into Tibet and the Himalaya continue to rise. To illustrate the magnitude of the uplift that has occurred during this collision, consider that marine fossils have been found in sandstone layers near Lhasa, Tibet, which has an altitude of 12,500 feet!

Mountains also can form along ocean–continent convergent boundaries. The Andes mountain range of South America formed as a result of rock deformation and volcanism caused by the subduction of the Nazca Plate under the South American Plate. The ocean trench marking the subduction zone is named the Peru-Chile Trench after the two countries most strongly affected by this plate boundary. When one oceanic plate descends beneath another at an ocean-ocean convergent plate boundary, a volcanic island arc forms behind the ocean trench. Volcanic arcs form as a result of partial melting of the subducting plate and mantle rock above the descending plate. The Japanese and Philippine Islands are examples of volcanic island arcs.

chapter content resources

Internet Resources
For additional content background, visit **bookf.msscience.com** to:
- access your book online
- find references to related articles in popular science magazines
- access Web links with related content background
- access current events with science journal topics

Print Resources

Global Tectonics by Philip Kearey & Frederick J. Vine, Blackwell Sciences Ltd, 1996

Plate Tectonics: Unraveling the Mysteries of the Earth by Jon Erickson, Replica Books, 2001

Moving Continents—Our Changing Earth by T.G. Aylesworth, Enslow, 1990

Dance of the Continents—Adventures With Rocks and Time by J.W. Harrington, J.P. Tarcher, 1983

Nicholas Parfitt/Stone

chapter 4

Chapter Vocabulary

continental drift, p. 98
Pangaea, p. 98
seafloor spreading, p. 103
plate tectonics, p. 106
plate, p. 106
lithosphere, p. 106
asthenosphere, p. 106
convection current, p. 111

Science Journal The continent shapes appear to match, plant and animal fossils from South America and Africa are very similar, and broad areas of rock in Africa and South America are the same type. All of this suggests that these continents were once joined.

INTERACTIVE CHALKBOARD with Image Bank

PowerPoint® Presentations

This CD-ROM is an editable Microsoft® PowerPoint® presentation that includes:

- a pre-made presentation for every chapter
- interactive graphics
- animations
- audio clips
- image bank
- all new section and chapter questions
- Standardized Test Practice
- transparencies
- pre-lab questions for all labs
- Foldables directions
- links to bookf.msscience.com

chapter 4

Plate Tectonics

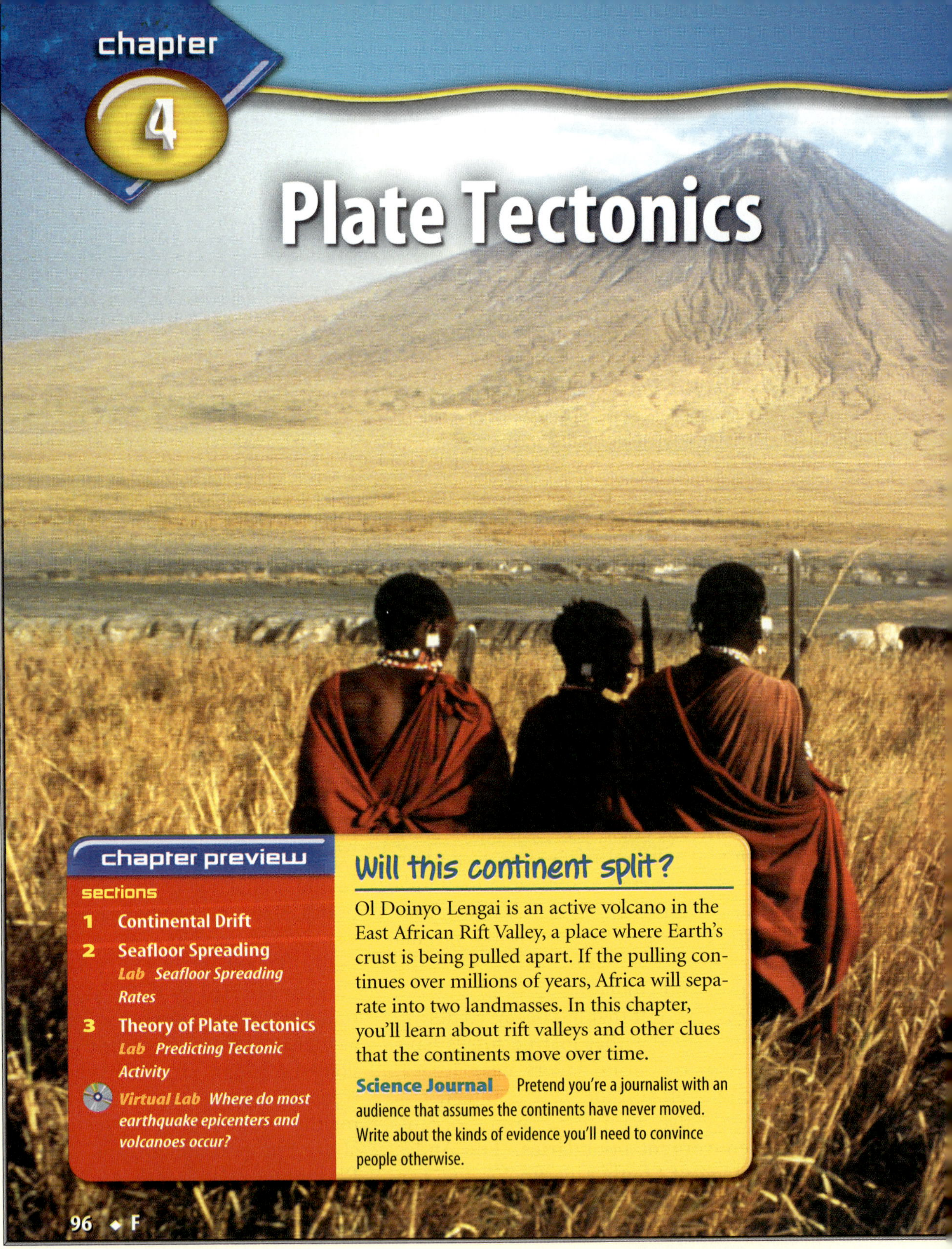

chapter preview

sections

1. Continental Drift
2. Seafloor Spreading
   Lab *Seafloor Spreading Rates*
3. Theory of Plate Tectonics
   Lab *Predicting Tectonic Activity*

Virtual Lab *Where do most earthquake epicenters and volcanoes occur?*

Will this continent split?

Ol Doinyo Lengai is an active volcano in the East African Rift Valley, a place where Earth's crust is being pulled apart. If the pulling continues over millions of years, Africa will separate into two landmasses. In this chapter, you'll learn about rift valleys and other clues that the continents move over time.

Science Journal Pretend you're a journalist with an audience that assumes the continents have never moved. Write about the kinds of evidence you'll need to convince people otherwise.

Theme Connection

Energy The transfer of energy in Earth's interior sets up massive convection currents in the mantle. These currents are thought to be the driving force that causes movement of Earth's plates.

About the Photo

Pulling Apart Africa This is a photo of Ol Doinyo Lengai, an active volcano in the East African Rift Valley. This valley is a location where Earth's crust is being pulled apart. Should this continue over millions of years, Africa will separate into two landmasses.

Start-Up Activities

Launch LAB

Reassemble an Image

Can you imagine a giant landmass that broke into many separate continents and Earth scientists working to reconstruct Earth's past? Do this lab to learn about clues that can be used to reassemble a supercontinent.

1. Collect interesting photographs from an old magazine.
2. You and a partner each select one photo, but don't show them to each other. Then each of you cut your photos into pieces no smaller than about 5 cm or 6 cm.
3. Trade your cut-up photo for your partner's.
4. Observe the pieces, and reassemble the photograph your partner has cut up.
5. **Think Critically** Write a paragraph describing the characteristics of the cut-up photograph that helped you put the image back together. Think of other examples in which characteristics of objects are used to match them up with other objects.

Science Online | Preview this chapter's content and activities at bookf.msscience.com

FOLDABLES Study Organizer

Plate Tectonics Make the following Foldable to help identify what you already know, what you want to know, and what you learned about plate tectonics.

STEP 1 **Fold** a vertical sheet of paper from side to side. Make the front edge about 1.25 cm shorter than the back edge.

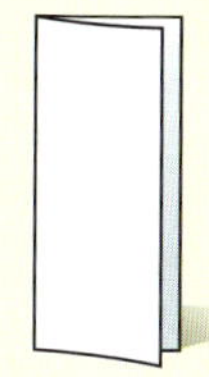

STEP 2 **Turn** lengthwise and fold into thirds.

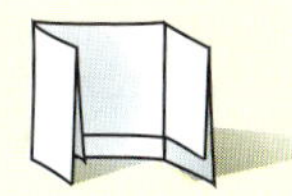

STEP 3 **Unfold and cut** only the layer along both folds to make three tabs.

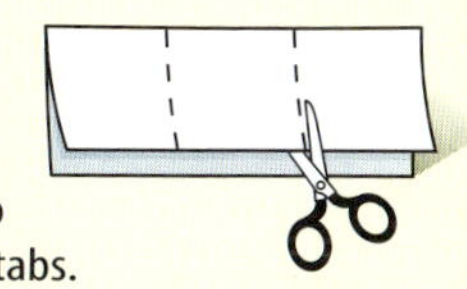

STEP 4 **Label** each tab.

Identify Questions Before you read the chapter, write what you already know about plate tectonics under the left tab of your Foldable, and write questions about what you'd like to know under the center tab. After you read the chapter, list what you learned under the right tab.

F ◆ 97

Launch LAB

Purpose Use the Launch Lab to introduce students to the idea that Earth's continents may have once been connected. L2 ELL

COOP LEARN IN **Linguistic**

Preparation Ask students to bring in old magazines.

Materials photos from discarded magazines, scissors

Teaching Strategy Have students compare their results with those of other students. Discuss differences and similarities.

Safety Precautions Remind students to handle scissors with care.

Think Critically

The shape and content of the pieces enable students to see how they fit together. Students may suggest jigsaw puzzles or broken artifacts.

Assessment

Process Ask students to write in their Science Journals the characteristics that make some photographs easier to fit together than others. Use **Performance Assessment in the Science Classroom,** p. 99.

FOLDABLES Study Organizer — Dinah Zike Study Fold

Student preparation materials for this Foldable are available in the Chapter *FAST FILE* Resources.

section 1 Continental Drift

1 Motivate

INTERACTIVE CHALKBOARD
PowerPoint® Presentations

Bellringer

Section Focus Transparencies also are available on the Interactive Chalkboard CD-ROM. L2 ELL

Section 1 Section Focus Transparency — A Cold Dig

If you were interested in the fossils of an animal that liked warm weather, would you think of digging in Antarctica? Archaeologists have found many interesting fossils there, including parts of a hadrosaur, a dinosaur previously found only in the Americas.

1. Antarctica has a very inhospitable climate. Why might fossils of warm-weather animals be found there?
2. What are some reasons that the climate of Antarctica might change in the future?

L2

Plate Tectonics

Tie to Prior Knowledge

Connecting Continents Have students recall the general shapes of Africa and South America. Ask if they can recall how the eastern coast of South America seems to fit into the western coast of Africa. Display a world map, asking again if students can make the connection.

Reading Check

Answer Alfred Wegener

---

as you read

What **You'll Learn**

- **Describe** the hypothesis of continental drift.
- **Identify** evidence supporting continental drift.

Why **It's Important**

The hypothesis of continental drift led to plate tectonics—a theory that explains many processes in Earth.

Review Vocabulary
continent: one of the six or seven great divisions of land on the globe

New Vocabulary
- continental drift
- Pangaea

Evidence for Continental Drift

If you look at a map of Earth's surface, you can see that the edges of some continents look as though they could fit together like a puzzle. Other people also have noticed this fact. For example, Dutch mapmaker Abraham Ortelius noted the fit between the coastlines of South America and Africa more than 400 years ago.

Pangaea German meteorologist Alfred Wegener (VEG nur) thought that the fit of the continents wasn't just a coincidence. He suggested that all the continents were joined together at some time in the past. In a 1912 lecture, he proposed the hypothesis of continental drift. According to the hypothesis of **continental drift,** continents have moved slowly to their current locations. Wegener suggested that all continents once were connected as one large landmass, shown in **Figure 1,** that broke apart about 200 million years ago. He called this large landmass **Pangaea** (pan JEE uh), which means "all land."

Reading Check *Who proposed continental drift?*

Figure 1 This illustration represents how the continents once were joined to form Pangaea. This fitting together of continents according to shape is not the only evidence supporting the past existence of Pangaea.

Section 1 Resource Manager

Chapter *Fast File* Resources
- Transparency Activity, p. 44
- Directed Reading for Content Mastery, pp. 19, 20
- Note-taking Worksheets, pp. 33–35
- MiniLAB, p. 3
- Enrichment, p. 30
- Reinforcement, p. 27
- Lab Activity, pp. 9–11

98 ◆ F CHAPTER 4 Plate Tectonics

A Controversial Idea Wegener's ideas about continental drift were controversial. It wasn't until long after Wegener's death in 1930 that his basic hypothesis was accepted. The evidence Wegener presented hadn't been enough to convince many people during his lifetime. He was unable to explain exactly how the continents drifted apart. He proposed that the continents plowed through the ocean floor, driven by the spin of Earth. Physicists and geologists of the time strongly disagreed with Wegener's explanation. They pointed out that continental drift would not be necessary to explain many of Wegener's observations. Other important observations that came later eventually supported Wegener's earlier evidence.

Science Online

Topic: Continental Drift
Visit bookf.msscience.com for Web links to information about the continental drift hypothesis.

Activity Research and write a brief report about the initial reactions, from the public and scientific communities, toward Wegener's continental drift hypothesis.

Fossil Clues Besides the puzzlelike fit of the continents, fossils provided support for continental drift. Fossils of the reptile *Mesosaurus* have been found in South America and Africa, as shown in **Figure 2.** This swimming reptile lived in freshwater and on land. How could fossils of *Mesosaurus* be found on land areas separated by a large ocean of salt water? It probably couldn't swim between the continents. Wegener hypothesized that this reptile lived on both continents when they were joined.

Reading Check *How do* Mesosaurus *fossils support the past existence of Pangaea?*

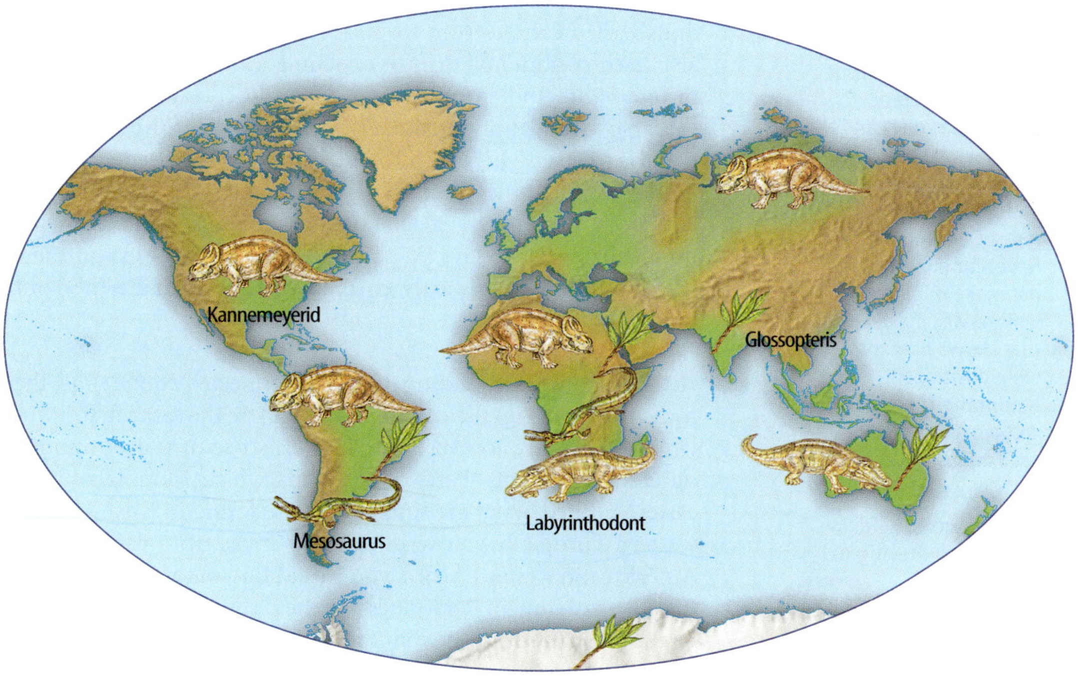

Figure 2 Fossil remains of plants and animals that lived in Pangaea have been found on more than one continent.
Evaluate *How do the locations of* Glossopteris, Mesosaurus, Kannemeyerid, Labyrinthodont, *and other fossils support Wegener's hypothesis of continental drift?*

2 Teach

Use an Analogy

Puzzle Pieces Have students recall that when putting together a jigsaw puzzle, they use the shapes of the puzzle pieces and the picture on the box as clues. When thinking about evidence of continental drift, the puzzle pieces are analogous to the shapes of the continents, and the picture clues are analogous to evidence found on each continent.

Discussion

Continental Drift Consider why Wegener's idea of continental drift was rejected? Scientists at the time thought Wegener's observations could be explained by something other than continental drift. Also, Wegener failed to provide a reasonable or believable mechanism to explain how continental drift might occur.

Reading Check

Answer Fossils of *Mesosaurus,* a freshwater and land animal, have been found in widespread areas separated by oceans of salt water, through which they could not swim.

Caption Answer

Figure 2 Matching fossils on widely separated continents provide evidence that these land masses were once joined.

Quick Demo

Paper "Plates"

Materials paper plate, red marker, blue marker, green marker, pair of scissors, 13″ × 9″ baking pan, access to water

Estimated Time 10 minutes

Procedure Draw a large red circle on the paper plate. Draw a blue circle inside the red circle. Draw a green circle inside the blue circle. Using scissors cut the paper plate in half. Fill the baking pan halfway with water. Place the two plate semicircles into the pan to demonstrate our floating continents.

Differentiated Instruction

Visually Impaired In order to help visually impaired students better understand how shapes that fit together can be used to help reconstruct Pangaea, make clay models of the continental masses that formed the supercontinent. Make sure the edges of the continents clearly match. Have students work with a partner to reconstruct the clay "Pangaea." L2

Fun Fact

Fossils of *Lystrosaurus,* a small reptile that lived about 200 million years ago, have been found in South Africa, Antarctica, and India.

Mini LAB

Purpose Students reaffirm that geologic clues can be used to show how continents that are now separate were once joined. L2 ELL LS **Kinesthetic**

Materials three colors of modeling clay or modeling dough; objects such as macaroni, small buttons, or peanuts to use as fossils; spatula for cutting landmasses apart

Teaching Strategy Review the procedure for making the landmasses with students in class before having them complete the activity at home.

Analysis

Possible answer: I looked for clues in the pattern of fossils and "mountain ranges."

Assessment

Oral Have students describe the characteristics they used to reconstruct the original landmass. Use **Performance Assessment in the Science Classroom,** p. 143.

Try at Home

Figure 3 This fossil plant, *Glossopteris,* grew in a temperate climate.

Mini LAB

Interpreting Fossil Data

Procedure

1. Build a three-layer landmass using **clay** or **modeling dough.**
2. Mold the clay into mountain ranges.
3. Place similar **"fossils"** into the clay at various locations around the landmass.
4. Form five continents from the one landmass. Also, form two smaller landmasses out of different clay with different mountain ranges and fossils.
5. Place the five continents and two smaller landmasses around the room.
6. Have someone who did not make or place the landmasses make a model that shows how they once were positioned.
7. Return the clay to its container so it can be used again.

Analysis

What clues were useful in reconstructing the original landmass?

Try at Home

A Widespread Plant Another fossil that supports the hypothesis of continental drift is *Glossopteris* (glahs AHP tur us). **Figure 3** shows this fossil plant, which has been found in Africa, Australia, India, South America, and Antarctica. The presence of *Glossopteris* in so many areas also supported Wegener's idea that all of these regions once were connected and had similar climates.

Climate Clues Wegener used continental drift to explain evidence of changing climates. For example, fossils of warm-weather plants were found on the island of Spitsbergen in the Arctic Ocean. To explain this, Wegener hypothesized that Spitsbergen drifted from tropical regions to the arctic. Wegener also used continental drift to explain evidence of glaciers found in temperate and tropical areas. Glacial deposits and rock surfaces scoured and polished by glaciers are found in South America, Africa, India, and Australia. This shows that parts of these continents were covered with glaciers in the past. How could you explain why glacial deposits are found in areas where no glaciers exist today? Wegener thought that these continents were connected and partly covered with ice near Earth's south pole long ago.

Rock Clues If the continents were connected at one time, then rocks that make up the continents should be the same in locations where they were joined. Similar rock structures are found on different continents. Parts of the Appalachian Mountains of the eastern United States are similar to those found in Greenland and western Europe. If you were to study rocks from eastern South America and western Africa, you would find other rock structures that also are similar. Rock clues like these support the idea that the continents were connected in the past.

Activity

Boundary Discussion Organize students into four groups and assign each group one of the following topics to master and present to the class: convergent boundaries, divergent boundaries, transform boundaries, and the driving mechanism of plate tectonics. L2 COOP LEARN LS **Interpersonal**

Teacher FYI

Ancient Antarctica In the 1960s, evidence was found in Antarctica that indicated a warm climate existed there 200 million years ago. *Glossopteris* fossils and rocks containing coal were found only 200 miles from the south pole.

Differentiated Instruction

Challenge New evidence suggests that mountains in southwest South America match up with the Appalachians. Have students research and write a report about how this came about. Possible answer: Before colliding with the African Plate, the North American Plate may have been positioned west of South America. L3 P

Figure 4 These computer models show the probable course the continents have taken. On the far left is their position 250 million years ago. In the middle is their position 135 million years ago. At right is their current position.

How could continents drift?

Although Wegener provided evidence to support his hypothesis of continental drift, he couldn't explain how, when, or why these changes, shown in **Figure 4,** took place. The idea suggested that lower-density, continental material somehow had to plow through higher-density, ocean-floor material. The force behind this plowing was thought to be the spin of Earth on its axis—a notion that was quickly rejected by physicists. Because other scientists could not provide explanations either, Wegener's idea of continental drift was initially rejected. The idea was so radically different at that time that most people closed their minds to it.

Rock, fossil, and climate clues were the main types of evidence for continental drift. After Wegener's death, more clues were found, largely because of advances in technology, and new ideas that related to continental drift were developed. You'll learn about a new idea, seafloor spreading, in the next section.

section 1 review

Summary

Evidence for Continental Drift

- Alfred Wegener proposed in his hypothesis of continental drift that all continents were once connected as one large landmass called Pangaea.
- Evidence of continental drift came from fossils, signs of climate change, and rock structures from different continents.

How could continents drift?

- During his lifetime, Wegener was unable to explain how, when, or why the continents drifted.
- After his death, advances in technology permitted new ideas to be developed to help explain his hypothesis.

Self Check

1. **Explain** how Wegener used climate clues to support his hypothesis of continental drift.
2. **Describe** how rock clues were used to support the hypothesis of continental drift.
3. **Summarize** the ways that fossils helped support the hypothesis of continental drift.
4. **Think Critically** Why would you expect to see similar rocks and rock structures on two landmasses that were connected at one time.

Applying Skills

5. **Compare and contrast** the locations of fossils of the temperate plant *Glossopteris,* as shown in **Figure 2,** with the climate that exists at each location today.

Science Online bookf.msscience.com/self_check_quiz

Visual Learning

Figure 4 What is happening to eastern Africa at present? It is splitting apart.

3 Assess

DAILY INTERVENTION

Check for Understanding

Visual-Spatial Make photocopies of the globe from 250 million years ago, 135 million years ago, and present day. Using these copies, refresh the students on the ideas of continental drift.

Reteach

Outlines Have students outline the section, including all of the important points. Have pairs exchange outlines and then use the outlines to quiz one another. L1 LS **Interpersonal**

Assessment

Content Have groups of students write and perform skits in which one student plays Wegener introducing his theory, and others play scientists debunking it. Encourage students to use visual props. Use **Performance Assessment in the Science Classroom,** p. 147.

section 1 review

1. Fossils of warm-weather plants found on islands in the Arctic Ocean and glacial features found in places such as Africa supported the idea that continents drift.
2. Rock structures on different continents are similar.
3. Fossils of the same terrestrial organism were found on widely separated continents.
4. The same rock structure they shared when attached would appear on both halves after the landmass split apart.
5. Possible answer: compare–all locations are on landmasses; contrast–some locations are in temperate climates, others are in arid, semiarid, or polar climates.

section 2 Seafloor Spreading

as you read

What **You'll Learn**

- **Explain** seafloor spreading.
- **Recognize** how age and magnetic clues support seafloor spreading.

Why **It's Important**

Seafloor spreading helps explain how continents moved apart.

Review Vocabulary

seafloor: portion of Earth's crust that lies beneath ocean waters

New Vocabulary

- seafloor spreading

Mapping the Ocean Floor

If you were to lower a rope from a boat until it reached the seafloor, you could record the depth of the ocean at that particular point. In how many different locations would you have to do this to create an accurate map of the seafloor? This is exactly how it was done until World War I, when the use of sound waves was introduced by German scientists to detect submarines. During the 1940s and 1950s, scientists began using sound waves on moving ships to map large areas of the ocean floor in detail. Sound waves echo off the ocean bottom—the longer the sound waves take to return to the ship, the deeper the water is.

Using sound waves, researchers discovered an underwater system of ridges, or mountains, and valleys like those found on the continents. In some of these underwater ridges are rather long rift valleys where volcanic eruptions and earthquakes occur from time to time. Some of these volcanoes actually are visible above the ocean surface. In the Atlantic, the Pacific, and in other oceans around the world, a system of ridges, called the mid-ocean ridges, is present. These underwater mountain ranges, shown in **Figure 5,** stretch along the center of much of Earth's ocean floor. This discovery raised the curiosity of many scientists. What formed these mid-ocean ridges?

Reading Check *How were mid-ocean ridges discovered?*

Figure 5 As the seafloor spreads apart at a mid-ocean ridge, new seafloor is created. The older seafloor moves away from the ridge in opposite directions.

1 Motivate

INTERACTIVE CHALKBOARD PowerPoint® Presentations

Bellringer

Section Focus Transparencies also are available on the Interactive Chalkboard CD-ROM. L2 ELL

Tie to Prior Knowledge

Ocean Floor Ask students if they ever have walked into an ocean, lake, or river. Have them describe how the floor of the body of water felt. Students may mention steep and gentle slopes, rocks, bars or ridges of sand, and depressions. Explain that the entire ocean floor has such features, and it took special tools to map it.

Reading Check

Answer using sound waves

Activity

Volcano Map Show students a map with the location of volcanoes indicated. Have students use it to determine subduction areas. L2 LS **Visual-Spatial**

Section 2 Resource Manager

Chapter *FAST FILE* Resources

Transparency Activity, p. 45
Directed Reading for Content Mastery, p. 20
Enrichment, p. 31
Reinforcement, p. 28
Lab Worksheets, pp. 5–6

Mathematics Skill Activities, p. 5

Science Inquiry Labs, pp. 23–24

102 ◆ F CHAPTER 4 Plate Tectonics

The Seafloor Moves In the early 1960s, Princeton University scientist Harry Hess suggested an explanation. His now-famous theory is known as **seafloor spreading.** Hess proposed that hot, less dense material below Earth's crust rises toward the surface at the mid-ocean ridges. Then, it flows sideways, carrying the seafloor away from the ridge in both directions, as seen in **Figure 5.**

As the seafloor spreads apart, magma is forced upward and flows from the cracks. It becomes solid as it cools and forms new seafloor. As new seafloor moves away from the mid-ocean ridge, it cools, contracts, and becomes denser. This denser, colder seafloor sinks, helping to form the ridge. The theory of seafloor spreading was later supported by the following observations.

Reading Check *How does new seafloor form at mid-ocean ridges?*

Figure 6 Many new discoveries have been made on the seafloor. These giant tube worms inhabit areas near hot water vents along mid-ocean ridges.

Evidence for Spreading

In 1968, scientists aboard the research ship *Glomar Challenger* began gathering information about the rocks on the seafloor. *Glomar Challenger* was equipped with a drilling rig that allowed scientists to drill into the seafloor to obtain rock samples. Scientists found that the youngest rocks are located at the mid-ocean ridges. The ages of the rocks become increasingly older in samples obtained farther from the ridges, adding to the evidence for seafloor spreading.

Using submersibles along mid-ocean ridges, new seafloor features and life-forms also were discovered there, as shown in **Figure 6.** As molten material is forced upward along the ridges, it brings heat and chemicals that support exotic life-forms in deep, ocean water. Among these are giant clams, mussels, and tube worms.

INTEGRATE Physics

Magnetic Clues Earth's magnetic field has a north and a south pole. Magnetic lines, or directions, of force leave Earth near the south pole and enter Earth near the north pole. During a magnetic reversal, the lines of magnetic force run the opposite way. Scientists have determined that Earth's magnetic field has reversed itself many times in the past. These reversals occur over intervals of thousands or even millions of years. The reversals are recorded in rocks forming along mid-ocean ridges.

INTEGRATE Chemistry

Curie Point Find out what the Curie point is and describe in your Science Journal what happens to iron-bearing minerals when they are heated to the Curie point. Explain how this is important to studies of seafloor spreading.

2 Teach

Reading Check

Answer Magma moves upward and out of cracks in the seafloor. As it solidifies on the surface, new seafloor forms. Older seafloor moves away from the ridge.

INTEGRATE Chemistry

Curie Point The Curie point is the temperature above which iron-bearing minerals lose their magnetism. As lava cools into rock on the ocean floor, the rock acquires a magnetic field like that of Earth's. If these rocks are reheated beyond the Curie point, the magnetic signature they acquired at the time they formed is lost.

Research Have students research and write about the scientist for whom the Curie point is named. French physicist Pierre Curie P

Teacher FYI

Mapping the Seafloor Harry Hess (1906–1969) was a geologist who collected important data about the seafloor during World War II, when he was captain of a navy transport vessel. His ship was fitted with a new device called a fathometer, an echo sounder that showed ocean-floor depth under the ship. The data was meant to allow troop ships to get close to shore, but Hess used it to map the floor of the Pacific Ocean.

Visual Learning

Figure 6 What allows the exotic life shown to exist? Heat from molten material forced upward along rift valley.

Differentiated Instruction

Challenge Have students research the scientific and technological advances that led to the theory of seafloor spreading. Have students display their discoveries on a time line. L3 LS **Logical-Mathematical**

Fun Fact

Certain species of green turtles swim from South America to Ascension Island in the South Atlantic to lay their eggs. The turtles may have started this trip when these land masses were much closer. As the seafloor spread, the instinctive trip became longer.

Caption Answer

Figure 7 It shows the rock continually formed and moved away from the ridge over time.

3 Assess

DAILY INTERVENTION

Check for Understanding

Kinesthetic Have students make a concept map that includes evidence for seafloor spreading using the following phrases: ages increase away from ridge, pattern of magnetic field reversals, mid-ocean ridge, pattern of ages, and reverses back and forth.

Reteach

Seafloor Model Have pairs of students construct three-dimensional models of the seafloor at a mid-ocean ridge.

L1 ELL COOP LEARN LS **Kinesthetic**

Assessment

Performance Assess students' understanding by challenging them to draw concept maps that contain the main ideas in this section. Use **Performance Assessment in the Science Classroom,** p. 161.

Discussion

Plate Boundaries and Earthquakes Why are earthquakes "associated" with plate boundaries? Plates do not move smoothly. Instead, they stick and catch on each other. When the plates are "stuck," strain, or potential energy, builds up in the rocks. When the plates move again, this energy is released as an earthquake.

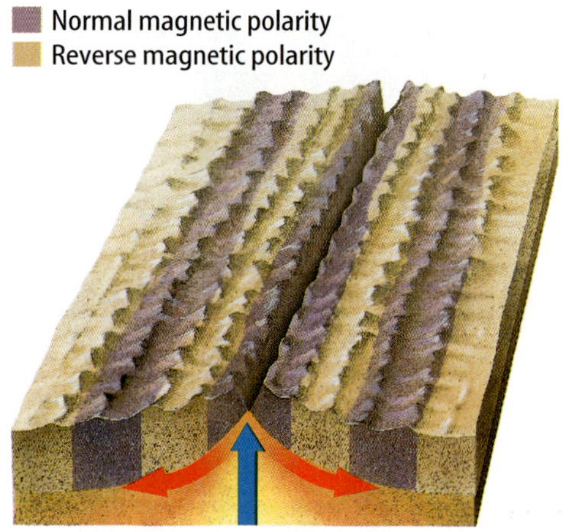

Figure 7 Changes in Earth's magnetic field are preserved in rock that forms on both sides of mid-ocean ridges.
Explain *why this is considered to be evidence of seafloor spreading.*

Magnetic Time Scale Iron-bearing minerals, such as magnetite, that are found in the rocks of the seafloor can record Earth's magnetic field direction when they form. Whenever Earth's magnetic field reverses, newly forming iron minerals will record the magnetic reversal.

Using a sensing device called a magnetometer (mag nuh TAH muh tur) to detect magnetic fields, scientists found that rocks on the ocean floor show many periods of magnetic reversal. The magnetic alignment in the rocks reverses back and forth over time in strips parallel to the mid-ocean ridges, as shown in **Figure 7.** A strong magnetic reading is recorded when the polarity of a rock is the same as the polarity of Earth's magnetic field today. Because of this, normal polarities in rocks show up as large peaks. This discovery provided strong support that seafloor spreading was indeed occurring. The magnetic reversals showed that new rock was being formed at the mid-ocean ridges. This helped explain how the crust could move—something that the continental drift hypothesis could not do.

section 2 review

Summary

Mapping the Ocean Floor

- Mid-ocean ridges, along the center of the ocean floor, have been found by using sound waves, the same method once used to detect submarines during World War I.
- Harry Hess suggested, in his seafloor spreading hypothesis, that the seafloor moves.

Evidence for Spreading

- Scientists aboard *Glomar Challenger* provided evidence of spreading by discovering that the youngest rocks are located at ridges and become increasingly older farther from the ridges.
- Magnetic alignment of rocks, in alternating strips that run parallel to ridges, indicates reversals in Earth's magnetic field and provides further evidence of seafloor spreading.

Self Check

1. **Summarize** What properties of iron-bearing minerals on the seafloor support the theory of seafloor spreading?
2. **Explain** how the ages of the rocks on the ocean floor support the theory of seafloor spreading.
3. **Summarize** How did Harry Hess's hypothesis explain seafloor movement?
4. **Explain** why some partly molten material rises toward Earth's surface.
5. **Think Critically** The ideas of Hess, Wegener, and others emphasize that Earth is a dynamic planet. How is seafloor spreading different from continental drift?

Applying Math

6. **Solve One-Step Equations** North America is moving about 1.25 cm per year away from a ridge in the middle of the Atlantic Ocean. Using this rate, how much farther apart will North America and the ridge be in 200 million years?

Science Online bookf.msscience.com/self_check_quiz

section 2 review

1. Magnetic reversals recorded in iron-bearing minerals show that new rock was being formed at the ridges over time.
2. The rocks get older as you move farther from the mid-ocean ridge.
3. Hot, dense material is forced upward at mid-ocean ridges. It then moves sideways, carrying the seafloor away from the ridge in both directions.
4. The molten material is less dense than the surrounding rock. So it is forced upward toward the surface.
5. Continental drift hypothesis provided no mechanism for movement; seafloor spreading explained how continents have separated over time as ocean basins enlarged.
6. If North America continues to move away from the Mid-Atlantic Ridge at 1.25 cm/y, these two features will separate by an additional 250 million centimeters or 2,500 kilometers in 200 million years.

104 ◆ F CHAPTER 4 Plate Tectonics

LAB

Seafloor Spreading Rates

How did scientists use their knowledge of seafloor spreading and magnetic field reversals to reconstruct Pangaea? Try this lab to see how you can determine where a continent may have been located in the past.

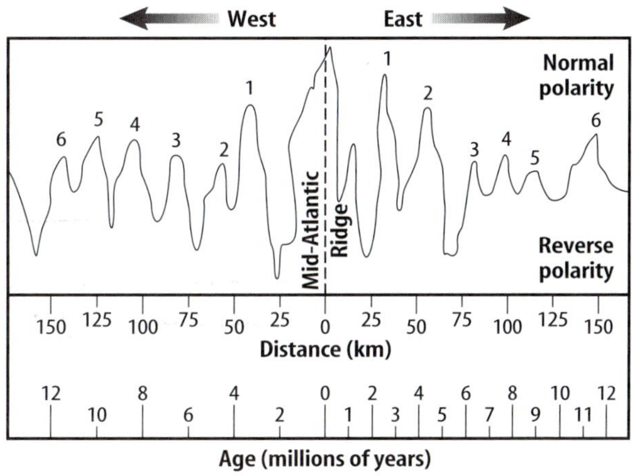

Real-World Question

Can you use clues, such as magnetic field reversals on Earth, to help reconstruct Pangaea?

Goals

- **Interpret** data about magnetic field reversals. Use these magnetic clues to reconstruct Pangaea.

Materials

metric ruler
pencil

Procedure

1. Study the magnetic field graph above. You will be working only with normal polarity readings, which are the peaks above the baseline in the top half of the graph.
2. Place the long edge of a ruler vertically on the graph. Slide the ruler so that it lines up with the center of peak 1 west of the Mid-Atlantic Ridge.
3. **Determine** and record the distance and age that line up with the center of peak 1 west. Repeat this process for peak 1 east of the ridge.
4. **Calculate** the average distance and age for this pair of peaks.
5. Repeat steps 2 through 4 for the remaining pairs of normal-polarity peaks.
6. **Calculate** the rate of movement in cm per year for the six pairs of peaks. Use the formula rate = distance/time. Convert kilometers to centimeters. For example, to calculate a rate using normal-polarity peak 5, west of the ridge:

$$\text{rate} = \frac{125 \text{ km}}{10 \text{ million years}} = \frac{12.5 \text{ km}}{\text{million years}} = \frac{1{,}250{,}000 \text{ cm}}{1{,}000{,}000 \text{ years}} = 1.25 \text{ cm/year}$$

Conclude and Apply

1. **Compare** the age of igneous rock found near the mid-ocean ridge with that of igneous rock found farther away from the ridge.
2. If the distance from a point on the coast of Africa to the Mid-Atlantic Ridge is approximately 2,400 km, calculate how long ago that point in Africa was at or near the Mid-Atlantic Ridge.
3. How could you use this method to reconstruct Pangaea?

BENCH TESTED

Real-World Question

Purpose Students can interpret magnetic field reversals in rock to determine the rate of seafloor spreading. L2

LS Logical-Mathematical

Process Skills make tables, use graphs, predict, observe, infer, use numbers, interpret data

Time Required 50 to 60 minutes

Teaching Strategy Note that the rate of movement (half the spreading rate) is about 1.25 cm/yr.

Conclude and Apply

1. The nearer rock is to the ridge, the younger it is.
2. about 192 million years ago, assuming a relatively constant rate of spreading
3. Students could determine when points on both coasts were at the ridge. This would mark when at least part of Pangaea was intact.

Assessment

Performance Have your students measure the distance between a point on the east coast of the United States and the Mid-Atlantic Ridge. Have them determine when that point was near the mid-ocean ridge. Use **Performance Assessment in the Science Classroom,** p. 99.

Communicating Your Data

Students can use a table such as this to record data.

Sample Data Table:

Peak	1	2	3	4	5	6
Distance west normal polarity	40	60	75	100	125	140
Distance east normal polarity	36	60	80	100	118	145
Average distance	38	60	78	100	122	142
Age from scale (millions of years)	3.5	4.5	6.7	8.0	9.0	10.5
Rate of movement (cm/yr)	1.1	1.2	1.2	1.3	1.3	1.4

Theory of Plate Tectonics

1 Motivate

INTERACTIVE CHALKBOARD
PowerPoint® Presentations

Bellringer

Section Focus Transparencies also are available on the Interactive Chalkboard CD-ROM. L2 ELL

SECTION 3 Section Focus Transparency **Valley of Ten Thousand Smokes**

One of the most massive volcanic eruptions ever investigated occurred in a valley in southern Alaska in 1912. The eruption covered over forty square miles with ash as deep as 210 meters and left thousands of vents (called fumaroles) in the valley spewing steam and gas.

1. How did this valley get its name, the Valley of Ten Thousand Smokes?
2. Why don't you see any smoke in the photograph?
3. Name some other places where there are volcanoes.

L2

Plate Tectonics

Tie to Prior Knowledge

Earthquakes Ask if anyone has ever experienced an earthquake. If so, have these students explain what happened. If no one has, explain that earthquakes cause the ground to shake, often causing great damage. Tell students that earthquakes often happen because of the movement of plates.

as you read

What **You'll Learn**

- **Compare and contrast** different types of plate boundaries.
- **Explain** how heat inside Earth causes plate tectonics.
- **Recognize** features caused by plate tectonics.

Why **It's Important**

Plate tectonics explains how many of Earth's features form.

Review Vocabulary
converge: to come together
diverge: to move apart
transform: to convert or change

New Vocabulary
- plate tectonics
- plate
- lithosphere
- asthenosphere
- convection current

Plate Tectonics

The idea of seafloor spreading showed that more than just continents were moving, as Wegener had thought. It was now clear to scientists that sections of the seafloor and continents move in relation to one another.

Plate Movements In the 1960s, scientists developed a new theory that combined continental drift and seafloor spreading. According to the theory of **plate tectonics,** Earth's crust and part of the upper mantle are broken into sections. These sections, called **plates,** move on a plasticlike layer of the mantle. The plates can be thought of as rafts that float and move on this layer.

Composition of Earth's Plates Plates are made of the crust and a part of the upper mantle, as shown in **Figure 8.** These two parts combined are the **lithosphere** (LIH thuh sfihr). This rigid layer is about 100 km thick and generally is less dense than material underneath. The plasticlike layer below the lithosphere is called the **asthenosphere** (as THE nuh sfihr). The rigid plates of the lithosphere float and move around on the asthenosphere.

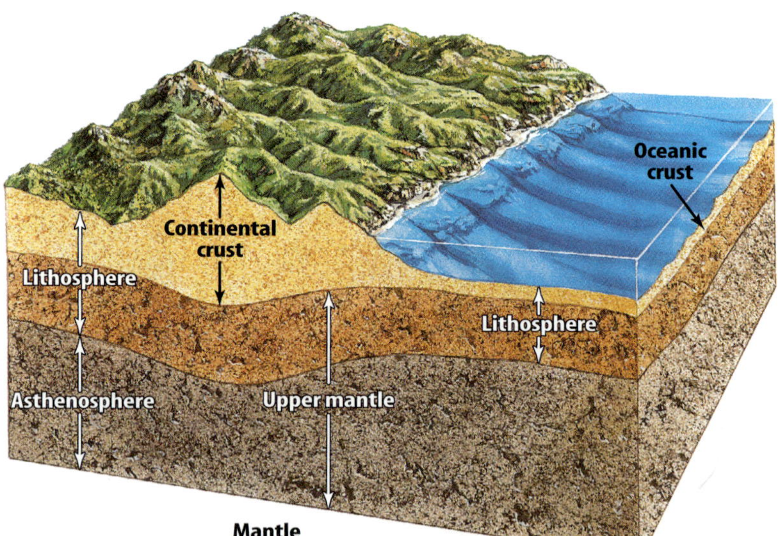

Figure 8 Plates of the lithosphere are composed of oceanic crust, continental crust, and rigid upper mantle.

Section 3 Resource Manager

Chapter *Fast File* Resources
- Transparency Activity, pp. 46, 48–49
- Directed Reading for Content Mastery, pp. 21, 22
- Lab Activity, pp. 13–15
- MiniLAB, p. 4
- Enrichment, p. 32
- Lab Worksheets, pp. 7–8
- Reinforcement, p. 29

Cultural Diversity, p. 47
Home and Community Involvement, p. 31
Earth Science Critical Thinking/Problem Solving, p. 8
Reading and Writing Skill Activities, p. 27
Performance Assessment in the Science Classroom, p. 42
Lab Management and Safety, p. 65

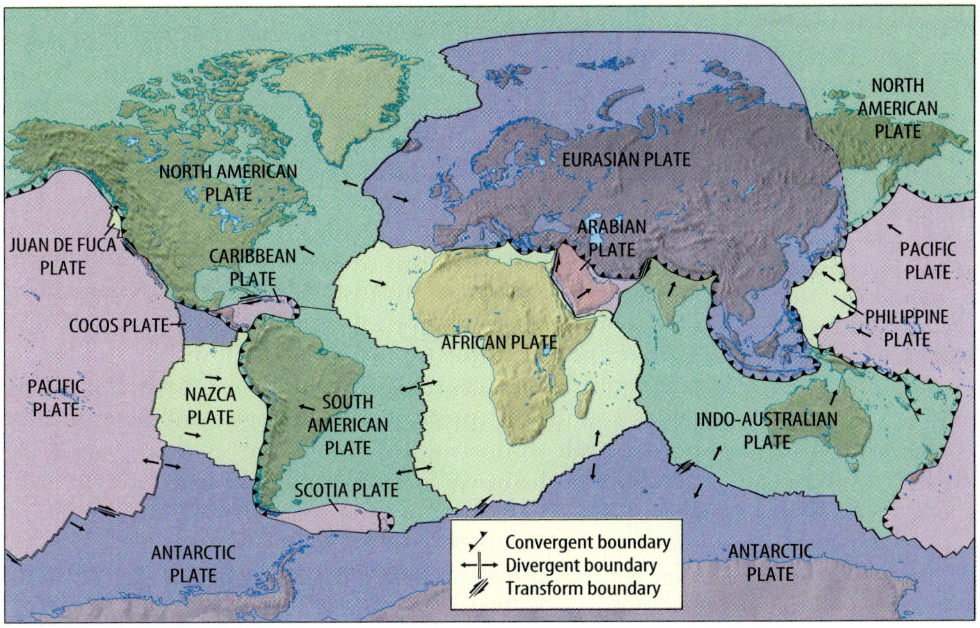

Figure 9 This diagram shows the major plates of the lithosphere, their direction of movement, and the type of boundary between them.

Analyze and Conclude *Based on what is shown in this figure, what is happening where the Nazca Plate meets the Pacific Plate?*

Plate Boundaries

When plates move, they can interact in several ways. They can move toward each other and converge, or collide. They also can pull apart or slide alongside one another. When the plates interact, the result of their movement is seen at the plate boundaries, as in **Figure 9.**

Reading Check *What are the general ways that plates interact?*

Movement along any plate boundary means that changes must happen at other boundaries. What is happening to the Atlantic Ocean floor between the North American and African Plates? Compare this with what is happening along the western margin of South America.

Plates Moving Apart The boundary between two plates that are moving apart is called a divergent boundary. You learned about divergent boundaries when you read about seafloor spreading. In the Atlantic Ocean, the North American Plate is moving away from the Eurasian and the African Plates, as shown in **Figure 9.** That divergent boundary is called the Mid-Atlantic Ridge. The Great Rift Valley in eastern Africa might become a divergent plate boundary. There, a valley has formed where a continental plate is being pulled apart. **Figure 10** shows a side view of what a rift valley might look like and illustrates how the hot material rises up where plates separate.

2 Teach

Quick Demo

Continental Movement

Materials globe or map

Estimated Time 10 minutes

Procedure Obtain a globe or make a map on which you can move continent pieces from a child's puzzle map. Use the globe or map to demonstrate continental movement.

Caption Answer

Figure 9 These plates are moving away from each other.

Reading Check

Answer Plates can collide, pull apart, or move past one another.

Fun Fact

The Indian Plate, which collided with Asia to form the Himalaya, continues to move at a rate of almost 5 cm per year. This massive plate is moving twice as fast as your fingernails grow!

Differentiated Instruction

English-Language Learners Pair English-language learning students with other students who can take them through the steps of this lab and visually assist them. After repeating steps 2 through 4 of this lab a few times, encourage English-language-learning students to try some graphing.

Science Journal

Theory Development Many scientists contributed ideas that led to plate tectonics theory. Have students select one from A.L. Du Toit, S.K. Runcorn, Bruce Heezen, Arthur Holmes, J. Tuzo Wilson, Jack Oliver, Lynn R. Sykes, Fred Vine, D.H. Matthews, and L.W. Morley and write a one-page report in their Science Journals about his contributions. L2 P

Use Science Words

Word Use Have students look up the words *diverge* and *converge* and use each word in a sentence. Then discuss how these meanings relate to plate boundaries. Possible answers: Two paths diverge at a fork in the road; traffic will converge in the center of the intersection. Plates converge, or come together, at some boundaries and diverge, or move apart, at others.

Applying Science

Answers

1. Yes, most fit together when continental shelves are included.
2. The continental shelves are the edges of continents. Present-day coastlines result from sea-level changes.

Science Online

Topic: Earthquakes and Volcanoes

Visit bookf.msscience.com for Web links to recent news or magazine articles about earthquakes and volcanic activity related to plate tectonics.

Activity Prepare a group demonstration about recent volcanic and earthquake events. Divide tasks among group members. Find and copy maps, diagrams, photographs, and charts to highlight your presentation. Emphasize the locations of events and the relationship to plate tectonics.

Plates Moving Together If new crust is being added at one location, why doesn't Earth's surface keep expanding? As new crust is added in one place, it disappears below the surface at another. The disappearance of crust can occur when seafloor cools, becomes denser, and sinks. This occurs where two plates move together at a convergent boundary.

When an oceanic plate converges with a less dense continental plate, the denser oceanic plate sinks under the continental plate. The area where an oceanic plate subducts, or goes down, into the mantle is called a subduction zone. Some volcanoes form above subduction zones. **Figure 10** shows how this type of convergent boundary creates a deep-sea trench where one plate bends and sinks beneath the other. High temperatures cause rock to melt around the subducting slab as it goes under the other plate. The newly formed magma is forced upward along these plate boundaries, forming volcanoes. The Andes mountain range of South America contains many volcanoes. They were formed at the convergent boundary of the Nazca and the South American Plates.

Applying Science

How well do the continents fit together?

Recall the Launch Lab you performed at the beginning of this chapter. While you were trying to fit pieces of a cut-up photograph together, what clues did you use?

Identifying the Problem

Take a copy of a map of the world and cut out each continent. Lay them on a tabletop and try to fit them together, using techniques you used in the Launch Lab. You will find that the pieces of your Earth puzzle—the continents—do not fit together well. Yet, several of the areas on some continents fit together extremely well.

Take out another world map—one that shows the continental shelves as well as the continents. Copy it and cut out the continents, this time including the continental shelves.

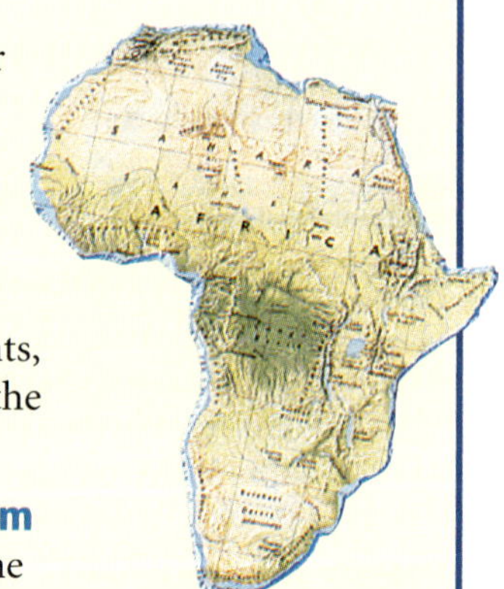

Solving the Problem

1. Does including the continental shelves solve the problem of fitting the continents together?
2. Why should continental shelves be included with maps of the continents?

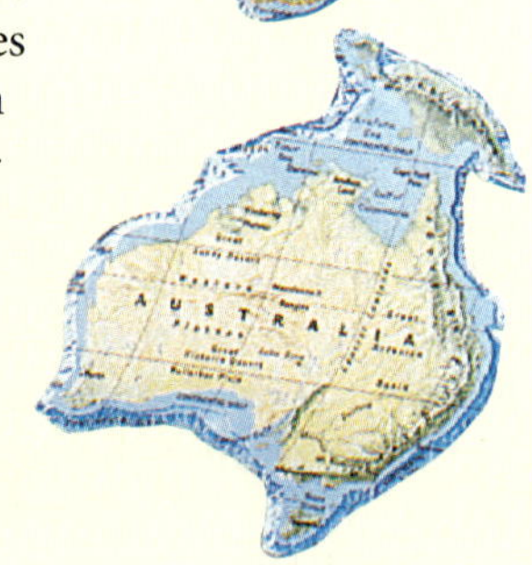

LAB DEMONSTRATION

Purpose to demonstrate compression forces that can form folded mountains

Materials two slabs of clay (5 cm thick and about 30 cm long), wax paper

Preparation Place the clay slabs on wax paper to make them easier to slide.

Procedure Lay the two clay pieces flat on a table. Have students predict what will happen when they are forced together. Push the two pieces together.

Expected Outcome Students will see folds and breaks form as the pieces of clay are pushed together.

Assessment

What landforms are the folds in the clay analogous to on Earth's surface? folded mountains

NATIONAL GEOGRAPHIC VISUALIZING PLATE BOUNDARIES

Figure 10

By diverging at some boundaries and converging at others, Earth's plates are continually—but gradually—reshaping the landscape around you. The Mid-Atlantic Ridge, for example, was formed when the North and South American Plates pulled apart from the Eurasian and African Plates (see globe). Some features that occur along plate boundaries—rift valleys, volcanoes, and mountain ranges—are shown on the right and below.

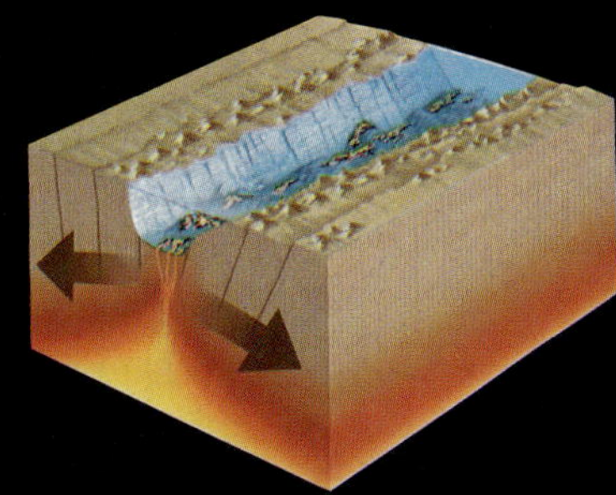

A RIFT VALLEY When continental plates pull apart, they can form rift valleys. The African continent is separating now along the East African Rift Valley.

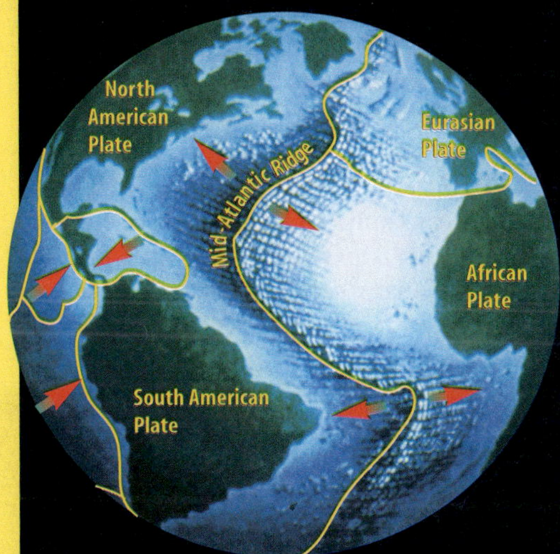

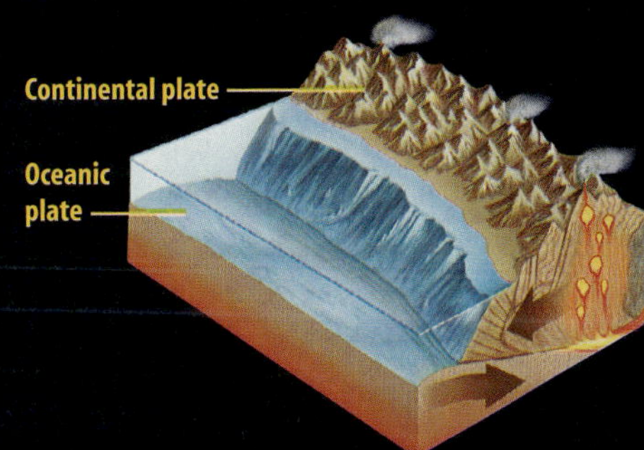

SUBDUCTION Where oceanic and continental plates collide, the oceanic plate plunges beneath the less dense continental plate. As the plate descends, molten rock (yellow) forms and rises toward the surface, creating volcanoes.

SEAFLOOR SPREADING A mid-ocean ridge, like the Mid-Atlantic Ridge, forms where oceanic plates continue to separate. As rising magma (yellow) cools, it forms new oceanic crust.

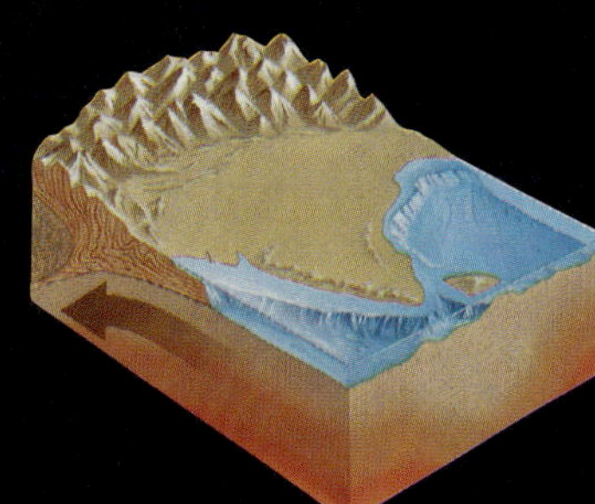

CONTINENTAL COLLISION Where two continental plates collide, they push up the crust to form mountain ranges such as the Himalaya.

NATIONAL GEOGRAPHIC

Visualizing Plate Boundaries

Have students examine the pictures and read the captions. Then ask the following questions.

How would you predict the size of the Atlantic Ocean will change over the next 100 million years? Why? The Atlantic Ocean will become larger because sea-floor spreading is occurring along the Mid-Atlantic Ridge.

The Andes mountains are found along the west coast of South America. How did this mountain chain form? The plate boundary along the west coast of South America is a convergent boundary, which results in the formation of mountains and volcanoes.

Activity

Surtsey Have small groups research the history of Surtsey, a small island in the North Atlantic Ocean. Ask them to draw a map of the island's location and write a summary of how the island formed, describing the type of plate boundary and the volcanic activity involved. L2

ELL COOP LEARN LS **Interpersonal**

Where Plates Collide A subduction zone also can form where two oceanic plates converge. In this case, the colder, older, denser oceanic plate bends and sinks down into the mantle. The Mariana Islands in the western Pacific are a chain of volcanic islands formed where two oceanic plates collide.

Usually, no subduction occurs when two continental plates collide, as shown in **Figure 10.** Because both of these plates are less dense than the material in the asthenosphere, the two plates collide and crumple up, forming mountain ranges. Earthquakes are common at these convergent boundaries. However, volcanoes do not form because there is no, or little, subduction. The Himalaya in Asia are forming where the Indo-Australian Plate collides with the Eurasian Plate.

Where Plates Slide Past Each Other The third type of plate boundary is called a transform boundary. Transform boundaries occur where two plates slide past one another. They move in opposite directions or in the same direction at different rates. When one plate slips past another suddenly, earthquakes occur. The Pacific Plate is sliding past the North American Plate, forming the famous San Andreas Fault in California, as seen in **Figure 11.** The San Andreas Fault is part of a transform plate boundary. It has been the site of many earthquakes.

Figure 11 The San Andreas Fault in California occurs along the transform plate boundary where the Pacific Plate is sliding past the North American Plate.

Overall, the two plates are moving in roughly the same direction. **Explain** *Why, then, do the red arrows show movement in opposite directions?*

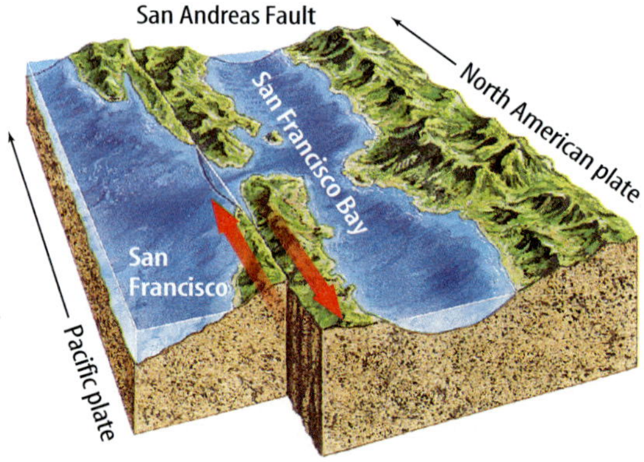

This photograph shows an aerial view of the San Andreas Fault.

IDENTIFYING Misconceptions

Rotation of Earth Because students feel as though they are standing perfectly still on Earth's surface, they often forget that they are moving in several ways at the same time. Remind students that, as they sit in class, Earth is rotating at 464 m/s, speeding around the sun at 29.8 km/s, and the tectonic plate on which they sit is sliding around on Earth's plastic-like mantle.

Visual Learning

Figure 11 Have students study the photograph of the San Andreas Fault and then describe evidence that shows the plates on either side of the fault are moving. Students should see that streams and other features that cross the fault are offset because of movement.

Caption Answer

Figure 11 The western side (Pacific Plate) of the fault is moving faster than the eastern side (North American Plate).

Curriculum Connection

Mathematics The deepest point on Earth's surface is the bottom of the Mariana Trench, 11.2 km below sea level. Have students find Earth's highest point. Mt. Everest is 8.8 km above sea level. After students determine which is bigger, have them draw a scale diagram showing Mt. Everest in the trench. Their drawings should show how many kilometers Mt. Everest's top would be below sea level. 2.4 km

Causes of Plate Tectonics

Many new discoveries have been made about Earth's crust since Wegener's day, but one question still remains. What causes the plates to move? Scientists now think they have a good idea. They think that plates move by the same basic process that occurs when you heat soup.

Convection Inside Earth Soup that is cooking in a pan on the stove contains currents caused by an unequal distribution of heat in the pan. Hot, less dense soup is forced upward by the surrounding, cooler, denser soup. As the hot soup reaches the surface, it cools and sinks back down into the pan. This entire cycle of heating, rising, cooling, and sinking is called a **convection current.** A version of this same process, occurring in the mantle, is thought to be the force behind plate tectonics. Scientists suggest that differences in density cause hot, plasticlike rock to be forced upward toward the surface.

Moving Mantle Material Wegener wasn't able to come up with an explanation for why plates move. Today, researchers who study the movement of heat in Earth's interior have proposed several possible explanations. All of the hypotheses use convection in one way or another. It is, therefore, the transfer of heat inside Earth that provides the energy to move plates and causes many of Earth's surface features. One hypothesis is shown in **Figure 12.** It relates plate motion directly to the movement of convection currents. According to this hypothesis, convection currents cause the movements of plates.

Mini LAB

Modeling Convection Currents

Procedure

1. Pour **water into a clear, colorless casserole dish** until it is 5 cm from the top.
2. Center the dish on a **hot plate** and heat it. **WARNING:** *Wear thermal mitts to protect your hands.*
3. Add a few drops of **food coloring** to the water above the center of the hot plate.
4. Looking from the side of the dish, observe what happens in the water.
5. Illustrate your observations in your **Science Journal.**

Analysis

1. Determine whether any currents form in the water.
2. Infer what causes the currents to form.

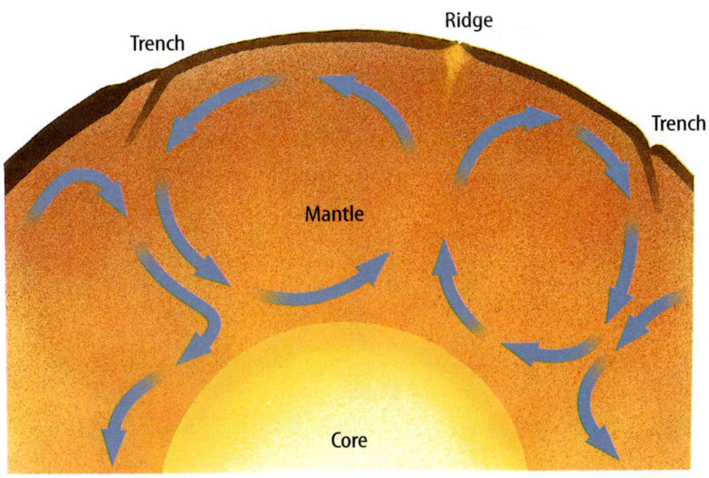

Figure 12 In one hypothesis, convection currents occur throughout the mantle. Such convection currents (see arrows) are the driving force of plate tectonics.

Mini LAB

Purpose Students model and observe currents. L2 ELL LS **Visual-Spatial**

Materials clear glass casserole dish, water, hot plate, food coloring, thermal mitts

Teaching Strategy Have students note any movement in the water.

Safety Precautions Students must wear thermal mitts. Be sure the dish is stove-top safe.

Analysis

1. Some students will observe currents; others won't.
2. The transfer of thermal energy from the burner to the dish warms the water near the bottom of the dish. The cooler, denser water at the top of the dish sinks, displacing the warmer, less dense water, which then moves toward the top of the dish. As the warmer water cools, it becomes denser and sinks to start the cycle again.

Assessment

Process Direct students to add informative labels to the drawings they made of their observations. The labels should be numbered and in sequence, explaining the steps in the formation and movement of convection currents. Use **Performance Assessment in the Science Classroom,** p. 127.

Active Reading

Write-Draw-Discuss This strategy encourages students to actively participate in reading and lectures, assimilating content creatively. Have students write about an idea, clarify it, then make an illustration or drawing. Ask students to share responses with the class and display several examples. Have students Write-Draw-Discuss about the causes of plate tectonics.

Cultural Diversity

Hawaiian Terms The Hawaiian Islands are volcanoes that formed as a result of magma rising through a "hot spot" in the middle of a plate. Some volcanic rocks have names that were made common in Hawaii. *Pahoehoe* (pa-hoe-ee-hoe-ee), from the Hawaiian word meaning "rope," forms in linear ridges. *Aa* (ah-ah) forms with sharp, jagged surfaces.

Caption Answer

Figure 13 tension

Make a Model

Seafloor Spreading Have students use paper to make a model of a divergent boundary with seafloor spreading. The model should be dynamic and show how spreading occurs. Students can draw parallel ridges on a long piece of paper and construct a mechanism for the paper to be drawn upward from both sides through a slot (the plate boundary), revealing "new" parallel ridges as it emerges. Accept any workable design. L2 ELL IS **Visual-Spatial**

Reading Check

Answer earthquakes

Features Caused by Plate Tectonics

Earth is a dynamic planet with a hot interior. This heat leads to convection, which powers the movement of plates. As the plates move, they interact. The interaction of plates produces forces that build mountains, create ocean basins, and cause volcanoes. When rocks in Earth's crust break and move, energy is released in the form of seismic waves. Humans feel this release as earthquakes. You can see some of the effects of plate tectonics in mountainous regions, where volcanoes erupt, or where landscapes have changed from past earthquake or volcanic activity.

Reading Check *What happens when seismic energy is released as rocks in Earth's crust break and move?*

Normal Faults and Rift Valleys Tension forces, which are forces that pull apart, can stretch Earth's crust. This causes large blocks of crust to break and tilt or slide down the broken surfaces of crust. When rocks break and move along surfaces, a fault forms. Faults interrupt rock layers by moving them out of place. Entire mountain ranges can form in the process, called fault-block mountains, as shown in **Figure 13.** Generally, the faults that form from pull-apart forces are normal faults—faults in which the rock layers above the fault move down when compared with rock layers below the fault.

Rift valleys and mid-ocean ridges can form where Earth's crust separates. Examples of rift valleys are the Great Rift Valley in Africa, and the valleys that occur in the middle of mid-ocean ridges. Examples of mid-ocean ridges include the Mid-Atlantic Ridge and the East Pacific Rise.

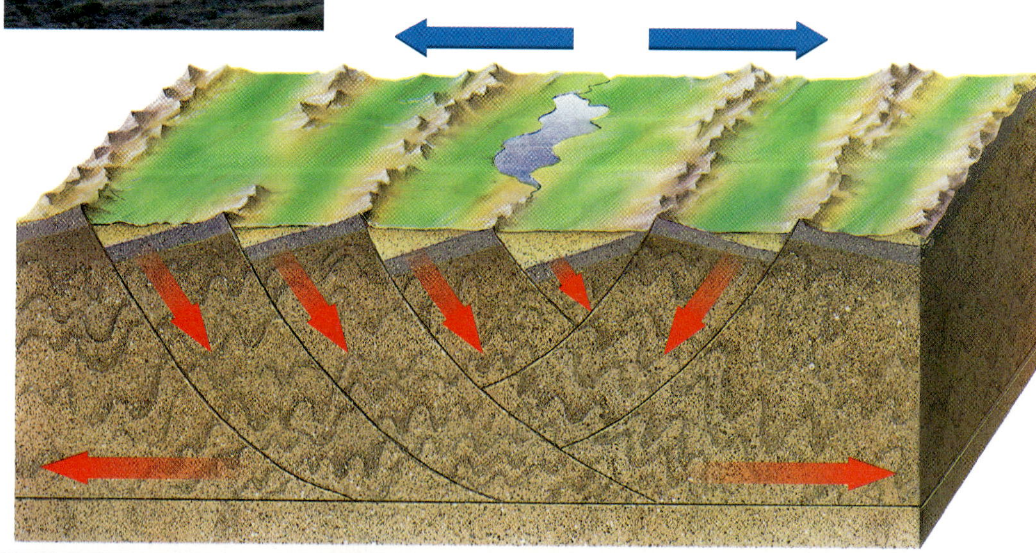

Figure 13 Fault-block mountains can form when Earth's crust is stretched by tectonic forces. The arrows indicate the directions of moving blocks.
Name *the type of force that occurs when Earth's crust is pulled in opposite directions.*

Differentiated Instruction

Challenge Have students place a piece of long, plastic tubing into the neck of a strong balloon and secure the tube so no air can escape from the balloon. Have them place the balloon at the bottom of an aquarium tank, with the tubing extending over the top of the tank. Then have students add alternating 2-mm layers of moist sand and dry sand to the bottom of the tank, covering the balloon. If they wish, students can build a small city on the top sand layer. Then have one student slowly blow up the balloon until the sand begins to crack to simulate an earthquake. Another student can videotape the "earthquake," and students can prepare a documentary showing how the quake affected the "city." L3 COOP LEARN IS **Kinesthetic**

Mountains and Volcanoes Compression forces squeeze objects together. Where plates come together, compression forces produce several effects. As continental plates collide, the forces that are generated cause massive folding and faulting of rock layers into mountain ranges such as the Himalaya, shown in **Figure 14,** or the Appalachian Mountains. The type of faulting produced is generally reverse faulting. Along a reverse fault, the rock layers above the fault surface move up relative to the rock layers below the fault.

Reading Check *What features occur where plates converge?*

As you learned earlier, when two oceanic plates converge, the denser plate is forced beneath the other plate. Curved chains of volcanic islands called island arcs form above the sinking plate. If an oceanic plate converges with a continental plate, the denser oceanic plate slides under the continental plate. Folding and faulting at the continental plate margin can thicken the continental crust to produce mountain ranges. Volcanoes also typically are formed at this type of convergent boundary.

INTEGRATE Career

Volcanologist This person's job is to study volcanoes in order to predict eruptions. Early warning of volcanic eruptions gives nearby residents time to evacuate. Volcanologists also educate the public about the hazards of volcanic eruptions and tell people who live near volcanoes what they can do to be safe in the event of an eruption. Volcanologists travel all over the world to study new volcanic sites.

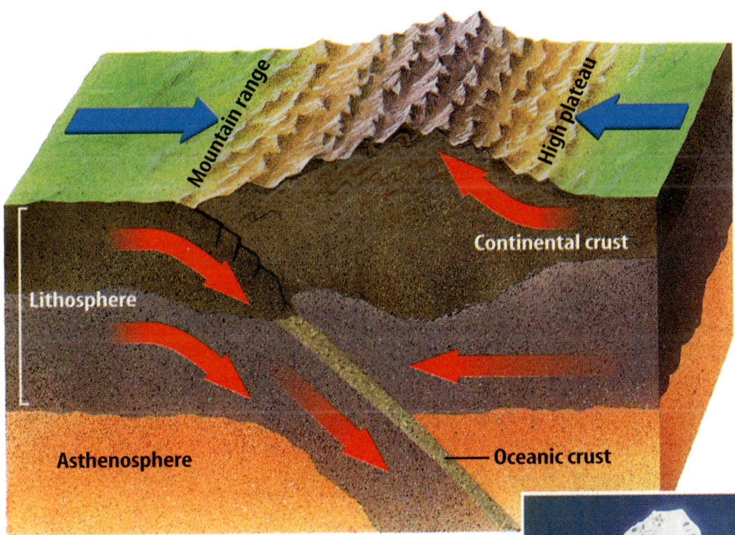

Figure 14 The Himalaya still are forming today as the Indo-Australian Plate collides with the Eurasian Plate.

Reading Check

Answer continental–continental: high mountain ranges; oceanic–oceanic: island arcs; oceanic–continental: mountains and volcanoes

Teacher FYI

Moving Mountains As the Indo-Australian Plate continues to push into and under the Eurasian Plate, the Eurasian Plate is thrust up over it. India moves into Asia at a rate of 3.7 to 5.4 cm/yr, thrusting the Himalaya about 1 cm higher each year. But because forces of erosion wear down the mountains by about the same amount annually, they remain about the same height.

Make a Model

Strike-Slip Faults Provide clay with which students can make models to show the movement of plates at a strike-slip fault. Have them include surface features offset by the movement. L1 LS **Kinesthetic**

INTEGRATE Career

Volcanologist Ask students to research to find out the type and level of degree required to become a volcanologist. Then have them research the course requirements and duration of such a degree at a university of their choice.

Curriculum Connection

Geography Have students search the Internet, newspapers, and magazines for stories about earthquakes and volcanic eruptions that have occurred as a result of plate movements along faults. Have students summarize their findings in written paragraphs to share in class. Then have them make a bulletin board display that includes a map showing the location of these tectonic events. L2 LS **Visual-Spatial**

Discussion

Living on a Fault Would you live or construct a building along a strike-slip fault? Explain. Most would not because of the danger of earthquakes and the likelihood of destruction. Why might so many people live along the San Andreas Fault? Possible answers: Some people were there before the danger was known; some think the danger is minimal, some might be unable to relocate; some may think that other factors (climate, economic advantages) outweigh the negatives of earthquake danger.

INTEGRATE Physics

Direction of Forces Convergent: toward each other; divergent: away from each other; transform: sliding past each other.

Teacher FYI

Moving Plates Creepmeters, lasers, and satellites are used to measure plate movements. They have shown that the Pacific Plate has been sliding past the North American Plate along the San Andreas Fault at a rate of 1.2 to 3.3 cm/yr. If that continues, the Pacific Plate will continue to move north relative to the North American Plate, bringing Los Angeles up next to San Francisco in about 27 million years.

Figure 15 Most of the movement along a strike-slip fault is parallel to Earth's surface. When movement occurs, human-built structures along a strike-slip fault are offset, as shown here in this road.

Strike-Slip Faults At transform boundaries, two plates slide past one another without converging or diverging. The plates stick and then slide, mostly in a horizontal direction, along large strike-slip faults. In a strike-slip fault, rocks on opposite sides of the fault move in opposite directions, or in the same direction at different rates. This type of fault movement is shown in **Figure 15.** One such example is the San Andreas Fault. When plates move suddenly, vibrations are generated inside Earth that are felt as an earthquake.

Earthquakes, volcanoes, and mountain ranges are evidence of plate motion. Plate tectonics explains how activity inside Earth can affect Earth's crust differently in different locations. You've seen how plates have moved since Pangaea separated. Is it possible to measure how far plates move each year?

Testing for Plate Tectonics

Until recently, the only tests scientists could use to check for plate movement were indirect. They could study the magnetic characteristics of rocks on the seafloor. They could study volcanoes and earthquakes. These methods supported the theory that the plates have moved and still are moving. However, they did not provide proof—only support—of the idea.

New methods had to be discovered to be able to measure the small amounts of movement of Earth's plates. One method, shown in **Figure 16,** uses lasers and a satellite. Now, scientists can measure exact movements of Earth's plates of as little as 1 cm per year.

INTEGRATE Physics

Direction of Forces In which directions do forces act at convergent, divergent, and transform boundaries? Demonstrate these forces using wooden blocks or your hands.

Differentiated Instruction

Challenge Challenge students to research the 1992 Landers, California, earthquake. In their Science Journals, have them write a summary explaining why this quake is important to the study of plate tectonics. It may indicate that a new fault is forming and that the Pacific–North American Plate boundary may be shifting eastward. L3 LS **Logical-Mathematical**

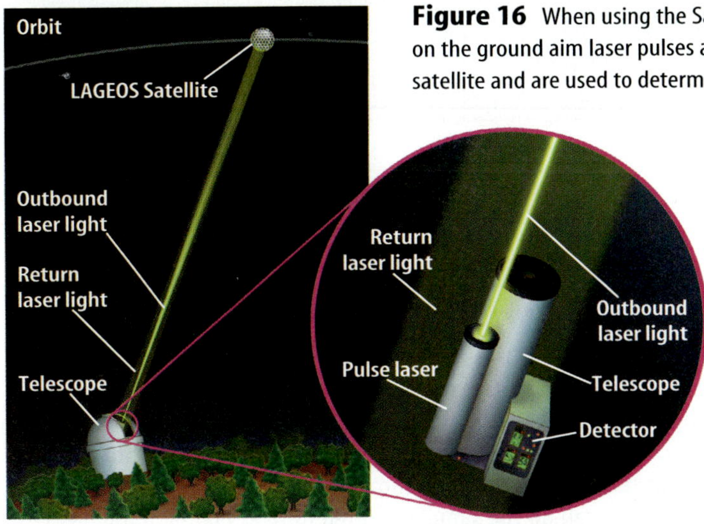

Figure 16 When using the Satellite Laser Ranging System, scientists on the ground aim laser pulses at a satellite. The pulses reflect off the satellite and are used to determine a precise location on the ground.

Current Data Satellite Laser Ranging System data show that Hawaii is moving toward Japan at a rate of about 8.3 cm per year. Maryland is moving away from England at a rate of 1.7 cm per year. Using such methods, scientists have observed that the plates move at rates ranging from about 1 cm to 12 cm per year.

section 3 review

Summary

Plate Tectonics

- The theory of plate tectonics states that sections of the seafloor and continents move as plates on a plasticlike layer of the mantle.

Plate Boundaries

- The boundary between two plates moving apart is called a divergent boundary.
- Plates move together at a convergent boundary.
- Transform boundaries occur where two plates slide past one another.

Causes of Plate Tectonics

- Convection currents are thought to cause the movement of Earth's plates.

Features Caused by Plate Tectonics

- Tension forces cause normal faults, rift valleys, and mid-ocean ridges at divergent boundaries.
- At convergent boundaries, compression forces cause folding, reverse faults, and mountains.
- At transform boundaries, two plates slide past one another along strike-slip faults.

Self Check

1. **Describe** what occurs at plate boundaries that are associated with seafloor spreading.
2. **Describe** three types of plate boundaries where volcanic eruptions can occur.
3. **Explain** how convection currents are related to plate tectonics.
4. **Think Critically** Using **Figure 9** and a world map, determine what natural disasters might occur in Iceland. Also determine what disasters might occur in Tibet. Explain why some Icelandic disasters are not expected to occur in Tibet.

Applying Skills

5. **Predict** Plate tectonic activity causes many events that can be dangerous to humans. One of these events is a seismic sea wave, or tsunami. Learn how scientists predict the arrival time of a tsunami in a coastal area.
6. **Use a Word Processor** Write three separate descriptions of the three basic types of plate boundaries—divergent boundaries, convergent boundaries, and transform boundaries. Then draw a sketch of an example of each boundary next to your description.

Science Online bookf.msscience.com/self_check_quiz

3 Assess

DAILY INTERVENTION

Check for Understanding

Interpersonal Students can take the answers they came up with in the section review and pair up with other students to discuss their answers.

Reteach

Conveyor Belt The movement of items on a supermarket conveyor belt is analogous to the movement of Earth's plates atop convection currents in the mantle. The belt is like the currents, and the moving groceries are like the plates moving on top of the currents. Have students draw diagrams showing these parallels. L1 LS **Visual-Spatial**

Assessment

Oral Copy portions of **Figure 10** showing different plates and the boundaries that they form. Have students identify the boundaries and the typical landforms formed by each type of interaction. Use **Performance Assessment in the Science Classroom** p. 89.

section 3 review

1. Plates move apart; rift-type volcanic eruptions.
2. Divergent; ocean–continent convergent; ocean–ocean convergent
3. Their movement in the mantle causes the movement of plates above them.
4. Iceland: volcanoes; Tibet: earthquakes; volcanoes are unlikely in Tibet because two continental plates converge there, with little or no subduction.
5. Answers should include the idea that scientists monitor undersea earthquakes, then estimate the potential arrival times of tsunamis in coastal areas.
6. Descriptions and sketches should match divergent boundaries, convergent boundaries, and transform boundaries as shown in **Figure 10** and **Figure 11.**

BENCH TESTED

Real-World Question

Purpose Students will obtain the latitudes and longitudes of recent earthquakes and volcanic eruptions. Data can be accessed through bookf.msscience.com/internet_lab. They will plot these locations on a map of the world. Once plotted, students will use the locations to determine tectonically active areas of Earth.

Process Skills Students will obtain maps of the world on which they will make their plots. They will use the plotted locations on the map to determine tectonically active areas of Earth.

Time Required two 35- to 40-minute class periods or one block of 70 to 80 minutes

Make a Plan

Internet Visit bookf.msscience.com/internet_lab to run through the steps that students will follow.

Non-Internet Obtain tissue paper or sheets of plastic for students to use to draw copies of maps showing the locations of recent earthquakes and volcanic eruptions. The locations of earthquake epicenters and erupting volcanoes can be obtained from the U.S. Geological Survey or from local newspapers.

LAB Use the Internet

Predicting Tectonic Activity

Goals

- **Research** the locations of earthquakes and volcanic eruptions around the world.
- **Plot** earthquake epicenters and the locations of volcanic eruptions.
- **Predict** locations that are tectonically active based on a plot of the locations of earthquake epicenters and active volcanoes.

Data Source

Science Online

Visit bookf.msscience.com/internet_lab for more information about earthquake and volcano sites, and data from other students.

Real-World Question

The movement of plates on Earth causes forces that build up energy in rocks. The release of this energy can produce vibrations in Earth that you know as earthquakes. Earthquakes occur every day. Many of them are too small to be felt by humans, but each event tells scientists something more about the planet. Active volcanoes can do the same and often form at plate boundaries.

Can you predict tectonically active areas by plotting locations of earthquake epicenters and volcanic eruptions?

Think about where earthquakes and volcanoes have occurred in the past. Make a hypothesis about whether the locations of earthquake epicenters and active volcanoes can be used to predict tectonically active areas.

Differentiated Instruction

Challenge Pair learning disabled students with other students who can assist them with working on the computer. Encourage learning disabled students to take the lead in recording data, with the assistance of their partners. L3

Alternative Inquiry Lab

Explore Through Research Have students research earthquake intensity based on the lab data. Ask students, through research, to hypothesize why earthquakes on some plate boundaries are more severe than others. This will help students with research ability and critical thinking, as well as acquiring a more detailed understanding of plate tectonics.

Using Scientific Methods

Make a Plan

1. Make a data table in your Science Journal like the one shown.
2. Collect data for earthquake epicenters and volcanic eruptions for at least the past two weeks. Your data should include the longitude and latitude for each location. For help, refer to the data sources given on the opposite page.

Locations of Epicenters and Eruptions

Earthquake Epicenter/ Volcanic Eruption	Longitude	Latitude
Answers will vary.		

Follow Your Plan

1. Make sure your teacher approves your plan before you start.
2. **Plot** the locations of earthquake epicenters and volcanic eruptions on a map of the world. Use an overlay of tissue paper or plastic.
3. After you have collected the necessary data, predict where the tectonically active areas on Earth are.
4. **Compare and contrast** the areas that you predicted to be tectonically active with the plate boundary map shown in **Figure 9.**

Analyze Your Data

1. What areas on Earth do you predict to be the locations of tectonic activity?
2. How close did your prediction come to the actual location of tectonically active areas?

Conclude and Apply

1. How could you make your predictions closer to the locations of actual tectonic activity?
2. Would data from a longer period of time help? Explain.
3. What types of plate boundaries were close to your locations of earthquake epicenters? Volcanic eruptions?
4. **Explain** which types of plate boundaries produce volcanic eruptions. Be specific.

Communicating Your Data

Find this lab using the link below. Post your data in the table provided. **Compare** your data to those of other students. Combine your data with those of other students and **plot** these combined data on a map to recognize the relationship between plate boundaries, volcanic eruptions, and earthquake epicenters.

Science Online
bookf.msscience.com/internet_lab

Teaching Strategy Encourage students to look for clusters of earthquake and volcanic activity, or for events that occur in a linear pattern.

Expected Outcome Students should see patterns in the locations of earthquake and volcanic activity.

Analyze Your Data

1. Predictions will likely match the locations of plate boundaries.
2. Answers will depend on data collected. Most occur along plate boundaries. Hot spot eruptions may not coincide with plate boundaries.

Conclude and Apply

1. by collecting more data points
2. Yes; it would provide more data, which would help to more closely pinpoint these areas.
3. Earthquakes: near any type with many near convergent and transform boundaries; volcanoes: near divergent boundaries and subduction zones
4. Convergent ocean-ocean and ocean-continental boundaries where one plate is subducted under the other produce magma that rises and forms volcanoes. Volcanoes also form along divergent boundaries where magma rises through cracks in the crust, either at mid-ocean ridges or on land in rift valleys.

Assessment

Oral Have pairs of students form a hypothesis that could explain the relationship between the locations of earthquake epicenters and active volcanoes and Earth's tectonic activity. Have each pair of students report to the class. Use **Performance Assessment in the Science Classroom,** p. 93.

Communicating Your Data

Plot all data from each student in the class on one large map. Lead students to the realization that as more and more data are placed on the map, the relationship of these data to the location of tectonically active areas on Earth becomes much more evident.

Science and Language Arts

Listening In

by Gordon Judge

I'm just a bit of seafloor on this mighty solid sphere.
With no mind to be broadened, I'm quite content
down here.
The mantle churns below me, and the sea's in turmoil, too;
But nothing much disturbs me, I'm rock solid through
and through.

I do pick up occasional low-frequency vibrations –
(I think, although I can't be sure, they're sperm whales'
conversations).
I know I shouldn't listen in, but what else can I do?
It seems they are all studying for degrees from the OU.

They've mentioned me in passing, as their minds begin
improving:

I think I've heard them say
"The theory says the sea-
floor's moving…".
They call it "Plate Tectonics", this
new theory in their noddle.
If they would only ask me, I
could tell them it's all
twaddle….

But, how can I be moving, when I know full well myself
That I'm quite firmly anchored to a continental shelf?
"Well, the continent is moving, too; you're *pushing* it,
you see,"
I hear those OU whales intone, hydro-acoustically….

Well, thank you very much, OU. You've upset my
composure.
Next time you send your student whales to look at
my exposure
I'll tell them it's a load of tosh: it's *they* who move,
not me,
Those arty-smarty blobs of blubber, clogging up the sea!

Understanding Literature

Point of View Point of view refers to the perspective from which an author writes. This poem begins, "I'm just a bit of seafloor…." Right away, you know that the poem, or story, is being told from the point of view of the speaker, or the "first person." What effect does the first-person narration have on the story?

Respond to the Reading

1. Who is narrating the poem?
2. Why might the narrator think he or she hasn't moved?
3. **Linking Science and Writing** Using the first-person point of view, write an account from the point of view of a living or nonliving thing.

INTEGRATE Earth Science Volcanoes can occur where two plates move toward each other. When an oceanic plate and a continental plate collide, a volcano will form. Subduction zones occur when one plate sinks under another plate. Rocks melt in the zones where these plates converge, causing magma to move upward and form volcanic mountains.

Understanding Literature

Point of View Answers may vary but might include that the use of "I" gives a personal tone or human touch to otherwise technical or scientific material. Giving inanimate objects human traits personalizes science and gives it warmth.

Respond to the Reading

1. a land, or continental, plate
2. Because the movement of continental, or land, plates is so slow.
3. Remind students to use the personal pronoun "I" in their accounts.

INTEGRATE Earth Science **Volcanoes** There are approximately 1,300 active volcanoes in the world. However, active volcanoes do not appear by chance. Most occur in belts, around mountain ranges that border the Pacific Ocean. The Pacific Plate pushes beneath the adjacent continental plates in a process known as subduction. This is where Japan and the Aleutian Islands converge. Approximately 80 percent of the world's active volcanoes that have been recorded occur in these subduction zones.

Resources for Teachers and Students

Volcano Cowboys: The Rocky Evolution of a Dangerous Science, by Dick Thompson, St. Martin's Press, July 2000

Melting the Earth: The History of Ideas on Volcanic Eruptions, by Haraldur Sigurdsson, Oxford University Press, June 1999

Reviewing Main Ideas

Section 1 Continental Drift

1. Alfred Wegener suggested that the continents were joined together at some point in the past in a large landmass he called Pangaea. Wegener proposed that continents have moved slowly, over millions of years, to their current locations.
2. The puzzlelike fit of the continents, fossils, climatic evidence, and similar rock structures support Wegener's idea of continental drift. However, Wegener could not explain what process could cause the movement of the landmasses.

Section 2 Seafloor Spreading

1. Detailed mapping of the ocean floor in the 1950s showed underwater mountains and rift valleys.
2. In the 1960s, Harry Hess suggested seafloor spreading as an explanation for the formation of mid-ocean ridges.
3. The theory of seafloor spreading is supported by magnetic evidence in rocks and by the ages of rocks on the ocean floor.

Section 3 Theory of Plate Tectonics

1. In the 1960s, scientists combined the ideas of continental drift and seafloor spreading to develop the theory of plate tectonics. The theory states that the surface of Earth is broken into sections called plates that move around on the asthenosphere.
2. Currents in Earth's mantle called convection currents transfer heat in Earth's interior. It is thought that this transfer of heat energy moves plates.
3. Earth is a dynamic planet. As the plates move, they interact, resulting in many of the features of Earth's surface.

Visualizing Main Ideas

Copy and complete the concept map below about continental drift, seafloor spreading, and plate tectonics.

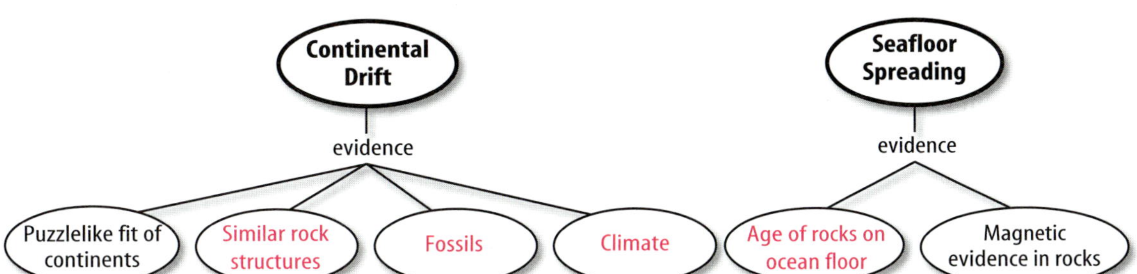

Reviewing Main Ideas

Summary statements can be used by students to review the major concepts of the chapter.

Visualizing Main Ideas

See student page.

Science Online
Visit bookf.msscience.com
/self_check_quiz
/interactive_tutor
/vocabulary_puzzlemaker
/chapter_review
/standardized_test

Assessment Transparency

For additional assessment questions, use the *Assessment Transparency* located in the transparency book.

Assessment

Assessment Transparency **Plate Tectonics**

Directions: *Carefully review the diagram and answer the following questions.*

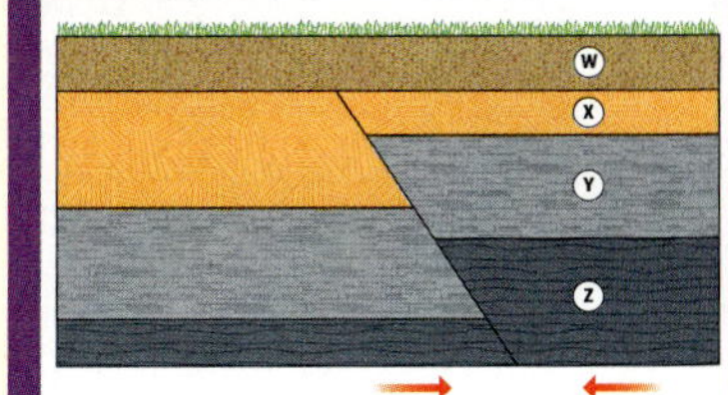

1. Which is the oldest rock layer in the picture?
   A W
   B X
   C Y
   D Z
2. The arrows indicate the directions the two plates are moving. What is this type of boundary called?
   F convergent boundary
   G divergent boundary
   H transform boundary
   J moving boundary
3. Which of the following is the danger most likely posed by the rock formation shown in the diagram?
   A flooding
   B earthquake
   C tornado
   D forest fire

L2

Plate Tectonics

CHAPTER STUDY GUIDE F ◆ 119

Using Vocabulary

1. asthenosphere
2. continental drift
3. Pangaea
4. lithosphere
5. plate tectonics
6. seafloor spreading

Checking Concepts

7. B
8. D
9. C
10. A
11. A
12. B
13. B
14. D

chapter 4 Review

Using Vocabulary

asthenosphere p. 106	Pangaea p. 98
continental drift p. 98	plate p. 106
convection current p. 111	plate tectonics p. 106
lithosphere p. 106	seafloor spreading p. 103

Each phrase below describes a vocabulary term from the list. Write the term that matches the phrase describing it.

1. plasticlike layer below the lithosphere
2. idea that continents move slowly across Earth's surface
3. large, ancient landmass that consisted of all the continents on Earth
4. composed of oceanic or continental crust and upper mantle
5. explains locations of mountains, trenches, and volcanoes
6. theory proposed by Harry Hess that includes processes along mid-ocean ridges

Checking Concepts

Choose the word or phrase that best answers the question.

7. Which layer of Earth contains the asthenosphere?
   A) crust **C)** outer core
   B) mantle **D)** inner core
8. What type of plate boundary is the San Andreas Fault part of?
   A) divergent **C)** convergent
   B) subduction **D)** transform
9. What hypothesis states that continents slowly moved to their present positions on Earth?
   A) subduction **C)** continental drift
   B) erosion **D)** seafloor spreading

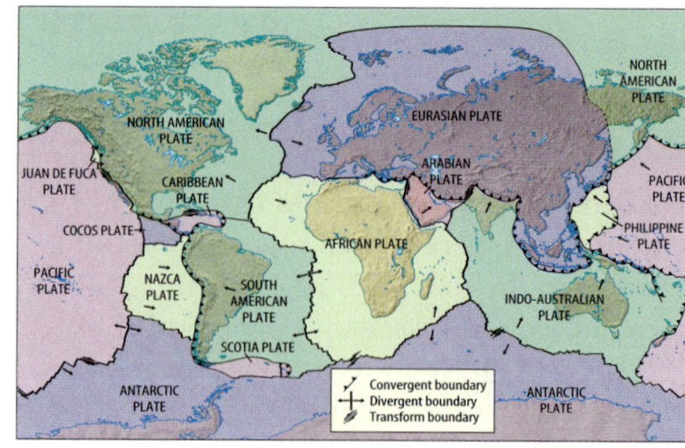

10. Which plate is subducting beneath the South American Plate to form the Andes mountain range?
    A) Nazca **C)** North American
    B) African **D)** Indo-Australian
11. Which of the following features are evidence that many continents were at one time near Earth's south pole?
    A) glacial deposits **C)** volcanoes
    B) earthquakes **D)** mid-ocean ridges
12. What evidence in rocks supports the theory of seafloor spreading?
    A) plate movement
    B) magnetic reversals
    C) subduction
    D) convergence
13. Which type of plate boundary is the Mid-Atlantic Ridge a part of?
    A) convergent **C)** transform
    B) divergent **D)** subduction
14. What theory states that plates move around on the asthenosphere?
    A) continental drift
    B) seafloor spreading
    C) subduction
    D) plate tectonics

Science Online bookf.msscience.com/vocabulary_puzzlemaker

Use the Exam*View*® Pro Testmaker CD-ROM to:
- create multiple versions of tests
- create modified tests with one mouse click for inclusion students
- edit existing questions and add your own questions
- build tests aligned with state standards using built-in State Curriculum Tags
- change English tests to Spanish with one mouse click and vice versa

Thinking Critically

15. Infer Why do many earthquakes but few volcanic eruptions occur in the Himalaya?

16. Explain Glacial deposits often form at high latitudes near the poles. Explain why glacial deposits have been found in Africa.

17. Describe how magnetism is used to support the theory of seafloor spreading.

18. Explain why volcanoes do not form along the San Andreas Fault.

19. Explain why the fossil of an ocean fish found on two different continents would not be good evidence of continental drift.

20. Form Hypotheses Mount St. Helens in the Cascade Range is a volcano. Use **Figure 9** and a U.S. map to hypothesize how it might have formed.

21. Concept Map Make an events-chain concept map that describes seafloor spreading along a divergent plate boundary. Choose from the following phrases: *magma cools to form new seafloor, convection currents circulate hot material along divergent boundary,* and *older seafloor is forced apart.*

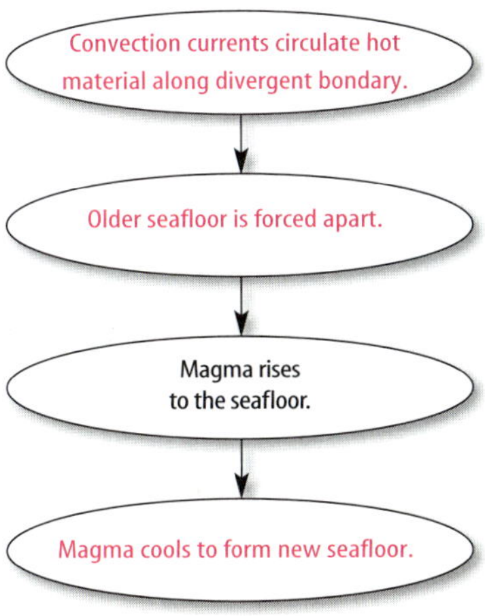

Performance Activities

22. Observe and Infer In the MiniLab called "Modeling Convection Currents," you observed convection currents produced in water as it was heated. Repeat the experiment, placing sequins, pieces of wood, or pieces of rubber bands into the water. How do their movements support your observations and inferences from the MiniLab?

Applying Math

23. A Growing Rift Movement along the African Rift Valley is about 2.1 cm per year. If plates continue to move apart at this rate, how much larger will the rift be (in meters) in 1,000 years? In 15,500 years?

Use the illustration below to answer questions 24 and 25.

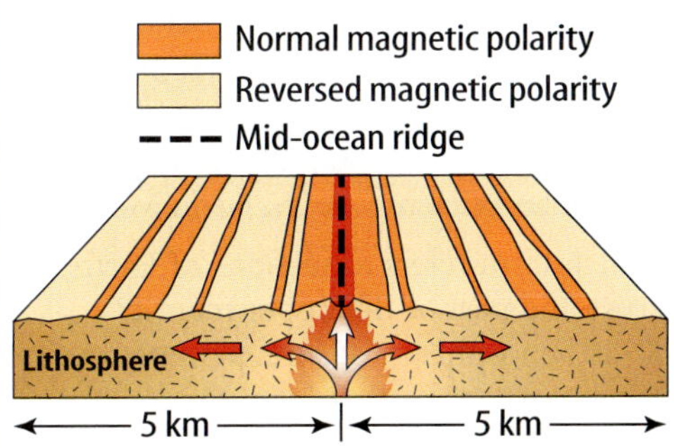

24. New Seafloor 10 km of new seafloor has been created in 50,000 years, with 5 km on each side of a mid-ocean ridge. What is the rate of movement, in km per year, of each plate? In cm per year?

25. Use a Ratio If 10 km of seafloor were created in 50,000 years, how many kilometers of seafloor were created in 10,000 years? How many years will it take to create a total of 30 km of seafloor?

ScienceOnline bookf.msscience.com/chapter_review

Assessment Resources

Reproducible Masters

Chapter *Fast File* Resources
Chapter Review, pp. 37–38
Chapter Tests, pp. 39–42
Assessment Transparency Activity, p. 49

Glencoe Science Web site
Chapter Review Test
Standardized Test Practice

Glencoe Technology
Assessment Transparency
Exam*View*® Pro Testmaker
MindJogger Videoquiz
Interactive Chalkboard

chapter 4 Review

Thinking Critically

15. The colliding continental plates cause earthquakes, but neither plate is forced deep into Earth, which would allow melting to occur and magma to rise through any cracks to form volcanoes.

16. Africa was located near the South Pole when all the continents were joined.

17. When molten material rises and cools at a ridge, magnetic rocks take on the orientation of Earth's magnetic field. Each time Earth's magnetic field reverses, new materials that form close to the ridge take on the new orientation.

18. It is a transform boundary. Without subduction no melting occurs and no magma is produced.

19. The fish could have moved through the oceans between continents.

20. The volcanoes of the CascadeRange formed as the Juan de Fuca Plate subducted beneath the North American Plate.

21. See student page.

Performance Activities

22. The objects move around because of convection processes going on inside the water. Use **Performance Assessment in the Science Classroom,** p. 89.

Applying Math

National Math Standards

1,9

23. 21 m; 325.5 m

24. 0.0001 km/year; 10 cm/year

25. 2 km; 150,000 years

CHAPTER REVIEW F ◆ 121

![FAST FILE]

Answer Sheet A practice answer sheet can be found at bookf.msscience.com/answer_sheet.

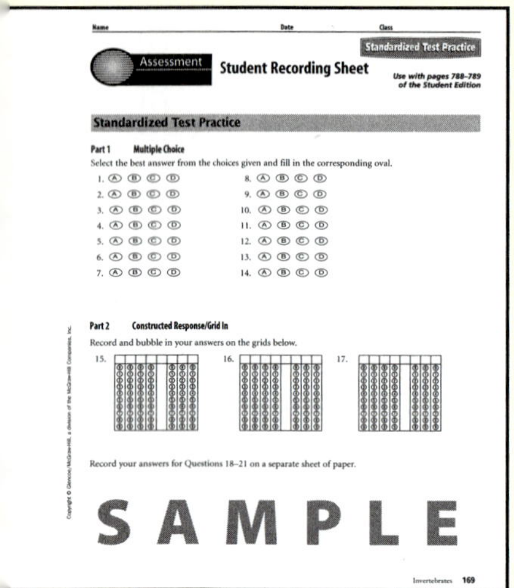

Part 1 Multiple Choice

1. A
2. C
3. B
4. B
5. C
6. D
7. C

Part 2 Short Response

8. an elongate, deep area on the ocean floor along a subduction zone
9. Island arcs form from volcanic eruptions above subduction zones.
10. Two plates meet at the San Andreas Fault. Sideways motion between the plates causes earthquakes to occur there.
11. A mid-ocean ridge is a mountain range on the ocean floor where two plates are separating. It is characterized by frequent, large eruptions of lava.
12. As oceanic lithosphere ages, it cools and becomes denser. When it becomes denser than the asthenosphere beneath, it sinks into the mantle.
13. 2.2 km; Student answers may vary. Accept reasonable responses.

chapter 4 Standardized Test Practice

Part 1 Multiple Choice

Record your answers on the answer sheet provided by your teacher or on a sheet of paper.

Use the illustration below to answer question 1.

1. What is the name of the ancient supercontinent shown above?
   A. Pangaea **C.** Laurasia
   B. Gondwanaland **D.** North America
2. Who developed the continental drift hypothesis?
   A. Harry Hess **C.** Alfred Wegener
   B. J. Tuzo Wilson **D.** W. Jason Morgan
3. Which term refers to sections of Earth's crust and part of the upper mantle?
   A. asthenosphere **C.** lithosphere
   B. plate **D.** core
4. About how fast do plates move?
   A. a few millimeters each year
   B. a few centimeters each year
   C. a few meters each year
   D. a few kilometers each year

Test-Taking Tip

Marking Answers Be sure to ask if it is okay to mark in the test booklet when taking the test, but make sure you mark all answers on your answer sheet.

5. Where do Earth's plates slide past each other?
   A. convergent boundaries
   B. divergent boundaries
   C. transform boundaries
   D. subduction zones

Study the diagram below before answering questions 6 and 7.

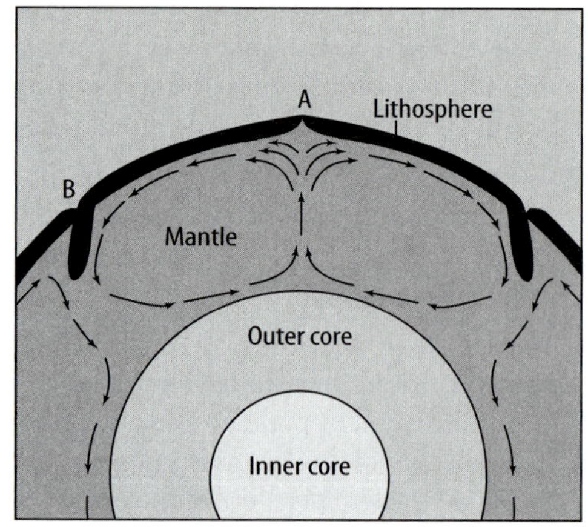

6. Suppose that the arrows in the diagram represent patterns of convection in Earth's mantle. Which type of plate boundary is most likely to occur along the region labeled "A"?
   A. transform
   B. reverse
   C. convergent
   D. divergent
7. Which statement is true of the region marked "B" on the diagram?
   A. Plates move past each other sideways.
   B. Plates move apart and volcanoes form.
   C. Plates move toward each other and volcanoes form.
   D. Plates are not moving.

14. 5.4 km; Accept reasonable responses.
15. The depth of ocean crust increases as the age increases. This occurs because the oceanic plate becomes denser as it cools.
16. a few centimeters each year
17. the aesthenophere
18. They use sound waves.

Part 3 Open Ended

19. Earth's magnetic field periodically reverses its polarity. At mid-ocean ridges, new seafloor is created when erupting lava cools. This seafloor then moves away from the ridge. The changing polarity of Earth's magnetic field is recorded in the volcanic rock.
20. Convection in Earth's mantle occurs because heat from Earth's interior causes some rock to be hotter and less dense than surrounding rock, forcing it upward. Denser rock sinks under gravity's influence.
21. states that the lithosphere is broken into sections called plates. These plates move on the asthenosphere.

Part 2 Short Response/Grid In

Record your answers on the answer sheet provided by your teacher or on a sheet of paper.

8. What is an ocean trench? Where do they occur?
9. How do island arcs form?
10. Why do earthquakes occur along the San Andreas Fault?
11. Describe a mid-ocean ridge.
12. Why do plates sometimes sink into the mantle?

Use the graph below to answer questions 13–15.

Relationship Between Depth and Age of Seafloor

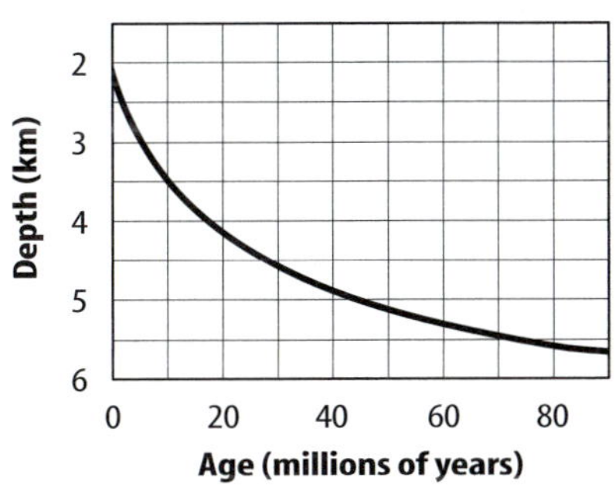

13. Use the graph to estimate the average depth below the ocean of ocean crust that has just formed.
14. Estimate the average depth of ocean crust that is 60 million years old.
15. Describe how the depth of ocean crust is related to the age of ocean crust.
16. On average, about how fast do plates move?
17. What layer in Earth's mantle do plates slide over?
18. Describe how scientists make maps of the ocean floor.

Part 3 Open Ended

Record your answers on a sheet of paper.

Use the illustration below to answer question 19.

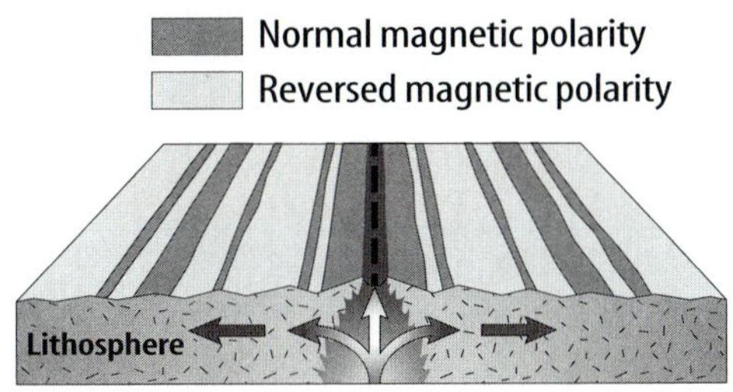

19. Examine the diagram above. Explain how the magnetic stripes form in rock that makes up the ocean crust.
20. What causes convection in Earth's mantle?
21. Explain the theory of plate tectonics.
22. What happened to the continents that made up Pangaea after it started to break up?
23. How does Earth's lithosphere differ from Earth's asthenosphere?
24. What types of life have been discovered near mid-ocean ridges?
25. What are the three types of motion that occur at plate boundaries? Describe each motion.
26. What forms when continents collide? Describe the process.
27. What occurs at the center of a mid-ocean ridge? What might you find there?
28. What evidence do we have that supports the hypothesis of continental drift?
29. Who proposed the first theories about plate tectonics? Explain why other scientists questioned these theories.

Rubrics

The following rubrics are sample scoring devices for short response and open-ended questions.

Short Response

Points	Description
2	The student demonstrates a thorough understanding of the science of the task. The response may contain minor flaws that do not detract from the demonstration of a thorough understanding.
1	The student has provided a response that is only partially correct.
0	The student has provided a completely incorrect solution or no response at all.

Open Ended

Points	Description
4	The student demonstrates a thorough understanding of the science of the task. The response may contain minor flaws that do not detract from the demonstration of a thorough understanding.
3	The student demonstrates an understanding of the science of the task. The response is essentially correct and demonstrates an essential but less than thorough understanding of the science.
2	The student demonstrates only a partial understanding of the science of the task. Although the student may have used the correct approach to a solution or may have provided a correct solution, the work lacks an essential understanding of the underlying science concepts.
1	The student demonstrates a very limited understanding of the science of the task. The response is incomplete and exhibits many flaws.
0	The student provides a completely incorrect solution or no response at all.

22. The continents gradually moved to their current positions as the North Atlantic Ocean and South Atlantic Ocean opened.
23. Earth's lithosphere is more brittle than the asthenosphere. The asthenosphere is hotter and is capable of plastic deformation.
24. giant clams, tube worms, and other strange creatures
25. Plates can separate, move toward each other, or slide past each other.
26. mountains; Continental plates collide forcing the crust upward.
27. a rift valley; volcanoes, new oceanic crust
28. Fossils of warm-weather plants found on islands in the Arctic Ocean and glacial features found in places such as Africa supported this idea.
29. many scientists in the 1960s; Controversial theories often are challenged by other scientists. The more challenges a theory can meet, the more reliable the theory becomes.

chapter 5 Organizer

Section/Objectives	Standards: National	Standards: State/Local	Labs/Features
Chapter Opener	See p. 9T–10T for a Key to Standards.		**Launch Lab:** Why do earthquakes occur?, p. 125 **Foldables,** p. 125
Section 1 Forces Inside Earth 2 sessions 1 block 1. **Explain** how earthquakes result from the buildup of energy in rocks. 2. **Describe** how compression, tension, and shear forces make rocks move along faults. 3. **Distinguish** among normal, reverse, and strike-slip faults.	National Content Standards: UCP.1–UCP.3, UCP.5, A.1, A.2, B.2, D.1, D.2, F.3		
Section 2 Features of Earthquakes 2 sessions 1 block 4. **Explain** how earthquake energy travels in seismic waves. 5. **Distinguish** among primary, secondary, and surface waves. 6. **Describe** the structure of Earth's interior.	National Content Standards: UCP.1–UCP.3, UCP.5, A.1, A.2, B.2, D.1, F.3, G.3		**Integrate Physics,** p. 131 **Visualizing Seismic Waves,** p. 132 **Science Online,** p. 133 **MiniLAB:** Interpreting Seismic Wave Data, p. 135 **Lab:** Epicenter Location, p. 138
Section 3 People and Earthquakes 3 sessions 1.5 blocks 7. **Explain** where most earthquakes in the United States occur. 8. **Describe** how scientists measure earthquakes. 9. **List** ways to make your classroom and home more earthquake-safe.	National Content Standards: UCP.1–UCP.3, UCP.5, A.1, A.2, D.1, F.3		**Integrate Career,** p. 141 **Science Online,** p. 142 **Applying Math:** Earthquake Energy, p. 143 **MiniLAB:** Modeling Seismic-Safe Structures, p. 144 **Lab:** Earthquake Depths, p. 146 **Science Stats:** Moving Earth!, p. 148

Glencoe Exclusive!
TeacherWorks™
All-In-One Planner and Resource Center

Lab Materials	Reproducible Resources	Section Assessment	Technology
Launch Lab: medium-grain sandpaper (2 sheets); tape; textbook; large, thick rubber bands (2)	**Chapter *FAST FILE* Resources** Foldables Worksheet, p. 19 Directed Reading Overview, p. 21 Note-taking Worksheets, pp. 35–37	GLENCOE'S ASSESSMENT ADVANTAGE	TeacherWorks includes: • Interactive Teacher Edition • Lesson Planner with calendar • Access to all program blacklines • Correlations to standards • Web links
Need materials? Contact Science Kit at 1-800-828-7777 or www.sciencekit.com on the Internet.	**Chapter *FAST FILE* Resources** Transparency Activity, p. 46 Enrichment, p. 32 Reinforcement, p. 29 Directed Reading, p. 22 Transparency Activity, pp. 49–50	**Performance** Activity, p. 128 **Content** Section Review, p. 129	Section Focus Transparency Teaching Transparency Guided Reading Audio Program Interactive Chalkboard CD-ROM
MiniLAB: Science Journal, **Figure 11** graph **Lab:** string, metric ruler, globe, chalk	**Chapter *FAST FILE* Resources** Transparency Activity, p. 47 MiniLAB, p. 3 Lab Activity, pp. 9–13 Enrichment, p. 33 Reinforcement, p. 30 Directed Reading, p. 23 Lab Worksheets, pp. 5–6 **Home and Community Involvement,** p. 40	**Portfolio** Science Journal, p. 133 **Performance** MiniLAB, p. 135 Applying Skills, p. 137 **Content** Section Review, p. 137	Section Focus Transparency Guided Reading Audio Program Interactive Chalkboard CD-ROM
MiniLAB: building blocks, rubber bands (medium and large sizes) **Lab:** graph paper, pencil	**Chapter *FAST FILE* Resources** Transparency Activity, p. 48 MiniLAB, p. 4 Lab Activity, pp. 15–18 Enrichment, p. 34 Reinforcement, p. 31 Directed Reading, pp. 23, 24 Lab Worksheets, pp. 7–8 **Lab Management and Safety,** p. 65	**Portfolio** Differentiated Instruction, p. 140 **Performance** Applying Math, p. 143 MiniLAB, p. 144 Applying Skills, p. 145 **Content** Section Review, p. 145	Section Focus Transparency Virtual Labs CD-ROM Guided Reading Audio Program Interactive Chalkboard CD-ROM Video Lab

End of Chapter Assessment

GLENCOE'S ASSESSMENT ADVANTAGE

Blackline Masters	Technology	Professional Series
Chapter *FAST FILE* Resources Chapter Review, pp. 39–40 Chapter Tests, pp. 41–44 **Standardized Test Practice,** pp. 23–26	MindJogger Videoquiz Virtual Labs CD-ROM Exam*View*® Pro Testmaker TeacherWorks CD-ROM Interactive Chalkboard CD-ROM	**Performance Assessment in the Science Classroom (PASC)**

chapter 5 Earthquakes

Transparencies

Section Focus

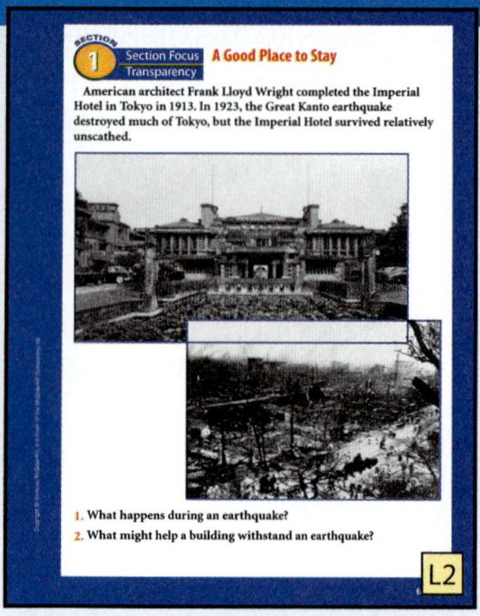

Section 1 Section Focus Transparency **A Good Place to Stay**

American architect Frank Lloyd Wright completed the Imperial Hotel in Tokyo in 1913. In 1923, the Great Kanto earthquake destroyed much of Tokyo, but the Imperial Hotel survived relatively unscathed.

1. What happens during an earthquake?
2. What might help a building withstand an earthquake?

L2

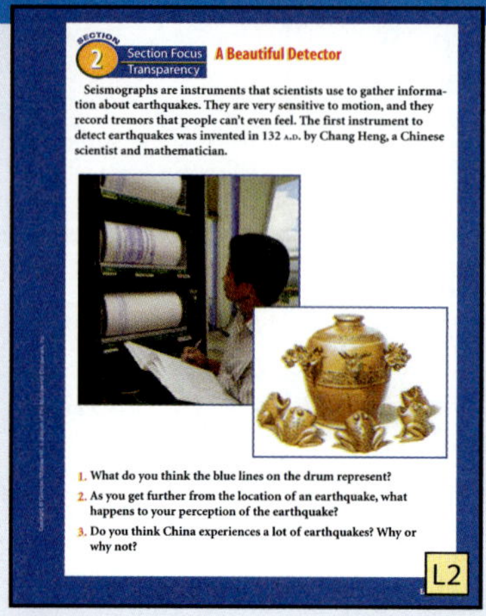

Section 2 Section Focus Transparency **A Beautiful Detector**

Seismographs are instruments that scientists use to gather information about earthquakes. They are very sensitive to motion, and they record tremors that people can't even feel. The first instrument to detect earthquakes was invented in 132 A.D. by Chang Heng, a Chinese scientist and mathematician.

1. What do you think the blue lines on the drum represent?
2. As you get further from the location of an earthquake, what happens to your perception of the earthquake?
3. Do you think China experiences a lot of earthquakes? Why or why not?

L2

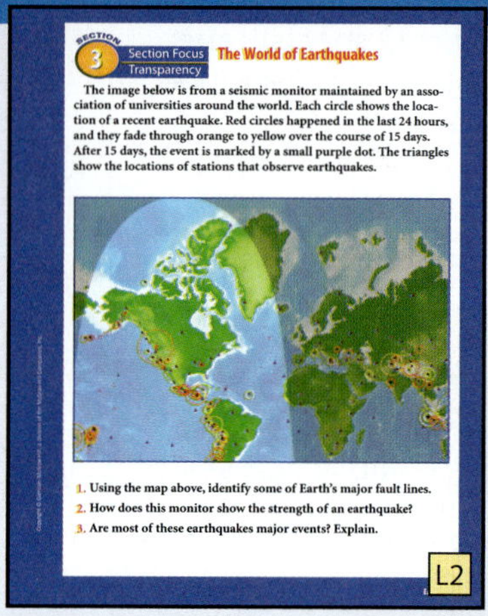

Section 3 Section Focus Transparency **The World of Earthquakes**

The image below is from a seismic monitor maintained by an association of universities around the world. Each circle shows the location of a recent earthquake. Red circles happened in the last 24 hours, and they fade through orange to yellow over the course of 15 days. After 15 days, the event is marked by a small purple dot. The triangles show the locations of stations that observe earthquakes.

1. Using the map above, identify some of Earth's major fault lines.
2. How does this monitor show the strength of an earthquake?
3. Are most of these earthquakes major events? Explain.

L2

This is a representation of key blackline masters available in the Teacher Classroom Resources. See Resource Manager boxes within the chapter for additional information.

Assessment

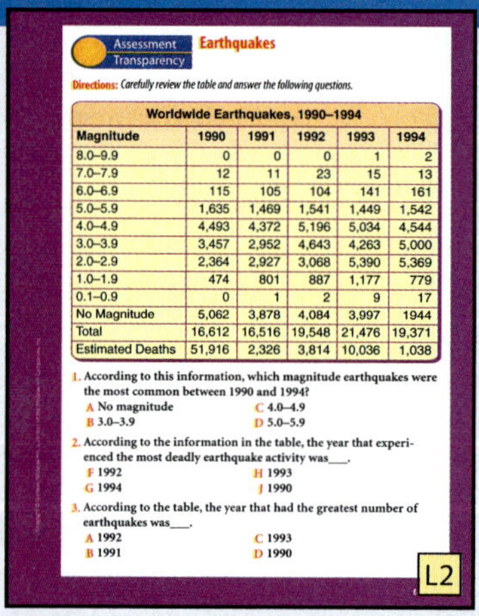

Assessment Transparency **Earthquakes**

Directions: *Carefully review the table and answer the following questions.*

Worldwide Earthquakes, 1990–1994					
Magnitude	1990	1991	1992	1993	1994
8.0–9.9	0	0	0	1	2
7.0–7.9	12	11	23	15	13
6.0–6.9	115	105	104	141	161
5.0–5.9	1,635	1,469	1,541	1,449	1,542
4.0–4.9	4,493	4,372	5,196	5,034	4,544
3.0–3.9	3,457	2,952	4,643	4,263	5,000
2.0–2.9	2,364	2,927	3,068	5,390	5,369
1.0–1.9	474	801	887	1,177	779
0.1–0.9	0	1	2	9	17
No Magnitude	5,062	3,878	4,084	3,997	1944
Total	16,612	16,516	19,548	21,476	19,371
Estimated Deaths	51,916	2,326	3,814	10,036	1,038

1. According to this information, which magnitude earthquakes were the most common between 1990 and 1994?
   A No magnitude C 4.0–4.9
   B 3.0–3.9 D 5.0–5.9
2. According to the information in the table, the year that experienced the most deadly earthquake activity was____.
   F 1992 H 1993
   G 1994 J 1990
3. According to the table, the year that had the greatest number of earthquakes was____.
   A 1992 C 1993
   B 1991 D 1990

L2

Teaching

Section 1 Teaching Transparency **Faults**

L2

Key to Teaching Strategies

The following designations will help you decide which activities are appropriate for your students.

L1 Level 1 activities should be appropriate for students with learning difficulties.

L2 Level 2 activities should be within the ability range of all students.

L3 Level 3 activities are designed for above-average students.

ELL ELL activities should be within the ability range of English Language Learners.

COOP LEARN Cooperative Learning activities are designed for small group work.

LS Multiple Learning Styles logos, as described on page 6T, are used throughout to indicate strategies that address different learning styles.

P These strategies represent student products that can be placed into a best-work portfolio.

PBL Problem-Based Learning activities apply real-world situations to learning.

Hands-on Activities

Student Text Lab Worksheet

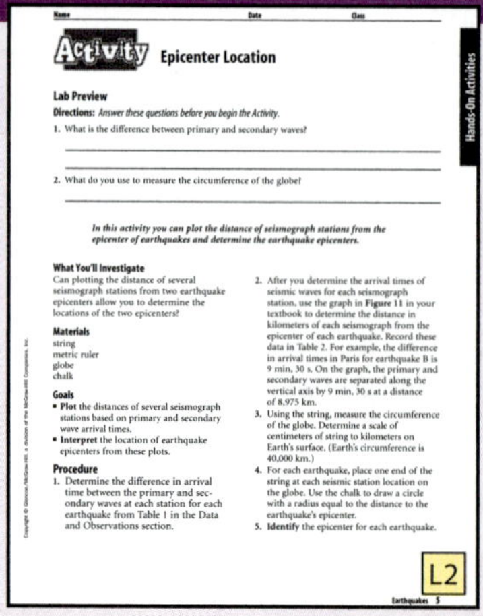

Name Date Class

Activity Epicenter Location

Hands-On Activities

Lab Preview

Directions: *Answer these questions before you begin the Activity.*

1. What is the difference between primary and secondary waves?
2. What do you use to measure the circumference of the globe?

In this activity you can plot the distance of seismograph stations from the epicenter of earthquakes and determine the earthquake epicenters.

What You'll Investigate

Can plotting the distance of several seismograph stations from two earthquake epicenters allow you to determine the locations of the two epicenters?

Materials

string
metric ruler
globe
chalk

Goals

- **Plot** the distances of several seismograph stations based on primary and secondary wave arrival times.
- **Interpret** the location of earthquake epicenters from these plots.

Procedure

1. Determine the difference in arrival time between the primary and secondary waves at each station for each earthquake from Table 1 in the Data and Observations section.
2. After you determine the arrival times of seismic waves for each seismograph station, use the graph in **Figure 11** in your textbook to determine the distance in kilometers of each seismograph from the epicenter of each earthquake. Record these data in Table 2. For example, the difference in arrival times in Paris for earthquake B is 9 min, 30 s. On the graph, the primary and secondary waves are separated along the vertical axis by 9 min, 30 s at a distance of 8,975 km.
3. Using the string, measure the circumference of the globe. Determine a scale of centimeters of string to kilometers on Earth's surface. (Earth's circumference is 40,000 km.)
4. For each earthquake, place one end of the string at each seismic station location on the globe. Use the chalk to draw a circle with a radius equal to the distance to the earthquake's epicenter.
5. **Identify** the epicenter for each earthquake.

L2

Earthquakes 5

Laboratory Activities

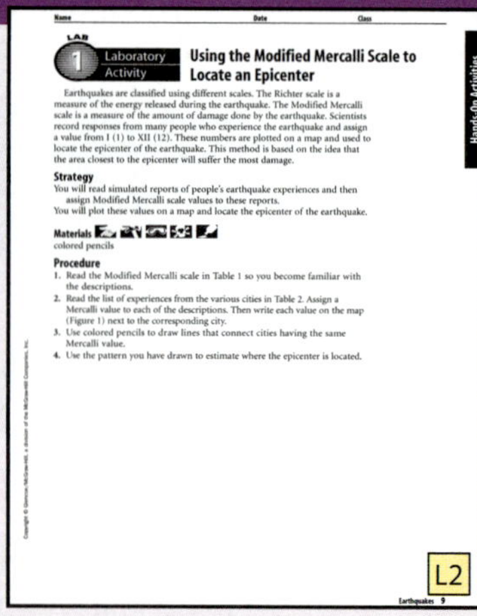

Name Date Class

Lab 1 Laboratory Activity **Using the Modified Mercalli Scale to Locate an Epicenter**

Hands-On Activities

Earthquakes are classified using different scales. The Richter scale is a measure of the energy released during the earthquake. The Modified Mercalli scale is a measure of the amount of damage done by the earthquake. Scientists record responses from many people who experience the earthquake and assign a value from I (1) to XII (12). These numbers are plotted on a map and used to locate the epicenter of the earthquake. This method is based on the idea that the area closest to the epicenter will suffer the most damage.

Strategy

You will read simulated reports of people's earthquake experiences and then assign Modified Mercalli scale values to these reports.
You will plot these values on a map and locate the epicenter of the earthquake.

Materials

colored pencils

Procedure

1. Read the Modified Mercalli scale in Table 1 so you become familiar with the descriptions.
2. Read the list of experiences from the various cities in Table 2. Assign a Mercalli value to each of the descriptions. Then write each value on the map (Figure 1) next to the corresponding city.
3. Use colored pencils to draw lines that connect cities having the same Mercalli value.
4. Use the pattern you have drawn to estimate where the epicenter is located.

L2

Earthquakes 9

Meeting Different Ability Levels

Content Outline

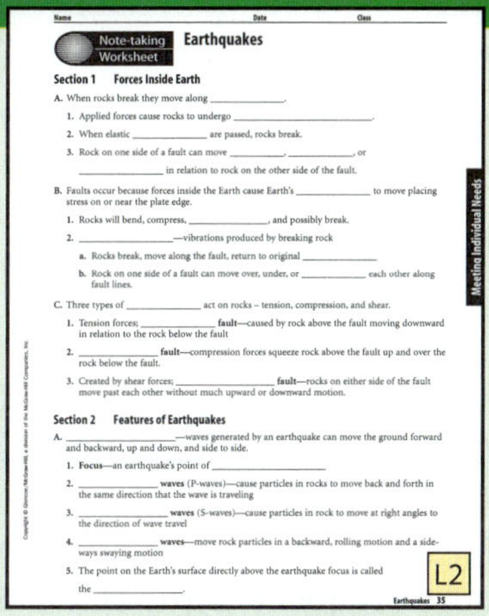

Reinforcement

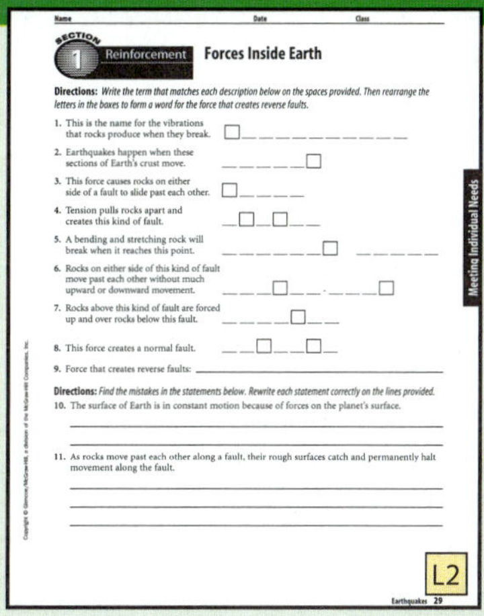

Enrichment

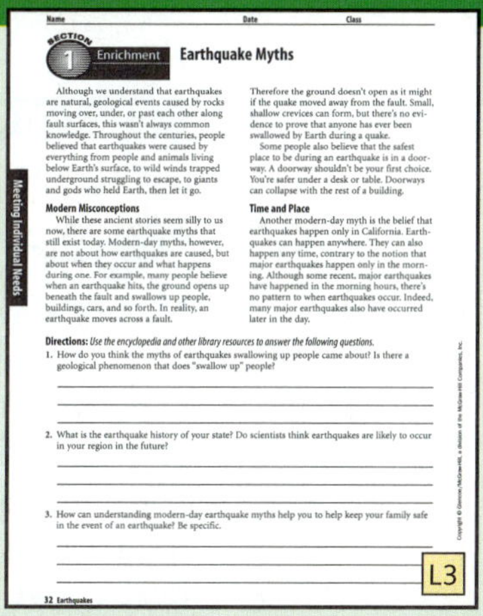

Directed Reading (English/Spanish)

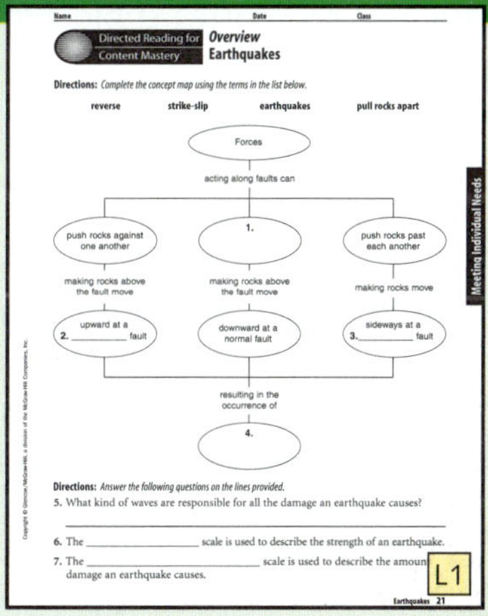

Study Guide

Reading Essentials

Assessment

Test Practice Workbook

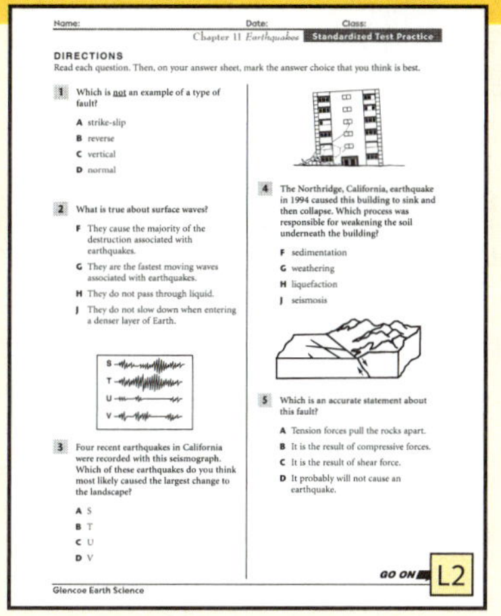

Chapter Review

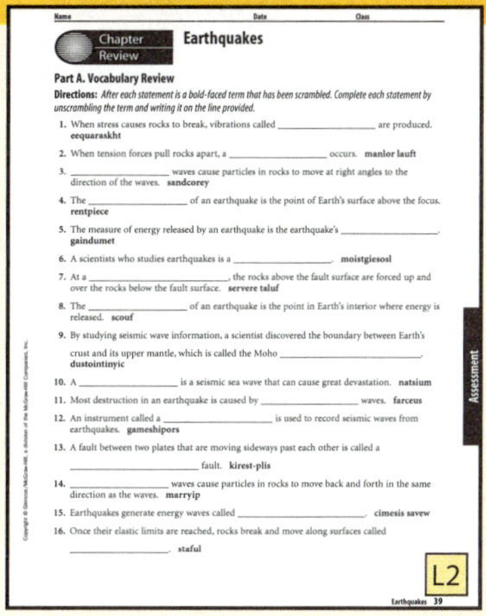

Chapter Tests

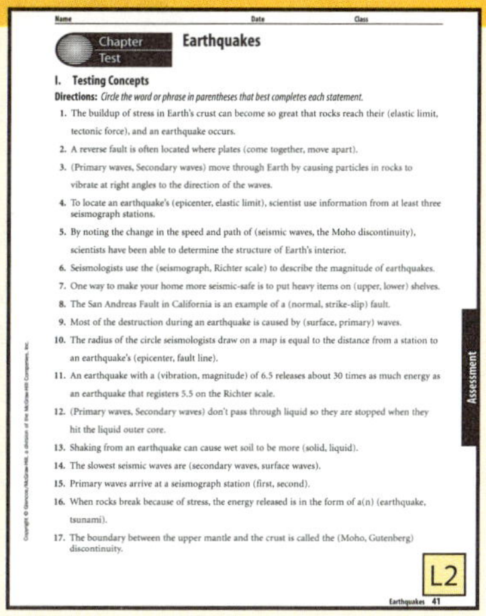

Science Content Background

section 1 Forces Inside Earth

Types of Earthquakes

Shallow-focus earthquakes occur in three places: where two of Earth's plates are moving apart, near the edges of plates that are converging, and where two plates are moving past each other. Intermediate and deep-focus earthquakes occur where one plate subducts under another. Subduction-zone quakes account for almost one-half of the destructive quakes.

section 2 Features of Earthquakes

Predicting Earthquakes

One method that is used to predict earthquakes is the study of seismic gaps. Seismic gaps are areas of a fault zone that have not produced a major earthquake during a specific time interval. When a seismic-gap area is located, researchers estimate how long it has been since the fault zone in question has experienced an earthquake. Using this information and knowledge of the average earthquake recurrence interval along the fault segment, scientists predict a time range over which the area could expect to experience a major earthquake.

Although much information is used in an attempt to predict earthquakes, success has been limited. Stress begins to build up in an area as soon as an earthquake has occurred. Without knowing how much stress can accumulate before rocks must shift, it is difficult to predict an earthquake with any amount of accuracy.

section 3 People and Earthquakes

Earthquake Magnitude

The extent of the damage caused by an earthquake depends on the distance to the epicenter, the type of bedrock in the area, the local soil or sediment types, and the number and type of structures subjected to the quake, among other factors.

Seismic-Safe Structures

Most loss of life in an earthquake occurs when people are trapped in and on crumbling structures, such as buildings, bridges, and highways. Making structures seismic-safe can reduce the loss of life in an earthquake. Seismic-safe structures are resistant to vibrations that occur during an earthquake.

The two main strategies for building seismic-safe structures are shock absorption and reinforcement. The goals of these building techniques are to prevent structures from collapsing and to minimize falling debris.

chapter content resources

Internet Resources

For additional content background, visit bookf.msscience.com to:

- access your book online
- find references to related articles in popular science magazines
- access Web links with related content background
- access current events with science journal topics

Print Resources

Geology Rocks, by Cindy Blobaum, Williamson Publishing Company, 1999

In Search of Pompeii, by Giovanni Caselli, McGraw-Hill, 1999

Earthquakes, by Janice VanCleave, John Wiley & Sons, 1993

IDENTIFYING Misconceptions

Find Out What Students Think

Students may think that . . .

Earthquakes occur only in California.
Much of the publicity found in the media features California earthquakes. As a result, many students think that the major location for earthquakes in the world is in California. Actually, earthquakes occur throughout the world, including many areas of the United States. Thus, it is important to feature earthquakes from a variety of areas of the world as students learn about this topic.

Activity

Use the Internet to find information on recent earthquakes. Useful links for earthquake information can be accessed through the link on the previous page. Have students plot on a world map earthquakes that have occurred over the last 30 days to begin to develop the idea that earthquakes occur worldwide. L2

Promote Understanding

Activity

Survival Kit After students understand that earthquakes are a world-wide phenomenon, reinforce the importance of having an "earthquake survival kit," especially if you live in an earthquake-prone area. Tell students that they have been hired to design the ideal kit. Explain that the kit must meet several minimum requirements:

- It should provide supplies for a family of four.
- It should provide enough supplies for three days.
- It should be able to be stored for long periods of time without spoiling.

Have students work in small groups to discuss what they would put in their kits. Have each group prepare a list of items they would include, giving both the item name and a quantity. Remind students that after a major earthquake many of the things we take for granted would not be available.

When students have made their lists, have groups compare their ideas. Are all items the same? What were some of the issues that different families might face? Prompt students to think of families with small children or infants. Discuss the types of foods that might be stored safely for long periods of time. Remind students that obtaining safe drinking water can be a major problem after an earthquake.

After students finish designing their kits, suggest they compare their ideas to those of the Red Cross or other disaster-management groups. What have they overlooked? What might your students recommend to these groups? L2

Assess

After completing the chapter, see *Identifying Misconceptions* in the Study Guide at the end of the chapter.

chapter 5

Chapter Vocabulary

fault, p. 126
earthquake, p. 127
normal fault, p. 128
reverse fault, p. 128
strike-slip fault, p. 129
seismic wave, p. 130
focus, p. 130
primary wave, p. 131
secondary wave, p. 131
surface wave, p. 131
epicenter, p. 131
seismograph, p. 133
magnitude, p. 140
liquefaction, p. 141
tsunami, p. 142

Science Journal Student responses will vary, but may include questions about personal background, experience, travel, and annual income.

INTERACTIVE CHALKBOARD with Image Bank

PowerPoint® Presentations

This CD-ROM is an editable Microsoft® PowerPoint® presentation that includes:

- a pre-made presentation for every chapter
- interactive graphics
- animations
- audio clips
- image bank
- all new section and chapter questions
- Standardized Test Practice
- transparencies
- pre-lab questions for all labs
- Foldables directions
- links to bookf.msscience.com

chapter 5

Earthquakes

chapter preview

sections

1. Forces Inside Earth
2. Features of Earthquakes
   Lab *Epicenter Location*
3. People and Earthquakes
   Lab *Earthquake Depths*

Virtual Lab *How do seismograph stations help determine an earthquake's epicenter?*

Was anyone hurt?

On October 17, 1989, the Loma Prieta earthquake rocked San Francisco, CA, leaving 62 dead and many more injured. Seismologists try to predict when and where earthquakes will occur so they can warn people of possible danger.

Science Journal Write *three* things that you would ask a scientist studying earthquakes.

Theme Connection

Energy The energy theme is highlighted as students learn how earthquakes unleash tremendous amounts of energy inside Earth's crust. That energy can change Earth's surface and affect both people and structures.

About the Photo

Earthquake! 5:04 P.M., October 17, 1989—62,000 fans are filling Candlestick Park for the third game of the World Series. The San Francisco Bay Area commute is at its heaviest flow. Suddenly, a Richter magnitude 7.1 earthquake strikes. The 20-second earthquake was followed 2.5 minutes later by a magnitude-5.2 aftershock. The San Andreas Fault has moved with devastating effects.

Start-Up Activities

Launch LAB

Why do earthquakes occur?

The bedrock beneath the soil can break to form cracks and move, forming faults. When blocks of bedrock move past each other along a fault, they cause the ground to shake. Why doesn't a block of bedrock move all the time, causing constant earthquakes? You'll find out during this activity.

1. Tape a sheet of medium-grain sandpaper to the tabletop.
2. Tape a second sheet of sandpaper to the cover of a textbook.
3. Place the book on the table so that both sheets of sandpaper meet.
4. Tie two large, thick rubber bands together and loop one of the rubber bands around the edge of the book so that it is not touching the sandpaper.
5. Pull on the free rubber band until the book moves. Record your observations.
6. **Think Critically** Write a paragraph that describes how the book moved. Using this model, predict why blocks of bedrock don't move all the time.

FOLDABLES™ Study Organizer

Earthquakes and Earth's Crust Make the following Foldable to help you understand the cause-and-effect relationship between earthquakes and movement in Earth's crust.

STEP 1 **Fold** a sheet of paper in half lengthwise.

STEP 2 **Fold** paper down 2.5 cm from the top. (Hint: From the tip of your index finger to your middle knuckle is about 2.5 cm.)

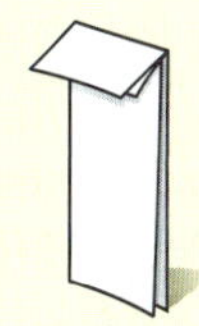

STEP 3 **Open and draw** lines along the 2.5 cm fold. **Label** as shown.

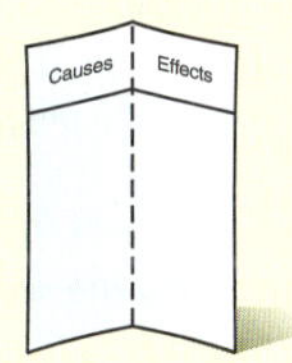

Read and Write As you read the chapter, write the causes and effects of earthquakes on your Foldable.

Science Online — Preview this chapter's content and activities at bookf.msscience.com

Launch LAB

Purpose Students explore the cause of earthquake activity at faults. LS **Kinesthetic** L2

Preparation Sandpaper can be bought at a hardware store.

Materials two sheets medium-grain sandpaper, textbook, two large rubber bands, tape

Teaching Strategy Have students work in pairs.

Safety Precautions Caution students to avoid scrapes or cuts when handling sandpaper.

Think Critically

The book moves with a quick jerk once the pulling force overcomes the friction between the two pieces of sandpaper. Rocks along fault lines build up stress in a similar manner until they rapidly slip past each other, causing an earthquake.

Assessment

Process Have students create a labeled diagram that illustrates what happened in the lab. Use **Performance Assessment in the Science Classroom**, p. 163.

FOLDABLES™ Study Organizer — Dinah Zike Study Fold

Student preparation materials for this Foldable are available in the Chapter *FAST FILE* Resources.

Differentiated Instruction

English-Language Learners Have students pair up with anyone needing assistance in understanding the Lab. Have those students visually demonstrate each step a few times so that it's understood why each step is important. Make sure your students also explain these steps using English, even if the English-language learner doesn't understand. The visual descriptions will help make connections to the English-language descriptions. ELL

1 Motivate

Bellringer

Section Focus Transparencies also are available on the Interactive Chalkboard CD-ROM. L2 ELL

Section 1 Section Focus Transparency — A Good Place to Stay

American architect Frank Lloyd Wright completed the Imperial Hotel in Tokyo in 1913. In 1923, the Great Kanto earthquake destroyed much of Tokyo, but the Imperial Hotel survived relatively unscathed.

1. What happens during an earthquake?
2. What might help a building withstand an earthquake?

L2

Tie to Prior Knowledge

Earthquakes in the News Ask students to recall any recent earthquakes they have heard about or experienced. Discuss with students where the earthquakes occurred and what happened as a result of them. Then tell students they will find out why earthquakes occur in this section. L2

section 1

Forces Inside Earth

as you read

What You'll Learn

- **Explain** how earthquakes result from the buildup of energy in rocks.
- **Describe** how compression, tension, and shear forces make rocks move along faults.
- **Distinguish** among normal, reverse, and strike-slip faults.

Why It's Important

Earthquakes cause billions of dollars in property damage and kill an average of 10,000 people every year.

Review Vocabulary
plate: a large section of Earth's crust and rigid upper mantle that moves around on the asthenosphere

New Vocabulary
- fault
- earthquake
- normal fault
- reverse fault
- strike-slip fault

Earthquake Causes

Recall the last time you used a rubber band. Rubber bands stretch when you pull them. Because they are elastic, they return to their original shape once the force is released. However, if you stretch a rubber band too far, it will break. A wooden craft stick behaves in a similar way. When a force is first applied to the stick, it will bend and change shape. The energy needed to bend the stick is stored inside the stick as potential energy. If the force keeping the stick bent is removed, the stick will return to its original shape, and the stored energy will be released as energy of motion.

Fault Formation There is a limit to how far a wooden craft stick can bend. This is called its elastic limit. Once its elastic limit is passed, the stick remains bent or breaks, as shown in **Figure 1.** Rocks behave in a similar way. Up to a point, applied forces cause rocks to bend and stretch, undergoing what is called elastic deformation. Once the elastic limit is passed, the rocks may break. When rocks break, they move along surfaces called **faults.** A tremendous amount of force is required to overcome the strength of rocks and to cause movement along a fault. Rock along one side of a fault can move up, down, or sideways in relation to rock along the other side of the fault.

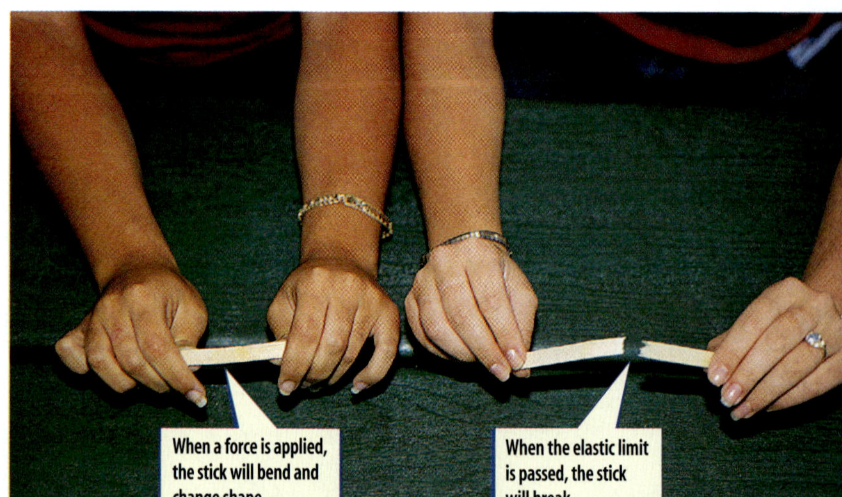

Figure 1 The bending and breaking of wooden craft sticks are similar to how rocks bend and break.

Section 1 Resource Manager

Chapter FAST FILE Resources
- Directed Reading for Content Mastery, pp. 21, 22
- Enrichment, p. 32
- Reinforcement, p. 29
- Transparency Activity, pp. 46, 49–50

Earth Science Critical Thinking/Problem Solving, pp. 1, 8, 10

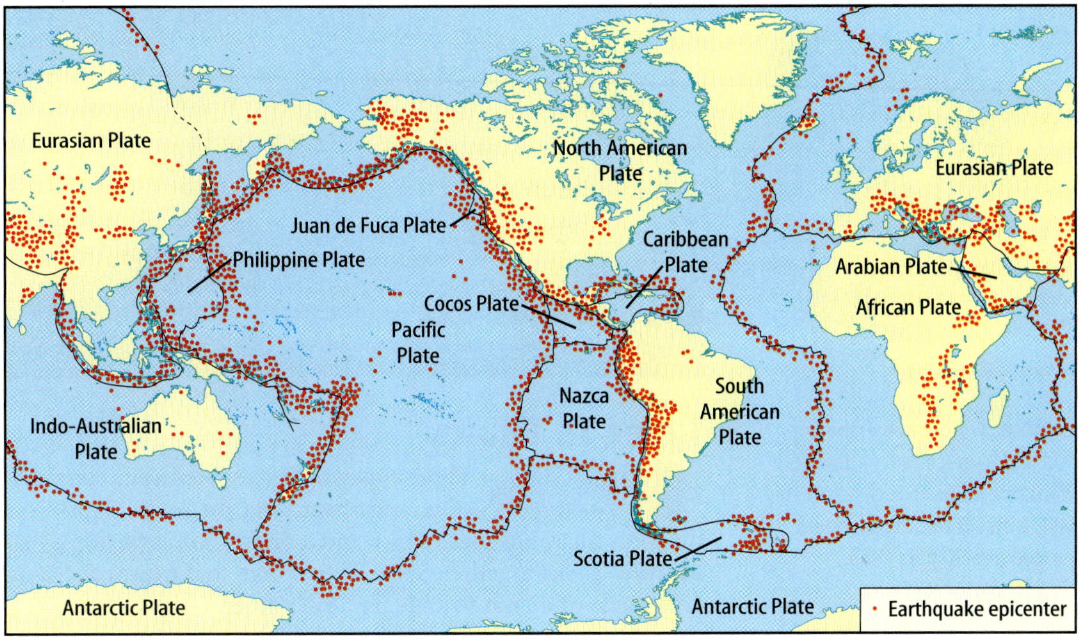

Figure 2 The dots represent the epicenters of major earthquakes over a ten-year period. Note that most earthquakes occur near plate boundaries.
Form a hypothesis *to explain why earthquakes rarely occur in the middle of a plate.*

What causes faults? What produces the forces that cause rocks to break and faults to form? The surface of Earth is in constant motion because of forces inside the planet. These forces cause sections of Earth's surface, called plates, to move. This movement puts stress on the rocks near the plate edges. To relieve this stress, the rocks tend to bend, compress, or stretch. If the force is great enough, the rocks will break. An **earthquake** is the vibrations produced by the breaking of rock. **Figure 2** shows how the locations of earthquakes outline the plates that make up Earth's surface.

Reading Check *Why do most earthquakes occur near plate boundaries?*

How Earthquakes Occur As rocks move past each other along a fault, their rough surfaces catch, temporarily halting movement along the fault. However, forces keep driving the rocks to move. This action builds up stress at the points where the rocks are stuck. The stress causes the rocks to bend and change shape. When the rocks are stressed beyond their elastic limit, they can break, move along the fault, and return to their original shapes. An earthquake results. Earthquakes range from unnoticeable vibrations to devastating waves of energy. Regardless of their intensity, most earthquakes result from rocks moving over, under, or past each other along fault surfaces.

2 Teach

Caption Answer

Figure 2 Because most of the stress on plates is at their edges, rock is more likely to move and cause quakes there than in central areas.

Activity

Elastic Limits Different materials have different elastic limits. Have students bend the following materials to see what happens once the elastic limit is reached: a piece of cardboard; a plastic drinking straw; a wooden tongue depressor; a thin, steel wire; some silicon putty; a piece of uncooked spaghetti. Have students wear goggles during this activity. L2 ELL LS **Kinesthetic**

Reading Check

Answer Most stress is at boundaries where plates meet.

Fun Fact

There is no spot on Earth that cannot experience an earthquake, but earthquakes are less likely to occur in Antarctica than in any other place.

Identifying Misconceptions

Locations Some students think that earthquakes occur only in California. Refer to page F at the beginning of this chapter for teaching strategies that address this misconception.

Teacher FYI

Quake Sounds Primary waves generated at the focus of an earthquake travel outward through Earth's interior. Sometimes these waves enter Earth's atmosphere and cause the loud noises associated with earthquakes.

Active Reading

Buddy Interviews This strategy helps students understand and clarify the reading. Have students interview one another to find out what helps them to understand what they are reading, how they find answers, and how they assimilate new vocabulary terms. Have students use Buddy Interviews to help them master the concept of what causes earthquakes. L2

Discussion

Hidden Faults Would it be possible to stand at a fault and not know it was there? Explain. Yes; the fault could be the site of hills or mountains, and the actual fault could have been covered by soil and rock as the result of erosion.

Activity

Edible Forces Have pairs of students apply compression, tension, and shear forces to bars of taffy. Tell students to make drawings that show what happens to the taffy in each instance. Have students write a paragraph relating this to what happens to rocks when forces are applied. Depending on the amount of force applied, students should see the taffy bend or break, which is what happens when force is applied to rock. L2 COOP LEARN LS **Interpersonal** P

Quick Demo

Edible Faults

Materials large, triple-decker peanut butter and jelly sandwich (crusts removed to show the layers)

Estimated Time 10 minutes

Procedure Construct a fault at about a 30° angle by cutting through the sandwich. Move the separate halves of the sandwich to demonstrate normal, reverse, and strike-slip faults. L2 **Visual-Kinesthetic**

Figure 3 Rock above the normal fault surface moves downward in relation to rock below the fault surface. This normal fault formed near Kanab, Utah.

Types of Faults

INTEGRATE Physics Three types of forces—tension, compression, and shear—act on rocks. Tension is the force that pulls rocks apart, and compression is the force that squeezes rocks together. Shear is the force that causes rocks on either side of a fault to slide past each other.

Normal Faults Tensional forces inside Earth cause rocks to be pulled apart. When rocks are stretched by these forces, a normal fault can form. Along a **normal fault,** rock above the fault surface moves downward in relation to rock below the fault surface. The motion along a normal fault is shown in **Figure 3.** Notice the normal fault shown in the photograph above.

Reverse Faults Reverse faults result from compression forces that squeeze rock. **Figure 4** shows the motion along a reverse fault. If rock breaks from forces pushing from opposite directions, rock above a **reverse fault** surface is forced up and over the rock below the fault surface. The photo below shows a large reverse fault in California.

Figure 4 The rock above the reverse fault surface moves upward in relation to the rock below the fault surface.

128 ◆ F CHAPTER 5 Earthquakes

Visual Learning

Figure 3 The blue arrows at the top of the figure illustrate how the tension forces are being applied. The red arrows beside the fault line illustrate the relationship between the two blocks of rock. Do the two pairs of arrows illustrate whether one block moved up or down? No, the movement is relative. Either or both blocks could have moved.

Differentiated Instruction

Challenge Have students do research and compare maps of the world's earthquakes and maps of the world's active volcanoes. What do they notice about the two areas? Most of the volcanoes in the world erupt where Earth's plates are bumping together or moving apart. Have students design an information mobile representing their findings and share their conclusions with the class. L3

Figure 5 Shear forces push on rock in opposite—but not directly opposite—horizontal directions. When they are strong enough, these forces split rock and create strike-slip faults.

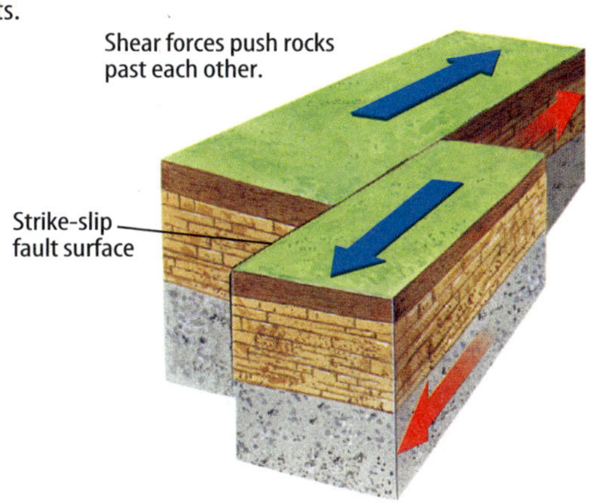

Strike-Slip Faults At a **strike-slip fault**, shown in **Figure 5**, rocks on either side of the fault are moving past each other without much upward or downward movement. The photo above shows the largest fault in California—the San Andreas Fault—which stretches more than 1,100 km through the state. The San Andreas Fault is the boundary between two of Earth's plates that are moving sideways past each other.

Reading Check *What is a strike-slip fault?*

section 1 review

Summary

Earthquake Causes

- Faults form when stressed rocks break along surfaces.
- Stresses on rock are created by plate movements.
- When rocks break along a fault, vibrations are created. This is an earthquake.

Types of Faults

- Normal faults can form when rocks undergo tension.
- Compression forces produce reverse faults.
- Strike-slip faults result when rocks move past each other without much upward or downward movement.

Self Check

1. **Infer** The Himalaya in Tibet formed when two of Earth's plates collided. What types of faults would you expect to find in these mountains? Why?
2. **State** In what direction do rocks move above a normal fault surface? What force causes this?
3. **Describe** how compression forces make rocks move along a reverse fault.
4. **Think Critically** Why is it easier to predict where an earthquake will occur than it is to predict when it will occur?

Applying Skills

5. **Infer** Why do the chances of an earthquake increase rather than decrease as time passes since the last earthquake?

Science Online bookf.msscience.com/self_check_quiz

Reading Check

Answer a fault where rocks on either side are moving past each other in a horizontal direction

3 Assess

DAILY INTERVENTION

Check for Understanding

Visual-Kinesthetic Have students use their hands to demonstrate the movement that occurs along each of the three kinds of faults: normal fault, reverse fault, and strike-slip fault. L1

Reteach

Demonstrating Faults Using a piece of wax paper, covered with frosting as magma and graham crackers as tectonic plates, demonstrate each type of fault. Start with two crackers placed on top of the frosting with edges touching and slide them apart to show a rift. Push two crackers together and move them to show a normal fault and then a reverse fault. Slide them past each other to model a strike-slip fault. L1

Assessment

Content Have students write a brief paragraph to explain why streambeds are offset at the San Andreas Fault. Rocks on either side of the fault move past each other, causing the streambeds to be offset. Use **Performance Assessment in the Science Classroom,** p. 159. L2

section 1 review

1. The overall stress regime in the Himalaya is compressive. Low-angle reverse faults, called thrust faults, occur at the front of the range. Other regions within the Himalaya have different stress regimes; strike-slip faults and normal faults also occur.
2. The rocks above a normal fault move down relative to the rocks below the fault surface. Tension promotes normal faulting.
3. The rocks above a reverse fault move up relative to the rocks below the fault surface.
4. One can use seismic history to predict whether more earthquakes are likely to occur in a given region. Predicting exactly when an earthquake will occur requires that the stress regime and rock strength be known precisely. This is not yet possible.
5. After an earthquake, stress and the amount of elastic energy stored in rock gradually increase.

section 2 Features of Earthquakes

1 Motivate

Bellringer

INTERACTIVE CHALKBOARD PowerPoint® Presentations

Section Focus Transparencies also are available on the Interactive Chalkboard CD-ROM. L2 ELL

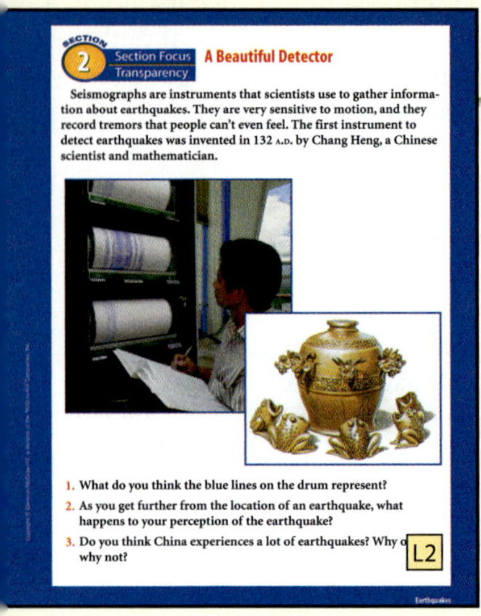

SECTION 2 Section Focus Transparency **A Beautiful Detector**

Seismographs are instruments that scientists use to gather information about earthquakes. They are very sensitive to motion, and they record tremors that people can't even feel. The first instrument to detect earthquakes was invented in 132 A.D. by Chang Heng, a Chinese scientist and mathematician.

1. What do you think the blue lines on the drum represent?
2. As you get further from the location of an earthquake, what happens to your perception of the earthquake?
3. Do you think China experiences a lot of earthquakes? Why or why not?

L2

Earthquakes

Tie to Prior Knowledge

Earthquake Energy Waves Tell students that the energy of earthquakes travels in waves. Ask students to recall other types of energy that travel in waves. Possible answers: sound, electromagnetic radiation such as visible light, energy in ocean waves. Tell students that in this section they will find out how earthquakes that happen inside Earth's crust can affect people and things on the surface.

as you read

What **You'll Learn**

- **Explain** how earthquake energy travels in seismic waves.
- **Distinguish** among primary, secondary, and surface waves.
- **Describe** the structure of Earth's interior.

Why **It's Important**

Seismic waves are responsible for most damage caused by earthquakes.

Review Vocabulary
wave: rhythmic movement that carries energy through matter and space

New Vocabulary
- seismic wave
- focus
- primary wave
- secondary wave
- surface wave
- epicenter
- seismograph

Seismic Waves

When two people hold opposite ends of a rope and shake one end, as shown in **Figure 6,** they send energy through the rope in the form of waves. Like the waves that travel through the rope, **seismic** (SIZE mihk) **waves** generated by an earthquake travel through Earth. During a strong earthquake, the ground moves forward and backward, heaves up and down, and shifts from side to side. The surface of the ground can ripple like waves do in water. Imagine trying to stand on ground that had waves traveling through it. This is what you might experience during a strong earthquake.

Origin of Seismic Waves You learned earlier that rocks move past each other along faults, creating stress at points where the rocks' irregular surfaces catch each other. The stress continues to build up until the elastic limit is exceeded and energy is released in the form of seismic waves. The point where this energy release first occurs is the **focus** (plural, *foci*) of the earthquake. The foci of most earthquakes are within 65 km of Earth's surface. A few have been recorded as deep as 700 km. Seismic waves are produced and travel outward from the earthquake focus.

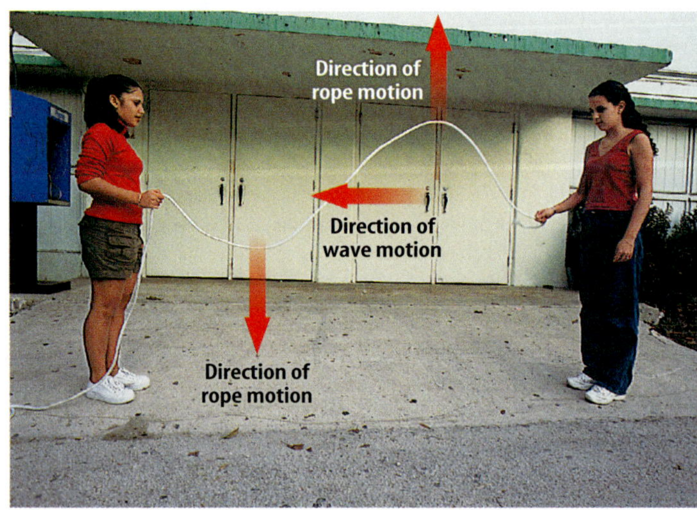

Figure 6 Some seismic waves are similar to the wave that is traveling through the rope. Note that the rope moves perpendicular to the wave direction.

Section 2 Resource Manager

Chapter *FAST FILE* Resources
- Directed Reading for Content Mastery, p. 23
- Enrichment, p. 33
- Lab Activity, pp. 9–13
- MiniLAB, p. 3
- Reinforcement, p. 30
- Transparency Activity, p. 47
- Lab Worksheets, pp. 5–6

Home and Community Involvement, p. 40

Primary Waves When earthquakes occur, three different types of seismic waves are produced. All of the waves are generated at the same time, but each behaves differently within Earth. **Primary waves** (P-waves) cause particles in rocks to move back and forth in the same direction that the wave is traveling. If you squeeze one end of a coiled spring and then release it, you cause it to compress and then stretch as the wave travels through the spring, as shown in **Figure 7.** Particles in rocks also compress and then stretch apart, transmitting primary waves through the rock.

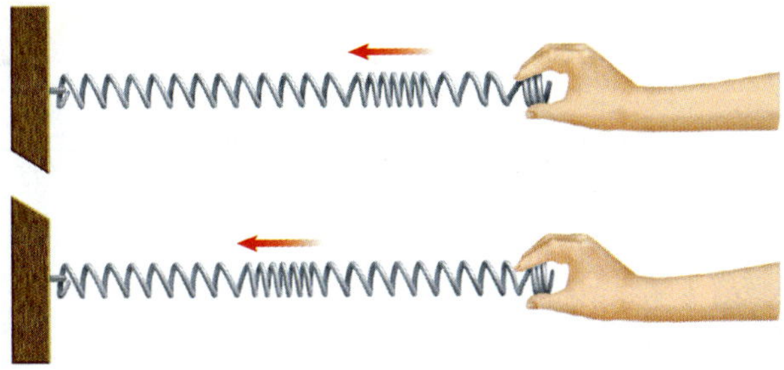

Figure 7 Primary waves move through Earth the same way that a wave travels through a coiled spring.

Secondary and Surface Waves **Secondary waves** (S-waves) move through Earth by causing particles in rocks to move at right angles to the direction of wave travel. The wave traveling through the rope shown in **Figure 6** is an example of a secondary wave.

Surface waves cause most of the destruction resulting from earthquakes. **Surface waves** move rock particles in a backward, rolling motion and a side-to-side, swaying motion, as shown in **Figure 8.** Many buildings are unable to withstand intense shaking because they are made with stiff materials. The buildings fall apart when surface waves cause different parts of the building to move in different directions.

INTEGRATE Physics

Sound Waves When sound is produced, waves move through air or some other material. Research sound waves to find out which type of seismic wave they are similar to.

Reading Check *Why do surface waves damage buildings?*

Surface waves are produced when earthquake energy reaches the surface of Earth. Surface waves travel outward from the epicenter. The earthquake **epicenter** (EH pih sen tur) is the point on Earth's surface directly above the earthquake focus. Find the focus and epicenter in **Figure 9.**

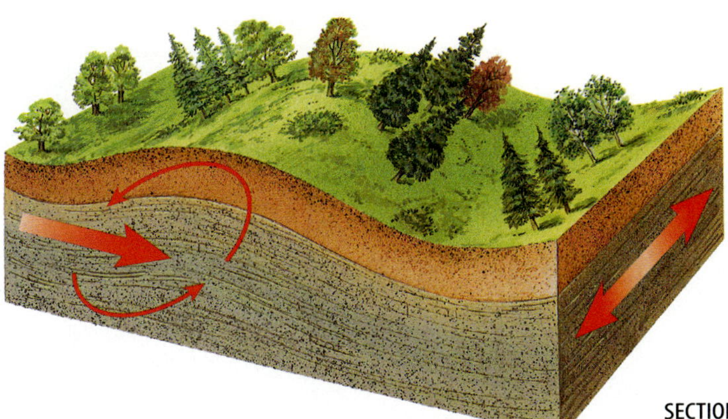

Figure 8 Surface waves move rock particles in a backward, rolling motion and a side-to-side, swaying motion.
Compare and contrast *surface waves and secondary waves.*

2 Teach

Discussion

Compression v. Tension Why is folding more likely with compression than with tension forces? Compression causes folding, as the plates press toward each other.

Use Science Words

Word Meaning Have students use a dictionary to determine the meaning of the word *focus*. Ask them to determine how an earthquake focus got its name. Focus is the point of origin of an earthquake. An earthquake is centered at its focus and waves diverge or move outward from there.

Reading Check

Answer Surface waves cause intense shaking of structures as they move rock at the surface in backward rolling and side-to-side swaying motions.

INTEGRATE Physics

Sound Waves Sound waves are similar to primary or P waves. In both cases, the wave is propagated by a forward and backward motion of particles moving in the same direction as the wave.

Caption Answer

Figure 8 Secondary waves move rock particles at right angles to the direction of the wave. Surface waves move rock particles in a backward rolling motion, which secondary waves do not.

Differentiated Instruction

Learning Disabled Have students prepare a three-circle Venn diagram comparing the three types of seismic waves: their direction of travel, destructive powers, and ability to travel through Earth's inner structure. It may be beneficial to include an example of each type of wave. L2

Teacher FYI

Quake Damage If the energy of an earthquake is strong enough to shake a house, light fixtures can fall, refrigerators and other large items can move, bookcases and TVs can topple over, and cabinet doors can fly open causing the contents to spill out.

SECTION 2 Features of Earthquakes F ◆ 131

NATIONAL GEOGRAPHIC

Visualizing Seismic Waves

Have students examine the pictures and read the captions. Then ask the following questions.

Which type of wave would a seismograph first record after an earthquake? A seismograph would first record a primary wave, then a secondary wave. Surface waves would be recorded last.

Why do some earthquakes cause more damage than others? Possible answers: Damage can be greater if the magnitude is higher or the focus is near Earth's surface. More damage also can occur if the epicenter is near populated areas or if buildings are not constructed to withstand earthquakes.

Activity

Demonstrating Waves Have students demonstrate seismic waves using a coiled spring toy, stretched 1–2 meters on a table. Have a student hold one end of the spring firmly. Have another student demonstrate P-waves by quickly pushing the other end of the spring toward the first student. S-waves can be demonstrated by moving the spring up and down. Gently moving one end of the spring side to side while at the same time moving it in a rolling motion will demonstrate surface waves. **WARNING:** *Students should wear goggles when performing this activity.* L2

NATIONAL GEOGRAPHIC **VISUALIZING SEISMIC WAVES**

Figure 9

As the plates that form Earth's lithosphere move, great stress is placed on rocks. They bend, stretch, and compress. Occasionally, rocks break, producing earthquakes that generate seismic waves. As shown here, different kinds of seismic waves—each with distinctive characteristics—move outward from the focus of the earthquake.

B Primary waves and secondary waves originate at the focus and travel outward in all directions. Primary waves travel about twice as fast as secondary waves.

C The point on Earth's surface directly above an earthquake's focus is known as the epicenter. Surface waves spread out from the epicenter like ripples in a pond.

D The amplitudes, or heights, of surface waves are greater than those of primary and secondary waves. Surface waves cause the most damage during an earthquake.

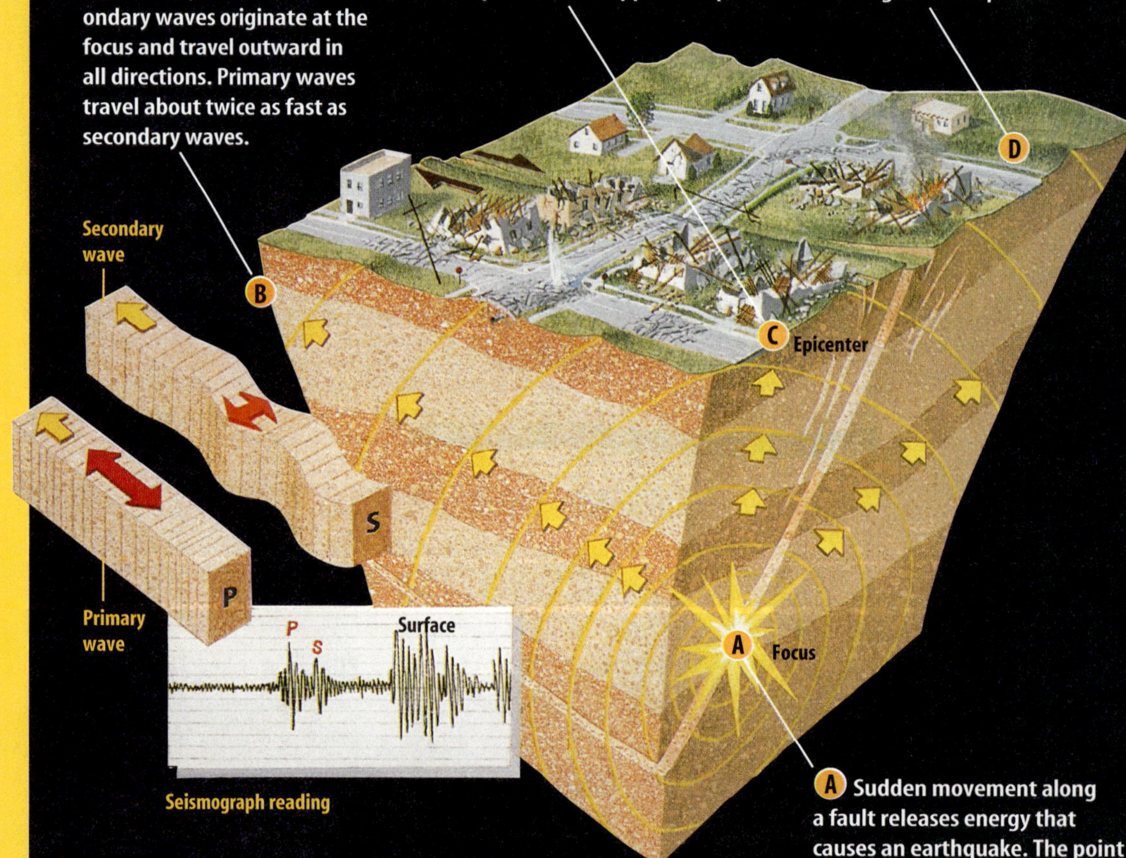

A Sudden movement along a fault releases energy that causes an earthquake. The point at which this movement begins is called the earthquake's focus.

Differentiated Instruction

Challenge Students can demonstrate types of waves in a bowl of water as they tap the surface with a pencil. Students investigate the two types of seismic motion: Raleigh waves, which move up and down, and Love waves, which move side-to-side. They also can investigate the scientists who first identified the movements and how these waves cause different types of damage. L3

Learning Disabled Drop a small rock into a tub of water. Have students observe the waves that are generated moving outward from the rock. Help students explain how the wave movement in the water is similar to the movement of certain types of earthquake waves at Earth's surface. L2 **Visual-Linguistic**

Locating an Epicenter

Different seismic waves travel through Earth at different speeds. Primary waves are the fastest, secondary waves are slower, and surface waves are the slowest. Can you think of a way this information could be used to determine how far away an earthquake epicenter is? Think of the last time you saw two people running in a race. You probably noticed that the faster person got further ahead as the race continued. Like runners in a race, seismic waves travel at different speeds.

Scientists have learned how to use the different speeds of seismic waves to determine the distance to an earthquake epicenter. When an epicenter is far from a location, the primary wave has more time to put distance between it and the secondary and surface waves, just like the fastest runner in a race.

Measuring Seismic Waves Seismic waves from earthquakes are measured with an instrument known as a **seismograph.** Seismographs register the waves and record the time that each arrived. Seismographs consist of a rotating drum of paper and a pendulum with an attached pen. When seismic waves reach the seismograph, the drum vibrates but the pendulum remains at rest. The stationary pen traces a record of the vibrations on the moving drum of paper. The paper record of the seismic event is called a seismogram. **Figure 10** shows two types of seismographs that measure either vertical or horizontal ground movement, depending on the orientation of the drum.

Science Online

Topic: Earthquake Data
Visit bookf.msscience.com for Web links to the National Earthquake Information Center and the World Data Center for Seismology.

Activity List the locations and distances of each reference that seigmograph stations used to determine the epicenter of the most recent earthquake.

Figure 10 Seismographs differ according to whether they are intended to measure horizontal or vertical seismic motions.
Infer *why one seismograph can't measure both horizontal and vertical motions.*

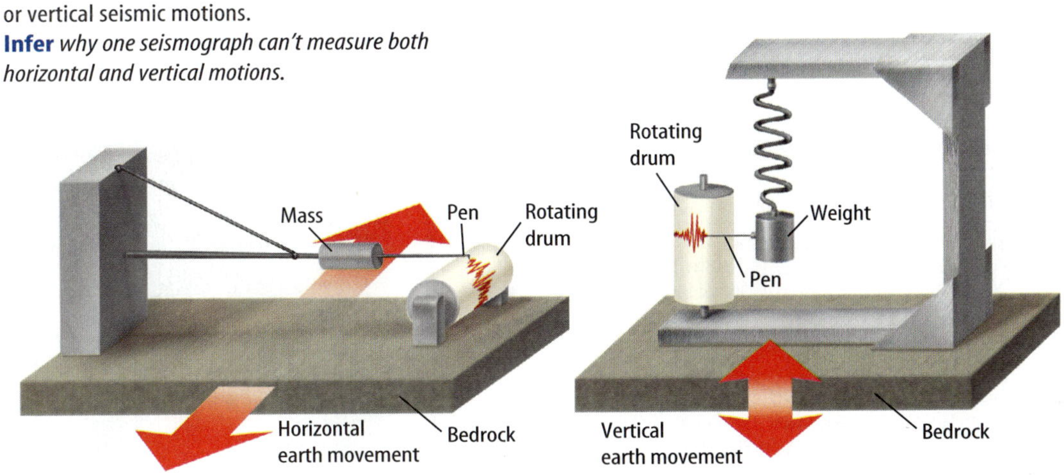

Fun Fact

Primary waves travel at about 6.0 km/s, secondary waves at about 3.5 km/s, and surface waves at about 3.0 km/s through granite crust.

IDENTIFYING Misconceptions

Prediction v. Probability Students may think that scientists can predict earthquakes. However, they can only estimate the probability that an earthquake will happen in a certain place in the future based on scientific data. For example, geologists can't say for certain when the next major earthquake in San Francisco will happen. But they can estimate that the probability of a major earthquake there in the next 30 years is 70 percent.

Caption Answer

Figure 10 The drum on the seismograph can be oriented to measure either horizontal or vertical motion, but not both.

Science Journal

Landers Quake Have students write a summary in their Science Journals after researching the 1992 earthquake in Landers, California. Have them explain why this event is so important to the study of faults and earthquakes. Data obtained from the Landers quake may indicate that a new fault is being formed in that area of the desert. P L2

SECTION 2 Features of Earthquakes F ◆ 133

Inquiry Lab

Observing Seismic Activity

Purpose To explore and observe seismic activity, have students design and construct simple seismographs.

Possible Materials bowls, variety of liquids, flashlights

Estimated Time 1 class session

Teaching Strategies

- Students can make a simple yet sensitive seismograph by reflecting light off a bowl containing a liquid onto a blank wall in a darkened room.
- Students can predict and then observe how the reflection is affected when they experiment with the intensity or location of the foci. (They may try clapping, or tapping the bowl, the liquid, or the table.)
- Students could also observe seismic activity in a liquid/solid mixture (like batter or gelatin) to model how different materials conduct seismic waves.
- Allow students to explore other questions that arise. L2

For additional inquiry activities, see *Science Inquiry Labs*.

Caption Answer

Figure 11 approximately 1,000 km from the earthquake epicenter

Seismograph Stations Each type of seismic wave reaches a seismograph station at a different time based on its speed. Primary waves arrive first at seismograph stations, and secondary waves, which travel slower, arrive second. Because surface waves travel slowest, they arrive at seismograph stations last. This difference in arrival times is used to calculate the distance from the seismograph station to the earthquake epicenter, as shown in **Figure 11.** If a seismograph station is located 4,000 km from an earthquake epicenter, primary waves will reach the station about 6 minutes before secondary waves.

If seismic waves reach three or more seismograph stations, the location of the epicenter can be determined. To locate an epicenter, scientists draw circles around each station on a map. The radius of each circle equals that station's distance from the earthquake epicenter. The point where all three circles intersect, shown in **Figure 12,** is the location of the earthquake epicenter.

Seismologists usually describe earthquakes based on their distances from the seismograph. Local events occur less than 100 km away. Regional events occur 100 km to 1,400 km away. Teleseismic events are those that occur at distances greater than 1,400 km.

Figure 11 Primary waves arrive at a seismograph station before secondary waves do.
Use Graphs *If primary waves reach a seismograph station two minutes before secondary waves, how far is the station from the epicenter?*

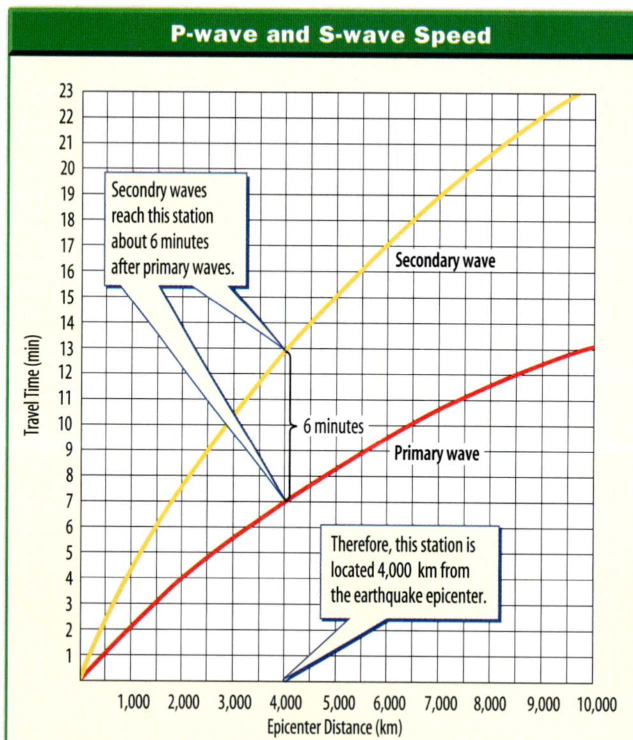

Figure 12 The radius of each circle is equal to the distance from the epicenter to each seismograph station. The intersection of the three circles is the location of the epicenter.

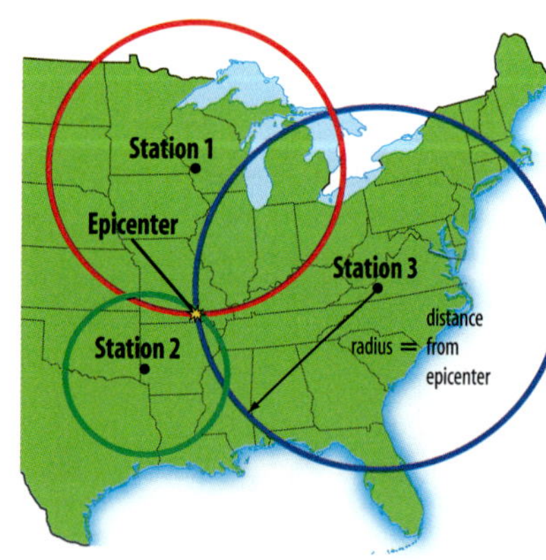

LAB DEMONSTRATION

Purpose to demonstrate each of the three types of seismic waves

Materials a coiled-spring toy

Preparation Tie pieces of string on the spring at five evenly spaced intervals.

Procedure Demonstrate primary waves by compressing about 20 coils together and releasing them, secondary waves by moving the spring from side to side, and surface waves by moving the spring in an elliptical path while also moving it from side to side.

Expected Outcome Students will see that each wave causes a different type of movement in the material through which it is moving.

Assessment

Ask students which type of wave would cause the most damage. Surface waves would cause the most damage because of the large amount of motion associated with them. L2

Basic Structure of Earth

Figure 13 shows Earth's internal structure. At the very center of Earth is a solid, dense inner core made mostly of iron with smaller amounts of nickel, oxygen, silicon, and sulfur. Pressure from the layers above causes the inner core to be solid. Above the solid inner core lies the liquid outer core, which also is made mainly of iron.

Reading Check *How do the inner and outer cores differ?*

Earth's mantle is the largest layer, lying directly above the outer core. It is made mostly of silicon, oxygen, magnesium, and iron. The mantle often is divided into an upper part and a lower part based on changing seismic wave speeds. A portion of the upper mantle, called the asthenosphere (as THE nuh sfihr), consists of weak rock that can flow slowly.

Earth's Crust The outermost layer of Earth is the crust. Together, the crust and a part of the mantle just beneath it make up Earth's lithosphere (LIH thuh sfihr). The lithosphere is broken into a number of plates that move over the asthenosphere beneath it.

The thickness of Earth's crust varies. It is more than 60 km thick in some mountainous regions and less than 5 km thick under some parts of the oceans. Compared to the mantle, the crust contains more silicon and aluminum and less magnesium and iron. Earth's crust generally is less dense than the mantle beneath it.

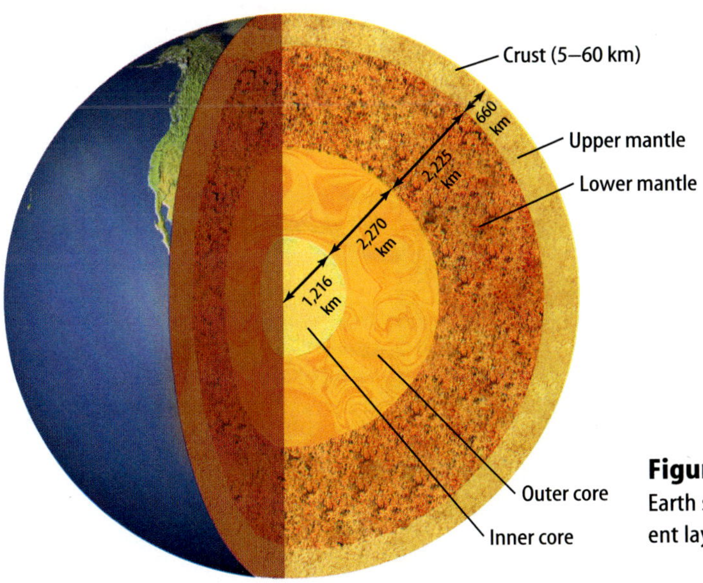

Figure 13 The internal structure of Earth shows that it is made of different layers.

Mini LAB

Interpreting Seismic Wave Data

Procedure

Copy the table below into your **Science Journal.** Use the **graph** in **Figure 11** to determine the difference in arrival times for primary and secondary waves at the distances listed in the data table below. Two examples are provided for you.

Wave Data

Distance (km)	Difference in Arrival Time
1,500	2 min, 50 s
2,250	3 min, 50 s
2,750	4 min, 30 s
3,000	4 min, 42 s
4,000	5 min, 55 s
7,000	8 min, 18 s
9,000	9 min, 42 s

Analysis

1. What happens to the difference in arrival times as the distance from the earthquake increases?
2. If the difference in arrival times at a seismograph station is 6 min, 30 s, how far away is the epicenter?

Mini LAB

Purpose Students will use seismic wave data to determine the distance to an earthquake epicenter. L2 LS **Logical-Mathematical**

Materials Figure 11 graph

Teaching Strategy Make sure students understand the graph before beginning.

Analysis

Wave Data

Distance (km)	Difference in Arrival Time
1,500	2 min, 50 s
2,250	3 min, 50 s
2,750	4 min, 30 s
3,000	4 min, 42 s
4,000	5 min, 55 s
7,000	8 min, 18 s
9,000	9 min, 42 s

Analysis

1. The difference in times increases with distance from the earthquake epicenter.
2. approximately 4,750 km

Assessment

Performance Have students determine distances to earthquakes whose primary and secondary wave arrival times are separated by 5 minutes and 7 minutes. 5 min—3,250 km; 7 min—5,400 km. Use **Performance Assessment in the Science Classroom,** p. 101.

Reading Check

Answer The inner core is solid whereas the outer core is liquid.

Curriculum Connection

Literature Have students do a library search to find science fiction books that have been written about journeys to Earth's center. Have students choose such a book and write a brief scientific critique comparing what the author imagined with the reality of Earth's interior. Have students read their analyses in class as part of a class discussion. L3

Differentiated Instruction

Challenge After reading a piece of literature about Earth's structure, earthquakes, plate tectonics, or another related topic, students may design a lunch bag book review similar to a kid's meal sack. Information displayed should include book summary, review, author biography, new vocabulary, puzzles, and illustrations created by the student. A related toy or treat may be placed inside; trade and enjoy. L3

Visual Learning

Figure 13 Which is the thinnest of Earth's layers? the crust

Quick Demo

Modeling Earth's Internal Layers

Materials tall, clear drinking glass; water; clear corn syrup; clear shampoo; a variety of sizes of coins

Estimated Time 15 minutes

Procedure Fill a glass one-fourth full with clear corn syrup. Slowly pour water into the glass until it is three-fourths full. Drop a coin from a height slightly above the surface of the water. Have students observe the path of the coin. How is the speed and direction of the path of the coin affected as it travels through each liquid and as it passes into the next liquid? How does the drop point differ from the final landing point? Explore the reaction using a variety of clear liquids. Does the coin react differently to different liquids? The path of coins, as well as that of earthquake energy waves, is most refracted when moving from one substance to a new substance of different density. Students may be interested in adding different substances or making a challenging target game for others to try. L2

Use an Analogy

Peach Analogy Explain to students that the internal structure of a peach is analogous to Earth's internal structure. The peach pit is like Earth's core. The meat of the peach, which is its thickest part, can be compared with the thickest part of Earth's interior, the mantle. The thin peach skin corresponds to Earth's crust, which is extremely thin compared with the planet's other layers.

Reading Check

Answer They change speed as the waves travel through rocks of different density in Earth's internal structure.

Mapping Earth's Internal Structure As shown in **Figure 14,** the speeds and paths of seismic waves change as they travel through materials with different densities. By studying seismic waves that have traveled through Earth, scientists have identified different layers with different densities. In general, the densities increase with depth as pressures increase. Studying seismic waves has allowed scientists to map Earth's internal structure without being there.

Early in the twentieth century, scientists discovered that large areas of Earth don't receive seismic waves from an earthquake. In the area on Earth between 105° and 140° from the earthquake focus, no waves are detected. This area, called the shadow zone, is shown in **Figure 14.** Secondary waves are not transmitted through a liquid, so they stop when they hit the liquid outer core. Primary waves are slowed and bent but not stopped by the liquid outer core. Because of this, scientists concluded that the outer core and mantle are made of different materials. Primary waves speed up again as they travel through the solid inner core. The bending of primary waves and the stopping of secondary waves create the shadow zone.

Reading Check *Why do seismic waves change speed as they travel through Earth?*

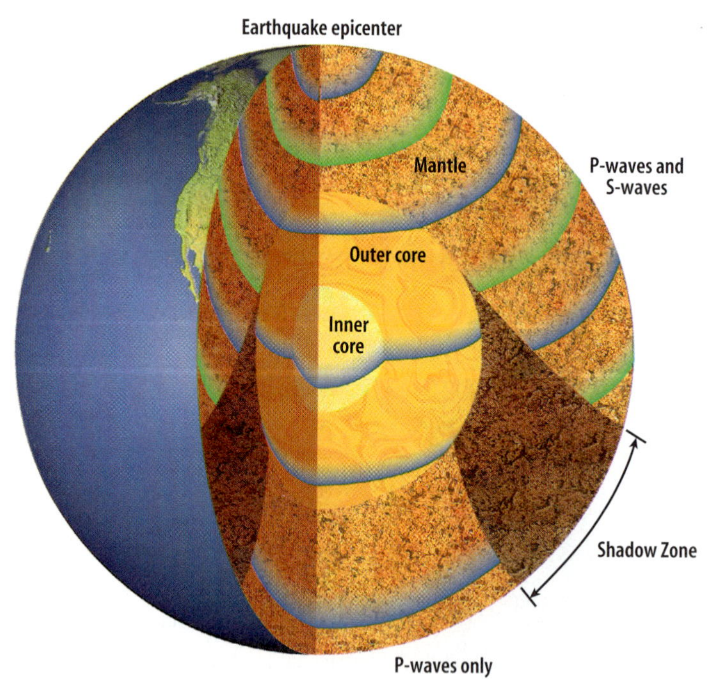

Figure 14 Seismic waves bend and change speed as the density of rock changes. Primary waves bend when they contact the outer core, and secondary waves are stopped completely. This creates a shadow zone where no seismic waves are received.

Differentiated Instruction

Learning Disabled To help these students, build a model of Earth's interior layers from modeling clay. First, make a ball of clay with a radius of 24 mm. Cut this ball in half and mold a 45-mm layer of different-colored clay around it. Around this second layer, mold a 58-mm layer of a third color of clay. Mold a very thin layer of a fourth color around this third layer. L1

Layer Boundaries **Figure 15** shows how seismic waves change speed as they pass through layers of Earth. Seismic waves speed up when they pass through the bottom of the crust and enter the upper mantle, shown on the far left of the graph. This boundary between the crust and upper mantle is called the Mohorovicic discontinuity (moh huh ROH vee chihch • dis kahn tuh NEW uh tee), or Moho.

The mantle is divided into layers based on changes in seismic wave speeds. For example, primary and secondary waves slow down again when they reach the asthenosphere. Then they generally speed up as they move through a more solid region of the mantle below the asthenosphere.

The core is divided into two layers based on how seismic waves travel through it. Secondary waves do not travel through the liquid core, as you can see in the graph. Primary waves slow down when they reach the outer core, but they speed up again upon reaching the solid inner core.

Seismic Wave Speeds

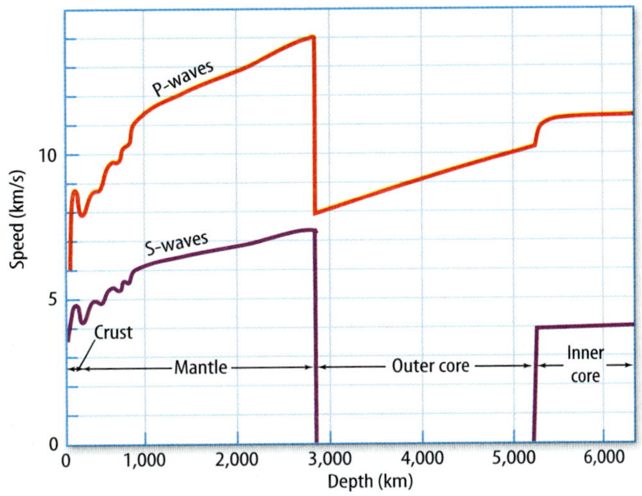

Figure 15 Changes in the speeds of seismic waves allowed scientists to detect boundaries between Earth's layers. S waves in the inner core form when P waves strike its surface.

section 2 review

Summary

Seismic Waves

- Stress builds up at the points where the surfaces of the fault touch.
- When the elastic limit of the rock is exceeded, it moves, producing seismic waves.
- There are three types of seismic waves—primary, secondary, and surface.

Locating an Epicenter

- A seismograph measures seismic waves.
- Three seismograph stations are needed to determine the location of an epicenter.

Basic Structure of Earth

- The inner core, the outer core, the lower mantle, the upper mantle, and the crust make up Earth.

Self Check

1. **Compare and contrast** the movement of rocks by primary waves, secondary waves, and surface waves.
2. **Explain** why surface waves cause the most damage to property.
3. **Describe** what makes up most of Earth's inner core.
4. **Explain** why three seismograph stations are needed to determine the location of an epicenter.
5. **Think Critically** Why do some seismograph stations receive both primary and secondary waves from an earthquake but other stations don't?

Applying Skills

6. **Simple Equations** Primary waves travel about 6 km/s through Earth's crust. The distance from Los Angeles to Phoenix is about 600 km. How long would it take primary waves to travel between the two cities?

3 Assess

DAILY INTERVENTION

Check for Understanding

Logical-Mathematical Working with a partner, have students draw a diagram of Earth's four layers, labeling each one. Ask them to choose a location for an earthquake epicenter and then demonstrate the P-waves, S-waves, and shadow zones by adding this information to their diagram. L2

Reteach

Seismic Waves Work with students to devise a large chart on the bulletin board or chalkboard that lists the properties and effects of the different types of seismic waves covered in this section. L1

Assessment

Performance Assess students' abilities to make and use graphs. Ask students to use **Figure 11** to determine travel times for primary waves at a distance of 3,000 km, 4,000 km, and 6,000 km. 5 minutes, 40 s; 7 minutes, 10 s; 9 minutes, 40 s L2

section 2 review

1. Primary waves cause rock to alternately compress and then expand. Secondary waves cause rock to vibrate in a direction that is perpendicular to wave travel. Surface waves cause rock to move in a side-to-side motion and a rolling up and down motion.
2. Surface waves combine the effects of primary and secondary waves, resulting in a backward, rolling motion and a side-to-side, swaying motion. The resulting amplitude of surface waves is greater than those of primary and secondary waves.
3. Iron is the most abundant element in Earth's inner core.
4. Three or more stations are needed to define a single location. Circles drawn around two stations intersect at two different points.
5. P-waves are able to travel through Earth's core; S-waves can not. Some seismograph stations would be in the shadow zone.
6. 100 seconds

BENCH TESTED

Real-World Question

Purpose Students will interpret data on an earthquake wave distance-time graph to determine the locations of earthquake epicenters. L2 ELL COOP LEARN

LS **Logical-Mathematical**

Process Skills use numbers, interpret data, make and use tables, make and use graphs, compare and contrast

Time Required 45 minutes

Procedure

Teaching Strategies Be sure students understand how to use **Figure 11** before beginning this lab.

Conclude and Apply

1. The difference in arrival time between P- and S-waves increases as the distance of the seismograph station from the earthquake increases. This time interval can be used to calculate the distance between the seismograph and the earthquake.
2. A: Mexico City, Mexico; B: San Francisco, California
3. a minimum of three
4. Those seismograph stations were probably within the outer core's shadow zone.

LAB

Epicenter Location

In this lab you can plot the distance of seismograph stations from the epicenters of earthquakes and determine the location of earthquake epicenters.

Real-World Question

How can plotting the distance of several seismograph stations from an earthquake epicenter allow you to determine the locations of the epicenter?

Goals

- **Plot** the distances from several seismograph stations based on primary and secondary wave arrival times.
- **Interpret** the location of earthquake epicenters from these plots.

Materials

string
metric ruler
globe
chalk

Earthquake Data

Location of Seismograph	Wave	Wave Arrival Times Earthquake A	Earthquake B
New York, New York	P	2:24:05 P.M.	1:19:42 P.M.
	S	2:29:15 P.M.	1:25:27 P.M.
Seattle, Washington	P	2:24:40 P.M.	1:14:37 P.M.
	S	2:30:10 P.M.	1:16:57 P.M.
Rio de Janeiro, Brazil	P	2:29:10 P.M.	—
	S	2:37:50 P.M.	—
Paris, France	P	2:30:30 P.M.	1:24:57 P.M.
	S	2:40:10 P.M.	1:34:27 P.M.
Tokyo, Japan	P	—	1:24:27 P.M.
	S	—	1:33:27 P.M.

Procedure

1. Determine the difference in arrival time between the primary and secondary waves at each station for each earthquake listed in the table.
2. After you determine the arrival time differences for each seismograph station, use the graph in **Figure 11** to determine the distance in kilometers of each seismograph from the epicenter of each earthquake. Record these data in a data table. For example, the difference in arrival times in Paris for earthquake B is 9 min, 30 s. On the graph, the primary and secondary waves are separated along the vertical axis by 9 min, 30 s at a distance of 8,975 km.
3. Using the string, measure the circumference of the globe. Determine a scale of centimeters of string to kilometers on Earth's surface. (Earth's circumference is 40,000 km.)
4. For each earthquake, place one end of the string at each seismic station location on the globe. Use the chalk to draw a circle with a radius equal to the distance to the earthquake's epicenter.
5. **Identify** the epicenter for each earthquake.

Conclude and Apply

1. How is the distance of a seismograph from the earthquake related to the arrival times of the waves?
2. **Identify** the location of the epicenter for each earthquake.
3. How many stations were needed to locate each epicenter accurately?
4. **Explain** why some seismographs didn't receive seismic waves from some quakes.

Assessment

Process Ask students to explain why data from two seismograph stations are not enough to locate an earthquake epicenter. Answers should state that data from just two seismograph stations would provide two possibilities for the epicenter. With three or more stations, there could only be one epicenter.

Differentiated Instruction

Physically Challenged Help these students by assigning each one a partner who will provide support as he or she measures distances on the globe. Encourage helpers to provide each physically challenged student with just enough assistance to accomplish the task. L2

section 3

People and Earthquakes

Earthquake Activity

Imagine waking up in the middle of the night with your bed shaking, windows shattering, and furniture crashing together. That's what many people in Northridge, California, experienced at 4:30 A.M. on January 17, 1994. The ground beneath Northridge shook violently—it was an earthquake.

Although the earthquake lasted only 15 s, it killed 51 people, injured more than 9,000 people, and caused $44 billion in damage. More than 22,000 people were left homeless. **Figure 16** shows some of the damage caused by the Northridge earthquake and a seismogram made by that quake.

Earthquakes are natural geological events that provide information about Earth. Unfortunately, they also cause billions of dollars in property damage and kill an average of 10,000 people every year. With so many lives lost and such destruction, it is important for scientists to learn as much as possible about earthquakes to try to reduce their impact on society.

Figure 16 The 1994 Northridge, California, earthquake was a costly disaster. Several major highways were damaged and 51 lives were lost.

Seismogram of the Northridge earthquake

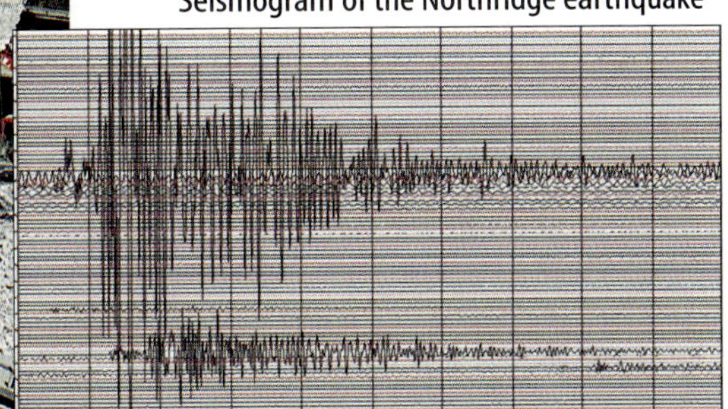

as you read

What You'll Learn

- **Explain** where most earthquakes in the United States occur.
- **Describe** how scientists measure earthquakes.
- **List** ways to make your classroom and home more earthquake-safe.

Why It's Important

Earthquake preparation can save lives and reduce damage.

Review Vocabulary

crest: the highest point of a wave

New Vocabulary

- magnitude
- liquefaction
- tsunami

1 Motivate

INTERACTIVE CHALKBOARD PowerPoint® Presentations

Bellringer

Section Focus Transparencies also are available on the Interactive Chalkboard CD-ROM. L2 ELL

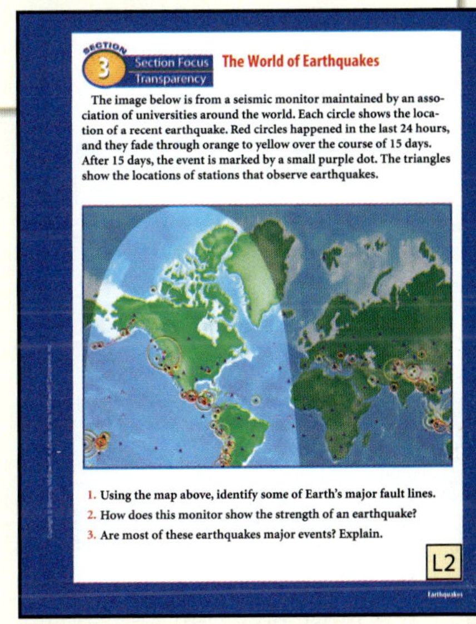
Section 3 Section Focus Transparency — The World of Earthquakes

The image below is from a seismic monitor maintained by an association of universities around the world. Each circle shows the location of a recent earthquake. Red circles happened in the last 24 hours, and they fade through orange to yellow over the course of 15 days. After 15 days, the event is marked by a small purple dot. The triangles show the locations of stations that observe earthquakes.

1. Using the map above, identify some of Earth's major fault lines.
2. How does this monitor show the strength of an earthquake?
3. Are most of these earthquakes major events? Explain.

L2

Tie to Prior Knowledge

Earthquakes in the News Help students recall photographs of the destruction caused by earthquakes they may have seen in newspapers, in magazines, or on television.

Section 3 Resource Manager

Chapter *FAST FILE* Resources

- Directed Reading, pp. 23, 24
- Enrichment, p. 34
- Lab Activity, pp. 15–18
- MiniLAB, p. 4
- Reinforcement, p. 31
- Transparency Activity, p. 48
- Lab Worksheets, pp. 7–8

Science Critical Thinking/Problem Solving, p. 9

Mathematical Skill Activities, p. 9

Cultural Diversity, p. 47

Performance Assessment in the Science Classroom, p. 42

2 Teach

Make a Model

Shoebox Seismograph Tape a stiff sheet of paper to the side of a closed shoe box. Have a volunteer slowly draw a straight line from the top of the sheet to the bottom. Then, have the volunteer attempt the same feat as another student bounces a small rubber ball on the top of the box. Use this model to explain how a seismograph works. LS **Visual-Spatial** L2

Discussion

Loss of Life Instruct students to look at **Table 1.** Why did some strong earthquakes cause so much loss of life while others caused little? Possible answers: Some areas were more populous than others; the buildings in some areas were more able to withstand earthquake vibrations. L2

Visual Learning

Table 1 Ask students to identify the most powerful earthquake listed in the table. the 1960 earthquake in Chile Which earthquake resulted in the most deaths? the 1556 earthquake in China L2

Virtual Labs

Seismograph Stations
How do seismograph stations help determine an earthquake's epicenter?

Figure 17 The 1999 earthquake in Turkey released about 32 times more energy than the 1994 Northridge earthquake did.

Studying Earthquakes Scientists who study earthquakes and seismic waves are seismologists. As you learned earlier, the instrument that is used to record primary, secondary, and surface waves from earthquakes all over the world is called a seismograph. Seismologists can use records from seismographs, called seismograms, to learn more than just where the epicenter of an earthquake is located.

Measuring Earthquake Magnitude The height of the lines traced on the paper of a seismograph is a measure of the energy that is released, or the **magnitude,** of the earthquake. The Richter magnitude scale is used to describe the strength of an earthquake and is based on the height of the lines on the seismogram. The Richter scale has no upper limit. However, scientists think that a value of about 9.5 would be the maximum strength an earthquake could register. For each increase of 1.0 on the Richter scale, the height of the line on a seismogram is ten times greater. However, about 32 times as much energy is released for every increase of 1.0 on the scale. For example, an earthquake with a magnitude of 8.5 releases about 32 times more energy than an earthquake with a magnitude of 7.5.

Table 1 Large-Magnitude Earthquakes

Year	Location	Magnitude	Deaths
1556	Shensi, China	?	830,000
1886	Charleston, SC	?	60
1906	San Francisco, CA	8.3	700 to 800
1923	Tokyo, Japan	9.2	143,000
1960	Chile	9.5	490 to 2,290
1975	Laoning Province, China	7.5	few
1976	Tangshan, China	8.2	242,000
1990	Iran	7.7	50,000
1994	Northridge, CA	6.8	51
2001	India	7.7	>20,000
2003	Northern Algeria	6.8	2,266

Past Earthquakes Damage from the 7.8-magnitude earthquake in Turkey in 1999 is shown in **Figure 17. Table 1** is a list of some large-magnitude earthquakes that have occurred around the world and the damage they have caused. Most of the earthquakes you hear about are large ones that cause great damage. However, of all the earthquakes detected throughout the world each year, most have magnitudes too low to be felt by humans. Scientists record thousands of earthquakes every day with magnitudes of less than 3.0. Each year, about 55,000 earthquakes are felt but cause little or no damage. These minor earthquakes have magnitudes that range from approximately 3.0 to 4.9 on the Richter scale.

Differentiated Instruction

Learning Disabled Have students use the information in **Table 1** to design and draw a time line of major earthquakes. Encourage students to illustrate the time line. Also encourage students to research other major earthquakes and add them to their time lines. L2 P

Describing Earthquake Intensity Earthquakes also can be described by the amount of damage they cause. The modified Mercalli intensity scale describes the intensity of an earthquake using the amount of structural and geologic damage in a specific location. The amount of damage done depends on the strength of the earthquake, the nature of surface material, the design of structures, and the distance from the epicenter.

Under ideal conditions, only a few people would feel an intensity-I earthquake, and it would cause no damage. An intensity-IV earthquake would be felt by everyone indoors during the day but would be felt by only a few people outdoors. Pictures might fall off walls and books might fall from shelves. However, an intensity-IX earthquake would cause considerable damage to buildings and would cause cracks in the ground. An intensity-XII earthquake would cause total destruction of buildings, and objects such as cars would be thrown upward into the air. The 1994 6.8-magnitude earthquake in Northridge, California, was listed at an intensity of IX because of the damage it caused.

Liquefaction Have you ever tried to drink a thick milkshake from a cup? Sometimes the milkshake is so thick that it won't flow. How do you make the milkshake flow? You shake it. Something similar can happen to very wet soil during an earthquake. Wet soil can be strong most of the time, but the shaking from an earthquake can cause it to act more like a liquid. This is called **liquefaction**. When liquefaction occurs in soil under buildings, the buildings can sink into the soil and collapse, as shown in **Figure 18.** People living in earthquake regions should avoid building on loose soils.

INTEGRATE Career

Magnetism In 1975, Chinese scientists successfully predicted an earthquake by measuring a slow tilt of Earth's surface and small changes in Earth's magnetism. Many lives were saved as a result of this prediction. Research the jobs that seismologists do and the types of organizations that they work for. Find out why most earthquakes have not been predicted.

Figure 18 San Francisco's Marina district suffered extensive damage from liquefaction in the 1989 Loma Prieta earthquake because it is built on a landfilled marsh.

Fun Fact

Three of the strongest earthquakes in the United States occurred in late 1811 and early 1812 along the New Madrid fault system centered in Missouri.

INTEGRATE Career

Magnetism Seismologists study earthquakes. They examine body waves and surface waves that are created from a sudden breaking of rock within Earth. Seismologists study these waves to understand how they interact, with the hope that they can better predict earthquake occurrence, as well as intensity. Most seismologists work for the United States Geological Survey, while others work for a branch of the USGS, the National Earthquake Information Center.

Research Have students research a variety of methods scientists employ to predict earthquakes. Some methods include technology, while other scientists are studying animals and their abilities to sense small vibrations in their environment and in their prey. Could animals also predict earthquakes? Students can design a brochure or earthquake prediction kit and attempt to "market" the idea to classmates. L2

Discussion

Liquefaction Why would the buildings in San Francisco's Marina district have been more susceptible to damage, having been built on a land-filled marsh? The soils of the filled-in marsh were probably not very compact and were saturated with water, causing them to be susceptible to liquefaction.

Teacher FYI

Moment Magnitude Moment magnitude, which is more precise than the Richter scale, is derived by multiplying the rigidity of the rock by the area of the fault rupture and then again by the amount of rock movement. This provides the seismic moment of the earthquake. The interaction of different segments of Earth on opposing sides of a fault set up internal torques that cause earthquakes. The magnitude usually first reported is Richter-scale magnitude modified for modern equipment. After further study, the moment magnitude can be determined and is applied to the earthquake.

SECTION 3 People and Earthquakes F ◆ 141

Quick Demo

Tsunami

Materials washtub, 2 bricks, toy boat, wax paper to cover bottom of washtub, string (2 pieces about 24 in long each)

Estimated Time 15 minutes

Procedure Lay a sheet of wax paper on the bottoom of the tub. Tie a piece of string around each end of one brick. With long sides touching, place both bricks on top of the wax paper. Fill the tub with water, just covering the bricks by about 3/4 inch (2 cm). Ask students to predict what will happen when the first brick is suddenly pulled away from the second brick. With a sharp tug, quickly pull the brick with the string away from the second brick. This simulates the movement of the ocean floor during an earthquake. Ask students to observe what happens at the surface of the water. A toy boat placed over the center of the two bricks will show the effects of a tsunami at the epicenter. Ask students to share their observations of the epicenter as well as the edges of the tub, which represent a shoreline. L2

Caption Answer

Figure 19 Flooding could occur as the huge wave comes ashore. Buildings might be destroyed, boats and other objects on shore could be swept out to sea, and people could be killed.

Science Online

Topic: Tsunamis

Visit bookf.msscience.com for Web links to information about tsunamis.

Activity Make a table that displays the location, date, earthquake magnitude, and maximum runup of the five most recent tsunamis.

Tsunamis Most earthquake damage occurs when surface waves cause buildings, bridges, and roads to collapse. People living near the seashore, however, have another problem. An earthquake under the ocean causes a sudden movement of the ocean floor. The movement pushes against the water, causing a powerful wave that can travel thousands of kilometers in all directions.

Ocean waves caused by earthquakes are called seismic sea waves, or **tsunamis** (soo NAH meez). Far from shore, a wave caused by an earthquake is so long that a large ship might ride over it without anyone noticing. But when one of these waves breaks on a shore, as shown in **Figure 19,** it forms a towering crest that can reach 30 m in height.

Tsunami Warnings Just before a tsunami crashes onto shore, the water along a shoreline might move rapidly toward the sea, exposing a large portion of land that normally is underwater. This should be taken as a warning sign that a tsunami could strike soon. You should head for higher ground immediately.

Because of the number of earthquakes that occur around the Pacific Ocean, the threat of tsunamis is constant. To protect lives and property, a warning system has been set up in coastal areas and for the Pacific Islands to alert people if a tsunami is likely to occur. The Pacific Tsunami Warning Center, located near Hilo, Hawaii, provides warning information including predicted tsunami arrival times at coastal areas.

However, even tsunami warnings can't prevent all loss of life. In the 1960 tsunami that struck Hawaii, 61 people died when they ignored the warning to move away from coastal areas.

Figure 19 A tsunami begins over the earthquake focus.
Infer *what might happen to towns located near the shore.*

Cultural Diversity

Tsunami! Tsunamis occur in many areas around the world and have been reported since ancient times. One of the earliest recorded tsunamis struck Syria around 2,000 B.C. Thera, one of the Cyclades Islands in the Mediterranean, may be the remnant of a volcano that erupted—causing tsunamis that ended the Minoan civilization on Crete. Tsunami is a Japanese word for "harbor wave." Many have struck the Japanese shore. Because Japan is an island nation, the threat of tsunamis is a national safety concern. Today, by using expected tsunami characteristics, the Japan Meteorological Agency can forecast tsunami heights for the Japanese coastline. This provides residents with the knowledge necessary to move a safe distance away from the shore.

142 ◆ F CHAPTER 5 Earthquakes

Earthquake Safety

Although earthquakes cannot be predicted reliably, **Figure 20** shows where earthquakes are most likely to occur in the United States. Knowing where earthquakes are likely to occur helps in long-term planning. Cities can take action to reduce damage and loss of life. Many buildings withstood the 1989 Loma Prieta earthquake because they were built with the expectation that such an earthquake would occur someday.

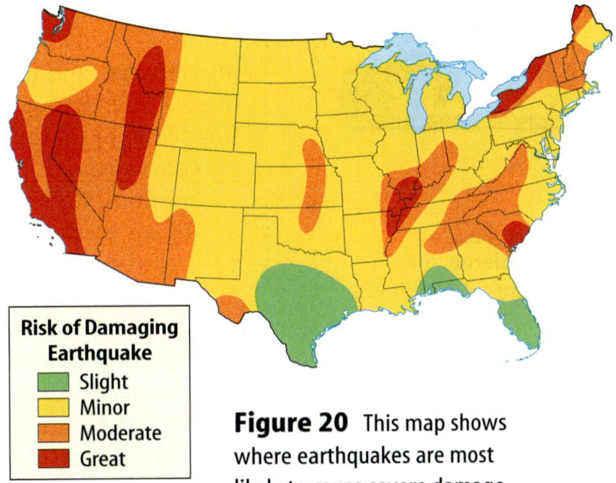

Figure 20 This map shows where earthquakes are most likely to cause severe damage.

Applying Math — Find a Ratio

EARTHQUAKE ENERGY An increase of one magnitude on the Richter scale for an earthquake means that 32 times more energy is released. How many times greater is the energy released by a magnitude-6 earthquake than the energy released by a magnitude-3 earthquake?

Solution

1 *This is what you know:*
- magnitude-6 earthquake, magnitude-3 earthquake
- energy increases 32 times per magnitude number

2 *This is what you need to find out:*
How many times greater is the energy of the larger earthquake than the energy of the smaller earthquake?

3 *This is the procedure you need to use:*
- Find the difference in magnitudes: 6 − 3 = 3.
- This is the number of times 32 is multiplied times itself. 32 × 32 × 32 = 32,768.
- The magnitude-6 earthquake releases 32,768 times more energy than the magnitude-3 earthquake.

4 *Check your answer:*
Count how many times you need to divide 32,768 by 32 to obtain 1. You should get 3.

Practice Problems

1. How many times greater is the energy released by a magnitude-7 earthquake than the energy released by a magnitude-2 earthquake?
2. How many times greater is the energy released by a magnitude-5 earthquake than the energy released by a magnitude-3 earthquake?

Science Online — For more practice, visit bookf.msscience.com/math_practice

Visual Learning

Figure 20 Have students determine whether your community has a high probability of having an earthquake. Answers will vary depending on your location. L2

Discussion

Earthquake Probability Of what use is a map that shows earthquake probability? It can alert people who live in certain areas of the country that there is a good chance they will experience an earthquake and that they should be prepared for one. How might some businesses be affected by being located in an earthquake zone? Answers might include: Construction companies might follow different building codes, rescue teams might have different training, and insurance companies might have different policy rates.

Applying Math

National Math Standards

Correlation to Mathematics Objectives

1, 2, 9

Teaching Strategy

Follow the steps in the example problem. The difference in magnitude numbers is 5. Multiply 32 times itself 5 times.

Answers to Practice Problems

1. 33,554,432 times more energy is released.
2. 1,024 times more energy is released.

Science Journal

Tsunami at Papua New Guinea Have students search newspaper articles online to write brief descriptions in their Science Journals of the damage done when a tsunami hit the northern coast of Papua New Guinea, on July 17, 1998. Trees and houses were swept away by the tsunami and at least 2,000 people died. L2

SECTION 3 People and Earthquakes F ◆ 143

Text Question Answer

Buildings can be designed to withstand the shaking of earthquakes.

Reading Check

Answer those built to resist damage from the shaking of earthquakes

Caption Answer

Figure 21 The rubber acts like a cushion to absorb movement.

Mini LAB

Purpose Students will model seismic-safe methods of construction. L2 ELL COOP LEARN

LS **Interpersonal**

Materials blocks, rubber bands of two sizes

Teaching Strategy Review the Earthquake-Resistant Structures text with students before having them complete the lab.

Analysis

1. Structures with rubber bands around them are more likely to withstand the "earthquake."
2. Concrete pillars could be wrapped with steel supports. The supports would decrease the chances of the pillars breaking during an earthquake.

Assessment

Performance Have students build a model seismic-safe highway. Use **Performance Assessment in the Science Classroom,** p. 123.

Try at Home

Mini LAB

Modeling Seismic-Safe Structures

Procedure

1. On a **tabletop,** build a structure out of **building blocks** by simply placing one block on top of another.
2. Build a second structure by wrapping sections of three blocks together with **rubber bands.** Then, wrap larger rubber bands around the entire completed structure.
3. Set the second structure on the tabletop next to the first one and pound on the side of the table with a slow, steady rhythm.

Analysis

1. Which of your two structures was better able to withstand the "earthquake" caused by pounding on the table?
2. How might the idea of wrapping the blocks with rubber bands be used in construction of supports for elevated highways?

Try at Home

Quake-Resistant Structures During earthquakes, buildings, bridges, and highways can be damaged or destroyed. Most loss of life during an earthquake occurs when people are trapped in or on these crumbling structures. What can be done to reduce loss of life?

Seismic-safe structures stand up to vibrations that occur during an earthquake. **Figure 21** shows how buildings can be built to resist earthquake damage. Today in California, some new buildings are supported by flexible, circular moorings placed under the buildings. The moorings are made of steel plates filled with alternating layers of rubber and steel. The rubber acts like a cushion to absorb earthquake waves. Tests have shown that buildings supported in this way should be able to withstand an earthquake measuring up to 8.3 on the Richter scale without major damage.

In older buildings, workers often install steel rods to reinforce building walls. Such measures protect buildings in areas that are likely to experience earthquakes.

Reading Check *What are seismic-safe structures?*

Figure 21 The rubber portions of this building's moorings absorb most of the wave motion of an earthquake. The building itself only sways gently.
Infer *what purpose the rubber serves.*

Curriculum Connection

Art Have students research what to do to keep safe during an earthquake. Then have them take one aspect of what they find and create a poster with one important "Earthquake Safety Tip." Hang the posters in class and use them as a foundation for a class discussion on earthquake safety. LS **Visual-Spatial** L2

Before an Earthquake To make your home as earthquake-safe as possible, certain steps can be taken. To reduce the danger of injuries from falling objects, move heavy objects from high shelves to lower shelves. Learn how to turn off the gas, water, and electricity in your home. To reduce the chance of fire from broken gas lines, make sure that water heaters and other gas appliances are held securely in place as shown in **Figure 22.** A newer method that is being used to minimize the danger of fire involves placing sensors on gas lines. The sensors automatically shut off the gas when earthquake vibrations are detected.

During an Earthquake If you're indoors, move away from windows and any objects that could fall on you. Seek shelter in a doorway or under a sturdy table or desk. If you're outdoors, stay in the open—away from power lines or anything that might fall. Stay away from chimneys or other parts of buildings that could fall on you.

After an Earthquake If water and gas lines are damaged, the valves should be shut off by an adult. If you smell gas, leave the building immediately and call authorities from a phone away from the leak area. Stay away from damaged buildings. Be careful around broken glass and rubble, and wear boots or sturdy shoes to keep from cutting your feet. Finally, stay away from beaches. Tsunamis sometimes hit after the ground has stopped shaking.

Figure 22 Sturdy metal straps on this gas water heater help reduce the danger of fires from broken gas lines during an earthquake.

section 3 review

Summary

Earthquake Activity

- The height of the lines traced on a seismogram can be used to determine an earthquake's magnitude.
- The intensity of an earthquake is determined by examining the amount of damage caused by the earthquake.

Earthquake Safety

- Knowing where large earthquakes are likely to occur helps people plan how to reduce damage.
- If you're ever in an earthquake, move away from windows or any object that might fall on you. Seek shelter in a doorway or under a sturdy table or desk.

Self Check

1. **Explain** how you can determine if you live in an area where an earthquake is likely to occur.
2. **Compare and contrast** the Richter and the Mercalli scales.
3. **Explain** what causes a tsunami.
4. **Describe** three ways an earthquake causes damage.
5. **Think Critically** How are shock absorbers on a car similar to the circular moorings used in modern earthquake-safe buildings? How do they absorb shock?

Applying Skills

6. **Infer** Seismographs around the world record the occurrence of thousands of earthquakes every day. Why are so few earthquakes in the news?

Science Online bookf.msscience.com/self_check_quiz

3 Assess

DAILY INTERVENTION

Check for Understanding

Logical-Mathematical Using a large, three-column chart, brainstorm with the class about what to do before, during, and after an earthquake. Discuss procedures for the community in public buildings and in private homes. L1

Reteach

Shaky Construction Using a collection of building blocks, small rectangular sponges, paper towel rolls, springs, or sugar cubes, provide time for students to explore and build earthquake-resistant building designs. Suggest bricks that have identical patterns on all sides, alternating patterns, long sides lying down, long sides standing up, or a combination of these methods. Test their designs when supported by a foundation of bricks, sand, or tubing while being shaken by a washing machine or other shaking device. L2

Assessment

Performance Have students make concept maps that show what people should do before, during, and after earthquakes to keep safe. Allow students to make large drawings on construction paper. Use **Performance Assessment in the Science Classroom,** p. 161. L2

section 3 review

1. Check maps or other sources to determine whether your region has a high probability for future earthquakes.
2. The Richter scale is a measure of earthquake magnitude. Its value is the same everywhere. The modified Mercalli scale is a measure of damage caused by the earthquake. Its value varies regionally.
3. Movement of rock or sediment beneath the ocean causes tsunamis. Earthquakes often cause this movement.
4. Shaking and liquefaction of sediment damage buildings, fires start when gas lines rupture, and tsunamis threaten coastal areas.
5. Both absorb vibrations. Shock absorbers and the building moorings will "give" to absorb shock. Rigid buildings are most susceptible to earthquake damage.
6. Most earthquakes cause no damage.

BENCH TESTED

Real-World Question

Purpose Students investigate whether there is a relationship between the depth of earthquake foci and epicenter locations and the movements of plates. L2

COOP LEARN LS

Logical-Mathematical

Process Skills observe and infer, communicate, use numbers, interpret data, hypothesize, use tables, make and use graphs, compare and contrast, separate and control variables

Time Required 45 minutes

Analyze Your Data

Teaching Strategies

- Review the definitions of focus and epicenter before starting the lab.
- Draw a blank version of the focus depth versus distance graph. Indicate the location of the coast on the graph. Help students begin by plotting the first two locations.

Expected Outcome Students should find that earthquake foci occur deeper the farther inland they are from the shore.

LAB

Earthquake Depths

Goals

- **Observe** any connection between earthquake-focus depth and epicenter location using the data provided on the next page.
- **Describe** any observed relationship between earthquake-focus depth and the movement of plates at Earth's surface.

Materials

graph paper
pencil

Real-World Question

You learned in this chapter that Earth's crust is broken into sections called plates. Stresses caused by movement of plates generate energy within rocks that must be released. When this release is sudden and rocks break, an earthquake occurs. Can a study of the foci of earthquakes tell you about plate movement in a particular region?

Analyze Your Data

1. Use graph paper and the data table on the right to make a graph plotting the depths of earthquake foci and the distances from the coast of a continent for each earthquake epicenter.
2. Use the graph below as a reference to draw your own graph. Place *Distance from the coast* and units on the *x*-axis. Begin labeling at the far left with 100 km west. To the right of it should be 0 km, then 100 km east, 200 km east, 300 km east, and so on through 700 km east. What point on your graph represents the coast?
3. Label the *y*-axis *Depth below Earth's surface.* Label the top of the graph *0 km* to represent Earth's surface. Label the bottom of the *y*-axis *–800 km.*
4. **Plot** the focus depths against the distance and direction from the coast for each earthquake in the table below.

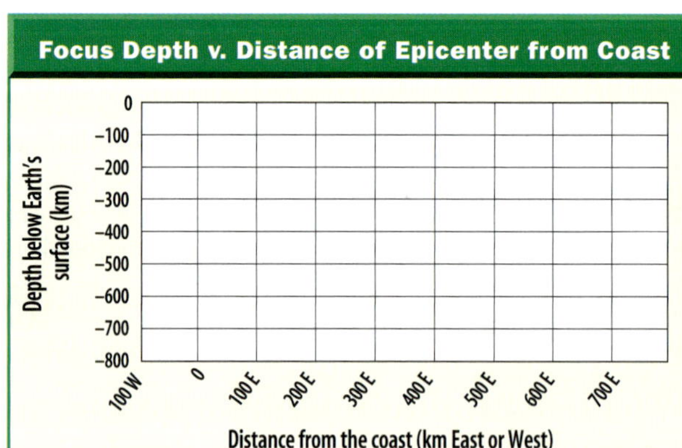

Differentiated Instruction

Visually Impaired Ask some of your students to prepare copies of the maps and charts being used in this lab in a manner useful to those who are visually impaired. Outline the locations under study with glue, and identify earthquake foci by raised bumps on the maps. Students who are able could produce maps and charts using Braille. L2

Alternative Inquiry Lab

Student Choice To make this Lab an Inquiry Lab, have students research global earthquake activity. Students can choose the geographic areas they want to study, plot the plate movement, and map the data they collect. By connecting the data collection to real-world earthquake activity, students will feel more personally involved.

Using Scientific Methods

Conclude and Apply

1. **Describe** any observed relationship between the location of earthquake epicenters and the depth of foci.
2. **Explain** why none of the plotted earthquakes occurred below 700 km.
3. Based on your graph, form a hypothesis to explain what is happening to the plates at Earth's surface in the vicinity of the plotted earthquake foci. In what direction are the plates moving relative to each other?
4. **Infer** what process is causing the earthquakes you plotted on your graph.
5. **Infer** whether these earthquakes are occurring along the eastern side of a continent or along the western side of a continent.
6. **Draw and label** a cross section of the Earth beneath this coast. Label the eastern plate, the western plate, and use arrows to show the directions the plates are moving.
7. **Form a hypothesis** to predict which continent these data might apply to. Apply what you have learned in this lab and the information in **Figure 2.** Explain your answer.

Focus and Epicenter Data

Earthquake	Focus Depth (km)	Distance of Epicenter from Coast (km)
A	−55	0
B	−295	100 east
C	−390	455 east
D	−60	75 east
E	−130	255 east
F	−195	65 east
G	−695	400 east
H	−20	40 west
I	−505	695 east
J	−520	390 east
K	−385	335 east
L	−45	95 east
M	−305	495 east
N	−480	285 east
O	−665	545 east
P	−85	90 west
Q	−525	205 east
R	−85	25 west
S	−445	595 east
T	−635	665 east
U	−55	95 west
V	−70	100 west

Communicating Your Data

Compare your graph with those of other members of your class. **For more help, refer to the Science Skill Handbook.**

Conclude and Apply

1. Earthquake foci become deeper as epicenters are plotted toward the East.
2. Below this depth, the rock in the subducting slab can no longer behave rigidly. It becomes too hot.
3. One plate is subducting beneath a second plate. The subducting plate is moving east relative to the overriding plate.
4. The earthquakes are occurring because of subduction. Earthquakes occur in the comparatively cool and rigid sinking plate.
5. They are occurring along the western side of a continent.
6. Drawings should show an oceanic plate subducting beneath a continental plate. The western plate (oceanic plate) should have an arrow pointing east. The eastern plate (continental plate) should have an arrow pointing west.
7. The west coast of South America or the northwest coast of the United States; subducting slabs sink into the mantle with relative easterly motion at both of these locations.

Error Analysis Students who do not get the correct outcome should check whether they have transferred data correctly from the table to the graph. Incorrect plotting of the graph itself also could introduce errors.

Communicating Your Data

Use a spreadsheet program to make a line graph to display the focus and epicenter data.

Assessment

Process Ask students to draw the plate boundary described by the data in this lab. Students should draw a convergent boundary where a sea plate meets a land plate and the sea plate is subducted. L2

Science Stats

Content Background

Students might have heard tsunamis referred to as tidal waves. This term is incorrect, as tsunamis have nothing to do with tides. The word *tsunami* actually means "harbor wave" in Japanese, and has become the accepted term for waves generated by seismic activity. Sometimes a drop in sea level precedes the arrival of a tsunami. People have been killed after moving out to inspect exposed sea life, and then suddenly being faced with a fast-moving wall of water.

Discussion

Tsunami Warning What preparations for a tsunami could the people of Hawaii make if given several hours notice? They could move ships out of harbors to the open sea and evacuate low-lying areas.

Duration Effects How might the duration of an earthquake contribute to its severity? The longer the shaking, the more opportunity for damage. The psychological toll exacted by an earthquake on residents is greater the longer the shaking continues.

Activity

Earthquake Reenactment Have the class act out the nearly five-minute duration of the Alaskan earthquake, imagining the panic and dangers involved. Use a stopwatch to give them accurate start and end times. Have them record their impressions in their Science Journals. LS **Kinesthetic and Linguistic** L2

Science Stats

Moving Earth!

Did you know...

... Tsunamis can travel as fast as commercial jets and can reach heights of 30 m. A wave that tall would knock over this lighthouse. Since 1945, more people have been killed by tsunamis than by the ground shaking from earthquakes.

... The most powerful earthquake to hit the United States in recorded history shook Alaska in 1964. At 8.5 on the Richter scale, the quake shook all of Alaska for nearly 5 min, which is a long time for an earthquake. Nearly 320 km of roads near Anchorage suffered damage, and almost half of the 204 bridges had to be rebuilt.

Applying Math How many 3.0-magnitude earthquakes would it take to equal the energy released by one 8.0-magnitude earthquake?

... Snakes can sense the vibrations made by a small rodent up to 23 m away. Does this mean that they can detect vibrations prior to major earthquakes? Unusual animal behavior was observed just before a 1969 earthquake in China—an event that was successfully predicted.

Write About It

Visit bookf.msscience.com/science_stats to research the history and effects of earthquakes in the United States. In a paragraph, describe how the San Francisco earthquake of 1906 affected earthquake research.

Applying Math

Answer It would take about 33,554,432 3.0-magnitude earthquakes to equal the energy released by one 8.0-magnitude earthquake.

Differentiated Instruction

Challenge Have students research causes and effects of earthquakes or tsunamis. Using the factual information, students write a simple, fictional biographical picture book depicting a survival story in a realistic setting. Characters, setting, and plot should reflect the environmental changes before, during, and after the catastrophic event. L3 **Linguistic**

Write About It

Organize Information Have students make charts to organize what they learn about various U.S. earthquakes. They can include columns to show the date, location, magnitude, and effects of each earthquake. L2

Reviewing Main Ideas

Section 1 Forces Inside Earth

1. Plate movements can cause rocks to bend and stretch. Rocks can break if the forces on them are beyond their elastic limit.
2. Earthquakes are vibrations produced when rocks break along a fault.
3. Normal faults form when rocks are under tension. Reverse faults form under compression and shearing forces produce strike-slip faults.

Section 2 Features of Earthquakes

1. Primary waves stretch and compress rock particles. Secondary waves move particles at right angles to the direction of wave travel.
2. Surface waves move rock particles in a backward, rolling motion and a side-to-side swaying motion.
3. Earthquake epicenters are located by recording seismic waves.
4. The boundaries between Earth's internal layers are determined by observing the speeds and paths of seismic waves.

Section 3 People and Earthquakes

1. A seismograph measures the magnitude of an earthquake.
2. The magnitude of an earthquake is related to the energy released by the earthquake.

Visualizing Main Ideas

Copy and complete the following concept map on earthquake damage.

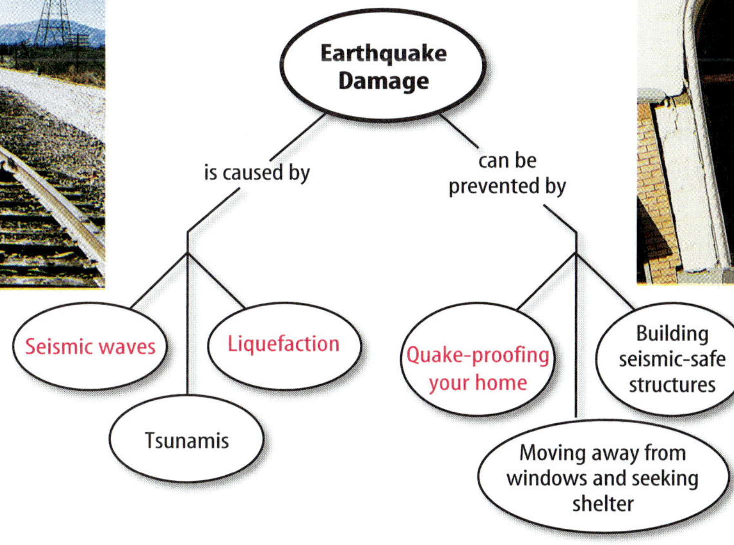

Identifying Misconceptions

Assess

Use the assessment as follow-up to page F at the beginning of this chapter.

Activity Have students bring in newspaper articles describing earthquakes around the world. Have the class work together to make a bulletin board display showing the location of different earthquakes, and describing their intensity and the damage done. Have students compare their findings with the map in **Figure 2.** Have students explain why earthquakes are found where they are, and why everyone should know how to respond to an earthquake. L2

chapter 5 Study Guide

Reviewing Main Ideas

Summary statements can be used by students to review the major concepts of the chapter.

Visualizing Main Ideas

See student page.

Science Online

Visit bookf.msscience.com
/self_check_quiz
/interactive_tutor
/vocabulary_puzzlemaker
/chapter_review
/standardized_test

Assessment Transparency

For additional assessment questions, use the *Assessment Transparency* located in the transparency book.

Assessment

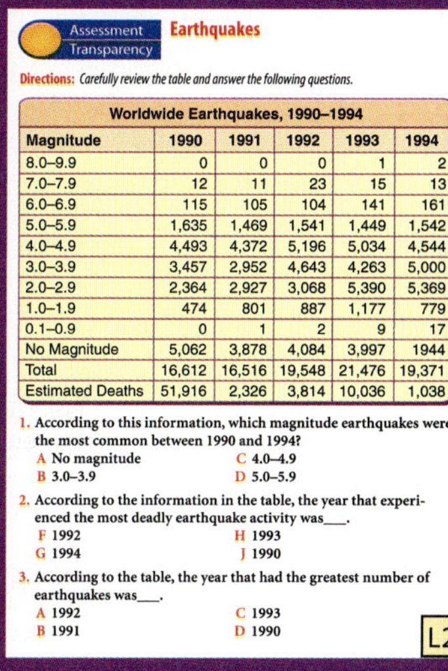
Assessment Transparency **Earthquakes**

Directions: *Carefully review the table and answer the following questions.*

Worldwide Earthquakes, 1990–1994					
Magnitude	1990	1991	1992	1993	1994
8.0–9.9	0	0	0	1	2
7.0–7.9	12	11	23	15	13
6.0–6.9	115	105	104	141	161
5.0–5.9	1,635	1,469	1,541	1,449	1,542
4.0–4.9	4,493	4,372	5,196	5,034	4,544
3.0–3.9	3,457	2,952	4,643	4,263	5,000
2.0–2.9	2,364	2,927	3,068	5,390	5,369
1.0–1.9	474	801	887	1,177	779
0.1–0.9	0	1	2	9	17
No Magnitude	5,062	3,878	4,084	3,997	1944
Total	16,612	16,516	19,548	21,476	19,371
Estimated Deaths	51,916	2,326	3,814	10,036	1,038

1. According to this information, which magnitude earthquakes were the most common between 1990 and 1994?
   A No magnitude C 4.0–4.9
   B 3.0–3.9 D 5.0–5.9
2. According to the information in the table, the year that experienced the most deadly earthquake activity was___.
   F 1992 H 1993
   G 1994 J 1990
3. According to the table, the year that had the greatest number of earthquakes was___.
   A 1992 C 1993
   B 1991 D 1990

L2

Earthquakes

Using Vocabulary

1. Surface waves
2. strike-slip fault
3. epicenter
4. magnitude
5. tsunami

Checking Concepts

6. C
7. A
8. D
9. B
10. B
11. A
12. B
13. B
14. D

chapter 5 Review

Using Vocabulary

earthquake p. 127	reverse fault p. 128
epicenter p. 131	secondary wave p. 131
fault p. 126	seismic wave p. 130
focus p. 130	seismograph p. 133
liquefaction p. 141	strike-slip fault p. 129
magnitude p. 140	surface wave p. 131
normal fault p. 128	tsunami p. 142
primary wave p. 131	

Fill in the blanks with the correct words.

1. ________ causes most of the damage in earthquakes because of the side to side swaying motion that many buildings are unable to withstand.
2. At a(n) ________, rocks move past each other without much upward or downward movement.
3. The point on Earth's surface directly above the earthquake focus is the ________.
4. The measure of the energy released during an earthquake is its ________.
5. An earthquake under the ocean can cause a(n) ________ that travels thousands of kilometers.

Checking Concepts

Choose the word or phrase that best answers the question.

6. Earthquakes can occur when which of the following is passed?
   A) tension limit C) elastic limit
   B) seismic unit D) shear limit
7. When the rock above the fault surface moves down relative to the rock below the fault surface, what kind of fault forms?
   A) normal C) reverse
   B) strike-slip D) shear

Use the illustration below to answer question 8.

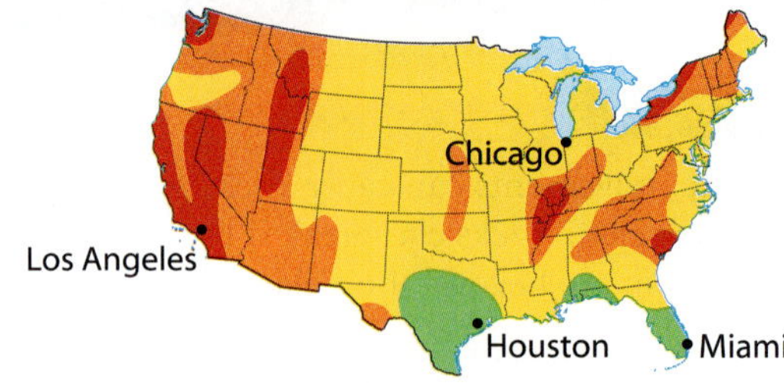

Risk of A Damaging Earthquake
Slight | Minor | Moderate | Great

8. Using the figure above, which city should be most prepared for an earthquake?
   A) Miami, FL C) Chicago, IL
   B) Houston, TX D) Los Angeles, CA
9. From which of the following do primary and secondary waves move outward?
   A) epicenter C) Moho
   B) focus D) tsunami
10. What kind of earthquake waves stretch and compress rocks?
   A) surface C) secondary
   B) primary D) shear
11. What are the slowest seismic waves?
   A) surface C) secondary
   B) primary D) pressure
12. What is the fewest number of seismograph stations that are needed to locate the epicenter of an earthquake?
   A) two C) four
   B) three D) five
13. What happens to primary waves when they pass from liquids into solids?
   A) slow down C) stay the same
   B) speed up D) stop
14. What part of a seismograph does not move during an earthquake?
   A) sheet of paper C) drum
   B) fixed frame D) pendulum

Science Online bookf.msscience.com/vocabulary_puzzlemaker

Use the ExamView® Pro Testmaker CD-ROM to:
- create multiple versions of tests
- create modified tests with one mouse click for inclusion students
- edit existing questions and add your own questions
- build tests aligned with state standards using built-in State Curriculum Tags
- change English tests to Spanish with one mouse click and vice versa

Thinking Critically

15. Infer The 1960 earthquake in the Pacific Ocean off the coast of Chile caused damage and loss of life in Chile, Hawaii, Japan, and other areas along the Pacific Ocean border. How could this earthquake do so much damage to areas thousands of kilometers from its epicenter?

16. Explain why a person who is standing outside in an open field is relatively safe during a strong earthquake.

17. Describe how a part of the seismograph remains at rest during an earthquake.

18. Explain why it is incorrect to call a tsunami a tidal wave.

19. Predict which is likely to be more stable during an earthquake—a single-story wood-frame house or a brick building. Explain.

20. Measure in SI Use an atlas and a metric ruler to answer the following question. Primary waves travel at about 6 km/s in continental crust. How long would it take a primary wave to travel from San Francisco, California, to Reno, Nevada?

Use the table below to answer question 21.

Seismograph Station Data

Station	Latitude	Longitude	Distance from Earthquake
1	45° N	120° W	1,300 km
2	35° N	105° W	1,200 km
3	40° N	115° W	790 km

21. Use Tables Use a map of the United States that has a distance scale, a compass for drawing circles, and the table above to determine the location of the earthquake epicenter.

Performance Activities

22. Model Use layers of different colors of clay to illustrate the three different kinds of faults. Label each model, explaining the forces involved and the rock movement.

Applying Math

23. Earthquake Magnitude An increase of one on the Richter scale corresponds to an increase of 10 in the size of the largest wave on a seismogram. How many times larger is the largest wave of a Richter magnitude-6 earthquake than a Richter magnitude-3 earthquake?

24. Tsunami Speed An underwater earthquake produces a tsunami 1,500 km away from Hawaii. If the tsunami travels at 600 km/h, how long will it take to reach Hawaii?

Use the graph below to answer question 25.

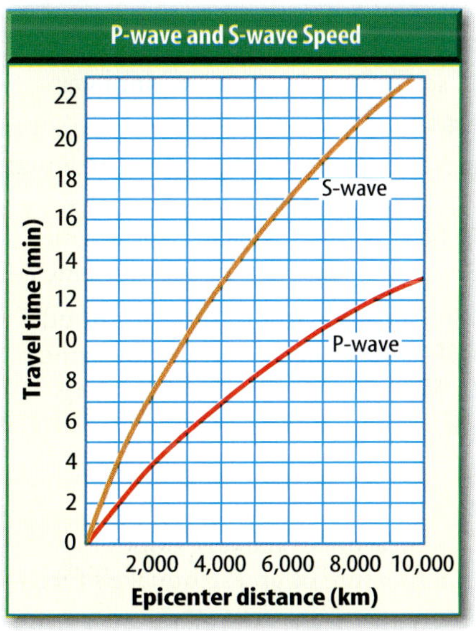

25. Earthquake Waves The graph above shows a P-wave and an S-wave plotted on a time-distance graph. According to the graph, which wave moves at the greater speed?

chapter Review 5

Thinking Critically

15. The earthquake caused tsunamis that traveled across the Pacific Ocean.

16. Buildings or other structures cannot fall on this person. Fire is unlikely in this area.

17. The heavy weight tends to remain at rest.

18. Tsunamis are not related to tides. Tides occur as a consequence of the Moon's gravity and the Sun's gravity.

19. A single-story wood-frame house would be more stable. This structure is less rigid than a brick building, so it can withstand more shaking.

20. about 61 s or 1 minute ($d = 367$ km)

21. The epicenter is just south of Los Angeles, California.

Performance Activities

22. Normal fault: block above fault moves down, tension; reverse fault: block above fault moves up, compression; strike-slip fault: blocks of rock slide past, shear. Use **Performance Assessment in the Science Classroom,** p. 123.

Applying Math

National Math Standards

1, 2, 5, 9

23. $10^3 = 1,000$

24. 1,500 km ÷ 600 km/h = 2.5 h

25. Primary waves move at higher speed. This can be determined by noticing that the primary wave curve covers more distance in unit time.

Assessment Resources

Reproducible Masters

Chapter *Fast File* Resources
- Chapter Review, pp. 39–40
- Chapter Tests, pp. 41–44
- Assessment Transparency Activity, p. 51

Glencoe Science Web site
- Chapter Review Test
- Standardized Test Practice

Glencoe Technology

- Assessment Transparency
- Exam*View*® Pro Testmaker
- MindJogger Videoquiz
- Interactive Chalkboard

FAST FILE

Answer Sheet A practice answer sheet can be found at bookf.msscience.com/answer_sheet.

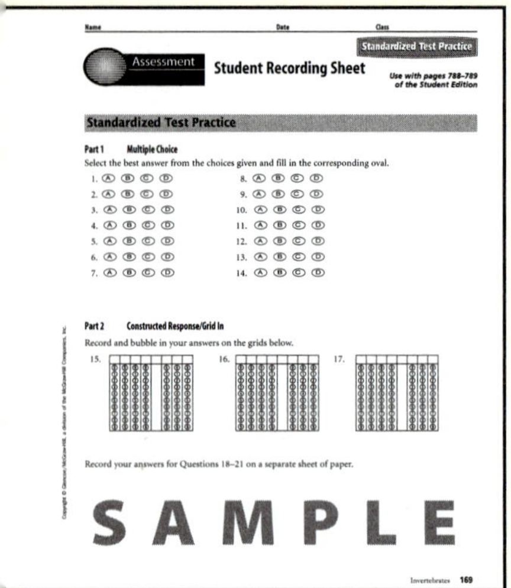
Name Date Class

Assessment **Student Recording Sheet** Standardized Test Practice

Use with pages 788–789 of the Student Edition

Standardized Test Practice

Part 1 Multiple Choice

Select the best answer from the choices given and fill in the corresponding oval.

Part 2 Constructed Response/Grid In

Record and bubble in your answers on the grids below.

Record your answers for Questions 18–21 on a separate sheet of paper.

SAMPLE

Part 1 Multiple Choice

1. B
2. C
3. A
4. C
5. C
6. C
7. C

Part 2 Short Response

8. The motion of Earth's plates causes rock to be under stress. This stress can cause elastic deformation of the rock. When the elastic limit of the rock is exceeded, movement can occur and the energy that was stored in the deformed rock is released as seismic waves.
9. reverse fault
10. compression
11. The plane that separates the two blocks is the fault plane.
12. Most motion occurs along plate boundaries. The interiors of plates are comparatively rigid.
13. $142 million

chapter 5 Standardized Test Practice

Part 1 Multiple Choice

Record your answers on the answer sheet provided by your teacher or on a sheet of paper.

Use the photo below to answer question 1.

1. The instrument above records seismic waves from an earthquake. Which of the following is the name of this instrument?
   A. seismogram **C.** tiltmeter
   B. seismograph **D.** strainmeter
2. Which of the following terms is used to indicate the region where no earthquake waves reach Earth's surface?
   A. light zone **C.** shadow zone
   B. waveless zone **D.** seismic zone
3. Which is used to measure magnitude?
   A. Richter scale **C.** shadow zone
   B. Mercalli scale **D.** seismic gap
4. What is earthquake intensity?
   A. a measure of energy released
   B. a measure of seismic risk
   C. a measure of damage done
   D. a measure of an earthquake's focus

Test-Taking Tip

Eliminate Incorrect Answers If you don't know the answer to a multiple choice question, try to eliminate as many incorrect answers as possible.

5. Which of the following describes liquefaction?
   A. the stopping of S-waves by Earth's molten outer core
   B. ice melting during an earthquake to cause flooding
   C. seismic waves shaking sediment, causing it to become more liquid like
   D. rivers diverted by the motion of earth-quake flooding
6. Which of the following describes the motion of secondary waves?
   A. a backward rolling and a side-to-side swaying motion
   B. a back-and-forth motion that is parallel to the direction of travel
   C. vibration in directions that are perpendicular to the direction of wave travel
   D. a forward rolling motion

Use the illustration below to answer question 7.

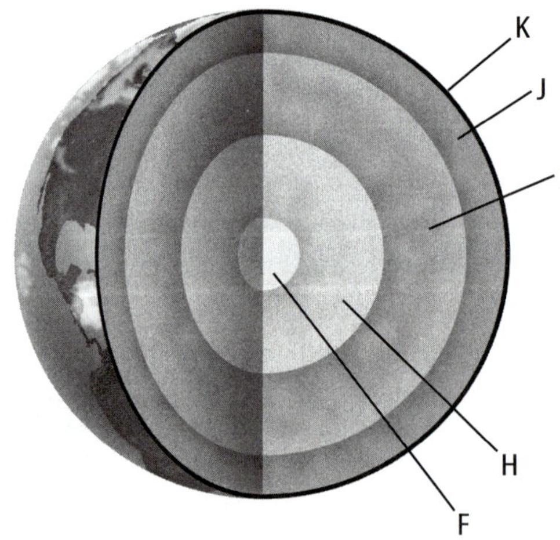

7. Which letter corresponds to the lower mantle?
   A. F **C.** L
   B. H **D.** J

Part 2 Short Response/Grid In

14. The amount of damage is affected by the following: distance from epicenter, type of rock and sediment, types of buildings, building codes, vulnerability of gas and electric lines. Other answers also might be correct.
15. Most earthquakes occur along plate boundaries.

Part 2 Short Response/Grid In

Record your answers on the answer sheet provided by your teacher or on a sheet of paper.

8. Explain how earthquakes occur. Include a description of how energy builds up in rocks and is later released.

Use the illustration below to answer questions 9–10.

9. What type of fault is shown?
10. What type of force caused this fault to form?
11. Where is the fault plane?
12. Explain why most earthquakes occur along plate boundaries.
13. An earthquake that occurred in San Fernando, California, in 1971 caused $500 million in damage. An earthquake that occurred in Whittier, California, in 1987 caused $358 million in damage. How much more damage was caused by the San Fernando earthquake?
14. What factors affect damage done by an earthquake?
15. Explain the relationship between worldwide earthquake distribution and tectonic boundaries.

Part 3 Open Ended

Record your answers on a sheet of paper.

16. The three types of faults are normal faults, reverse faults, and strike-slip faults. Draw each type of fault including arrows that show which way the rocks move.
17. Describe how a person should prepare for an earthquake and how a person should react if an earthquake occurs.

Use the map below to answer question 18.

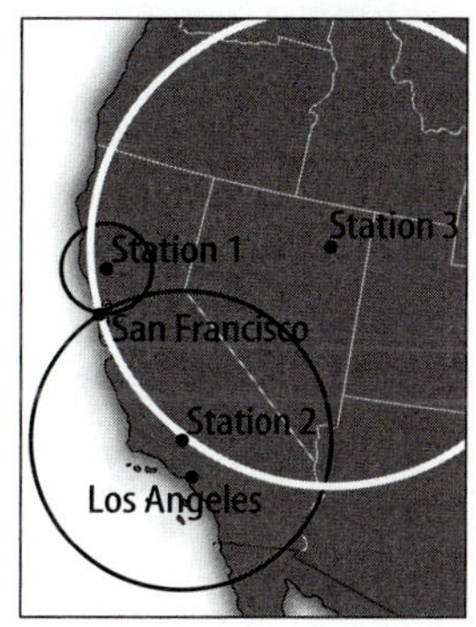

18. The map above shows three circles drawn around three different seismograph stations. The circles have radii equal to the distance between the seismograph station and the earthquake's epicenter. Which labeled point on the map represents the earthquake's epicenter? How do you know?
19. Compare and contrast the three types of seismic waves.
20. Draw three diagrams to show how each type of seismic wave moves through rocks.
21. Describe what happens to S-waves when they contact Earth's outer core. Decribe what happens to P-waves when they reach Earth's outer core. What is the *shadow zone?*

Part 3 Open Ended

16. Drawings should show fault surfaces and should indicate relative sense of motion.
17. To prepare, a person should gather emergency supplies, and should practice emergency procedures. During an earthquake, take cover and avoid potential hazards, such as glass windows.
18. San Francisco, all three circles intersect at this point
19. Primary waves are fastest and consist of alternating compressions and rarefactions. Secondary waves have intermediate speed and consist of vibrations perpendicular to wave travel. Surface waves are slowest, cause the most damage, and move rock in a side-to-side and rolling motion.
20. Students should use arrows to indicate the types of motion that occur with passage of each type of seismic wave.
21. S-waves cannot travel in Earth's outer core. P-waves are refracted upon entering Earth's outer core. A shadow zone is a zone where seismic waves are not detected.

Rubrics

The following rubrics are sample scoring devices for short response and open-ended questions.

Short Response

Points	Description
2	The student demonstrates a thorough understanding of the science of the task. The response may contain minor flaws that do not detract from the demonstration of a thorough understanding.
1	The student has provided a response that is only partially correct.
0	The student has provided a completely incorrect solution or no response at all.

Open Ended

Points	Description
4	The student demonstrates a thorough understanding of the science of the task. The response may contain minor flaws that do not detract from the demonstration of a thorough understanding.
3	The student demonstrates an understanding of the science of the task. The response is essentially correct and demonstrates an essential but less than thorough understanding of the science.
2	The student demonstrates only a partial understanding of the science of the task. Although the student may have used the correct approach to a solution or may have provided a correct solution, the work lacks an essential understanding of the underlying science concepts.
1	The student demonstrates a very limited understanding of the science of the task. The response is incomplete and exhibits many flaws.
0	The student provides a completely incorrect solution or no response at all.

chapter 6 Organizer

<table>
<tr><th rowspan="2">Section/Objectives</th><th colspan="2">Standards</th><th rowspan="2">Labs/Features</th></tr>
<tr><th>National</th><th>State/Local</th></tr>
<tr><td>Chapter Opener</td><td>See pp. 9T–10T for a Key to Standards.</td><td></td><td>Launch Labs: Map a volcano, p. 155
Foldables, p. 155</td></tr>
<tr><td>Section 1 Volcanoes and Earth's Moving Plates
2 sessions
1 block
1. Describe how volcanoes can affect people.
2. List conditions that cause volcanoes to form.
3. Identify the relationship between volcanoes and Earth's moving plates.</td><td>National Content Standards: UCP.1, UCP.2, UCP.3, UCP.5, A.1, A.2, B.2, D.1, D.2, F.3</td><td></td><td>Integrate Career, p. 158
MiniLAB: Modeling Magma Movement, p. 160</td></tr>
<tr><td>Section 2 Types of Volcanoes
2 sessions
1 block
4. Explain how the explosiveness of a volcanic eruption is related to the silica and water vapor content of its magma.
5. List three forms of volcanoes.</td><td>National Content Standards: UCP.1 UCP.2, UCP.3, UCP.5, A.1, A.2, B.2, D.1, F.3</td><td></td><td>Science Online, p. 163
Visualizing Lava, p. 164
Integrate Health, p. 165
MiniLAB: Modeling Volcanic Cones, p. 166
Lab: Identifying Types of Volcanoes, p. 170</td></tr>
<tr><td>Section 3 Igneous Rock Features
3 sessions
1.5 blocks
6. Describe intrusive igneous rock features and how they form.
7. Explain how a volcanic neck and a caldera form.</td><td>National Content Standards: UCP.1 UCP.2, UCP.3, UCP.5, A.1, A.2, B.2, D.1, F.3, G.2</td><td></td><td>Applying Math: Classifying Igneous Rocks, p. 172
Science Online, p. 173
Lab: How do calderas form?, p. 176
Oops! Accidents in Science: Buried in Ash, p. 178</td></tr>
</table>

Glencoe Exclusive! TeacherWorks™ All-In-One Planner and Resource Center

Lab Materials	Reproducible Resources	Section Assessment	Technology
Launch Lab: one-half of a foam ball, metric ruler, permanent marker	**Chapter *FAST FILE* Resources** Foldables Worksheet, p. 15 Directed Reading Overview, p. 17 Note-taking Worksheets, pp. 31–33	Glencoe's Assessment Advantage	TeacherWorks includes: • Interactive Teacher Edition • Lesson Planner with calendar • Access to all program blacklines • Correlations to standards • Web links
MiniLAB: water, 2 clear-plastic cups, olive oil, dropper *Need materials?* Contact Science Kit at 1-800-828-7777 or www.sciencekit.com on the Internet.	**Chapter *FAST FILE* Resources** Transparency Activity, p. 42 MiniLAB, p. 3 Enrichment, p. 28 Reinforcement, p. 25 Directed Reading, p. 18 Transparency Activity, pp. 45–46 Lab Activity, pp. 9–10 **Cultural Diversity,** p. 47	**Portfolio** Science Journal, p. 159 **Performance** MiniLab, p. 160 Applying Skills, p. 161 **Content** Section Review, p. 161	Section Focus Transparency Teaching Transparency Guided Reading Audio Program Interactive Chalkboard CD-ROM
MiniLAB: dry sand or sugar, 2 paper plates, plaster of paris, protractor **Lab:** Table 1, paper, pencil	**Chapter *FAST FILE* Resources** Transparency Activity, p. 43 MiniLAB, p. 4 Enrichment, p. 29 Reinforcement, p. 26 Directed Reading, p. 18 Lab Activity, pp. 11–14 Lab Worksheets, pp. 5–6 **Reading and Writing Skill Activities,** p. 27	**Portfolio** Curriculum Connection, p. 165 **Performance** MiniLAB, p. 166 Applying Skills, p. 169 **Content** Section Review, p. 169	Section Focus Transparency Virtual Labs CD-ROM Guided Reading Audio Program Interactive Chalkboard CD-ROM
Lab: small box, small balloon, paper, newspaper, flour, plastic tubing, clamp for tubing, tape, scissors	**Chapter *FAST FILE* Resources** Transparency Activity, p. 44 Enrichment, p. 30 Reinforcement, p. 27 Directed Reading, pp. 19, 20 Lab Worksheets, pp. 7–8 **Mathematics Skill Activities,** p. 43 **Lab Management and Safety,** p. 38	**Portfolio** Challenge, p. 174 **Performance** Applying Math, p. 172 Applying Math, p. 175 **Content** Section Review, p. 175	Section Focus Transparency Guided Reading Audio Program Interactive Chalkboard CD-ROM Video Lab

End of Chapter Assessment

Glencoe's Assessment Advantage

Blackline Masters	Technology	Professional Series
Chapter Fast File Resources Chapter Review, pp. 35–36 Chapter Tests, pp. 37–40 **Standardized Test Practice,** pp. 27–30	MindJogger Videoquiz Virtual Labs CD-ROM Exam*View*® Pro Testmaker TeacherWorks CD-ROM Interactive Chalkboard CD-ROM	**Performance Assessment in the Science Classroom (PASC)**

chapter 6 Volcanoes

Transparencies

Section Focus

Section 1 Section Focus Transparency **River Ablaze**

If you saw a black-and-white picture of this stream, it probably wouldn't impress you. But in living color, it's astonishing. The surface temperature is about 1,000°C (1,832°F).

1. What is this fiery, flowing substance?
2. Where did this glowing river originate?
3. What will become of this river of fire?

L2

Section 2 Section Focus Transparency **An Island Escape**

Trapped gases under pressure sometimes are released by volcanic eruptions, but there are other ways that Earth vents gases, too. One example is Dominica's Boiling Lake. Located on an island with dormant volcanoes, the lake is heated by an underwater vent that releases hot gases.

1. Why is Boiling Lake shrouded in steam?
2. What might happen if the underwater vents closed?

L2

Section 3 Section Focus Transparency **Ancient Volcanic Rock**

Over many years, erosion by weather and water can create exotic landscapes from the layers of material deposited by volcanoes. Examples of this are the rock homes shown below, which are in an area of Turkey called Cappadocia.

1. Where did the rock formations you see here come from?
2. How is the action of weather and water on a landscape similar to the action of a sculptor working marble?
3. Name some other places that volcanic activity has shaped the landscape.

L2

This is a representation of key blackline masters available in the Teacher Classroom Resources. See Resource Manager boxes within the chapter for additional information.

Assessment

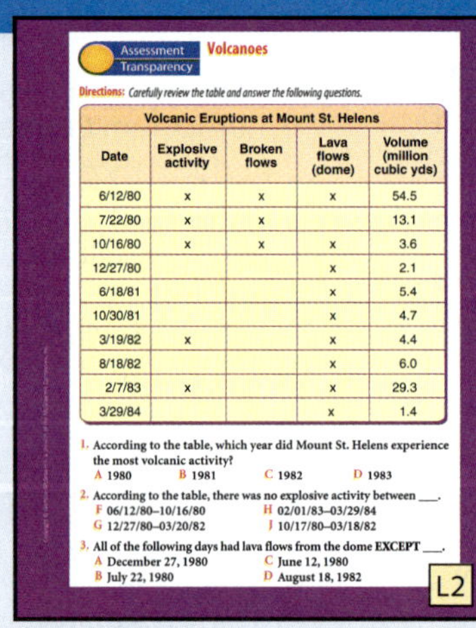
Assessment Transparency **Volcanoes**

Directions: *Carefully review the table and answer the following questions.*

Volcanic Eruptions at Mount St. Helens

Date	Explosive activity	Broken flows	Lava flows (dome)	Volume (million cubic yds)
6/12/80	x	x	x	54.5
7/22/80	x	x		13.1
10/16/80	x	x	x	3.6
12/27/80			x	2.1
6/18/81			x	5.4
10/30/81			x	4.7
3/19/82	x		x	4.4
8/18/82			x	6.0
2/7/83	x		x	29.3
3/29/84			x	1.4

1. According to the table, which year did Mount St. Helens experience the most volcanic activity?
A 1980 B 1981 C 1982 D 1983
2. According to the table, there was no explosive activity between ___.
F 06/12/80–10/16/80 H 02/01/83–03/29/84
G 12/27/80–03/20/82 J 10/17/80–03/18/82
3. All of the following days had lava flows from the dome EXCEPT ___.
A December 27, 1980 C June 12, 1980
B July 22, 1980 D August 18, 1982

L2

Teaching

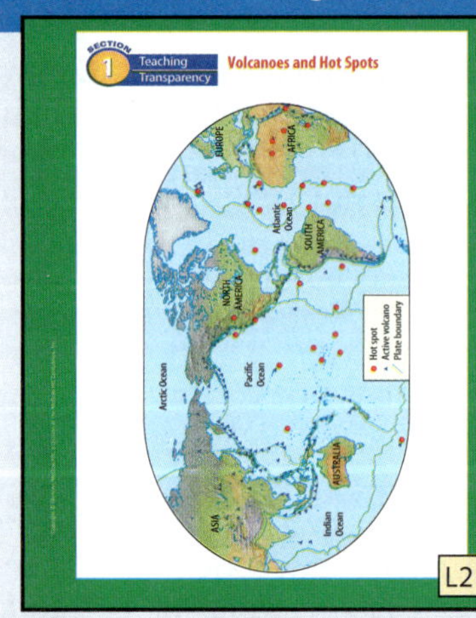
Section 1 Teaching Transparency **Volcanoes and Hot Spots**

L2

Key to Teaching Strategies

The following designations will help you decide which activities are appropriate for your students.

- L1 Level 1 activities should be appropriate for students with learning difficulties.
- L2 Level 2 activities should be within the ability range of all students.
- L3 Level 3 activities are designed for above-average students.
- ELL ELL activities should be within the ability range of English Language Learners.
- COOP LEARN Cooperative Learning activities are designed for small group work.
- LS Multiple Learning Styles logos, as described on page 6T, are used throughout to indicate strategies that address different learning styles.
- P These strategies represent student products that can be placed into a best-work portfolio.
- PBL Problem-Based Learning activities apply real-world situations to learning.

Hands-on Activities

Student Text Lab Worksheet

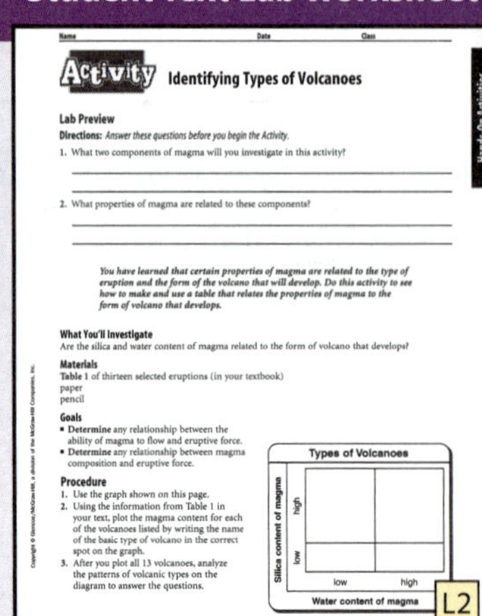
Name Date Class

Activity **Identifying Types of Volcanoes**

Lab Preview

Directions: *Answer these questions before you begin the Activity.*

1. What two components of magma will you investigate in this activity?
2. What properties of magma are related to these components?

You have learned that certain properties of magma are related to the type of eruption and the form of the volcano that will develop. Do this activity to see how to make and use a table that relates the properties of magma to the form of volcano that develops.

What You'll Investigate

Are the silica and water content of magmas related to the form of volcano that develops?

Materials

Table 1 of thirteen selected eruptions (in your textbook)
paper
pencil

Goals

- Determine any relationship between the ability of magma to flow and eruptive force.
- Determine any relationship between magma composition and eruptive force.

Procedure

1. Use the graph shown on this page.
2. Using the information from Table 1 in your text, plot the magma content for each of the volcanoes listed by writing the name of the basic type of volcano in the correct spot on the graph.
3. After you plot all 13 volcanoes, analyze the patterns of volcanic types on the diagram to answer the questions.

L2

Laboratory Activities

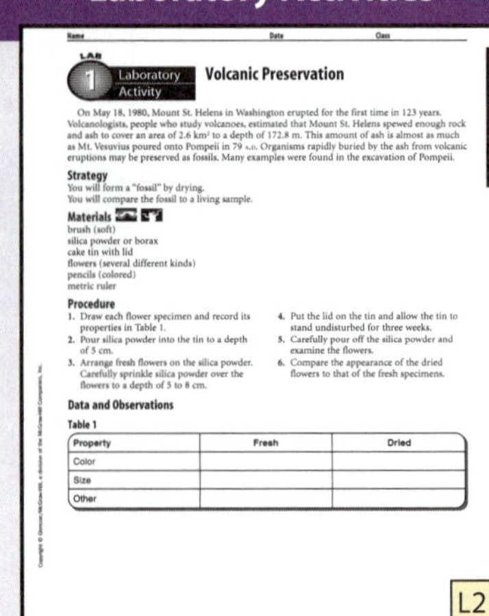
Name Date Class

Lab 1 Laboratory Activity **Volcanic Preservation**

On May 18, 1980, Mount St. Helens in Washington erupted for the first time in 123 years. Volcanologists, people who study volcanoes, estimated that Mount St. Helens spewed enough rock and ash to cover an area of 2.6 km² to a depth of 172.8 m. This amount of ash is almost as much as Mt. Vesuvius poured onto Pompeii in 79 A.D. Organisms rapidly buried by the ash from volcanic eruptions may be preserved as fossils. Many examples were found in the excavation of Pompeii.

Strategy

You will form a "fossil" by drying.
You will compare the fossil to a living sample.

Materials

brush (soft)
silica powder or borax
cake tin with lid
flowers (several different kinds)
pencils (colored)
metric ruler

Procedure

1. Draw each flower specimen and record its properties in Table 1.
2. Pour silica powder into the tin to a depth of 3 cm.
3. Arrange fresh flowers on the silica powder. Carefully sprinkle silica powder over the flowers to a depth of 3 to 4 cm.
4. Put the lid on the tin and allow the tin to stand undisturbed for three weeks.
5. Carefully pour off the silica powder and examine the flowers.
6. Compare the appearance of the dried flowers to that of the fresh specimens.

Data and Observations

Table 1

Property	Fresh	Dried
Color		
Size		
Other		

L2

Meeting Different Ability Levels

Content Outline

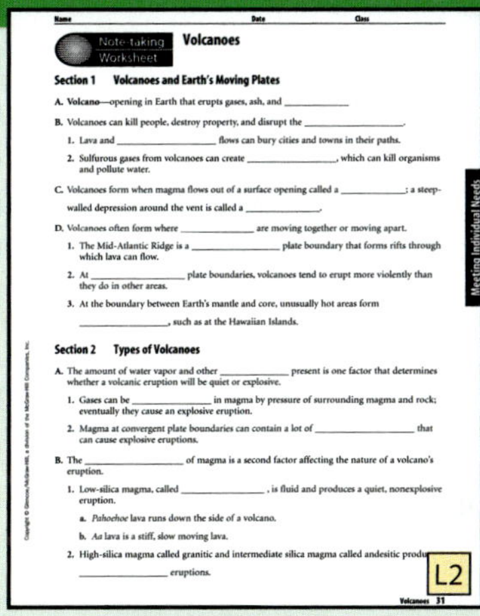

Reinforcement

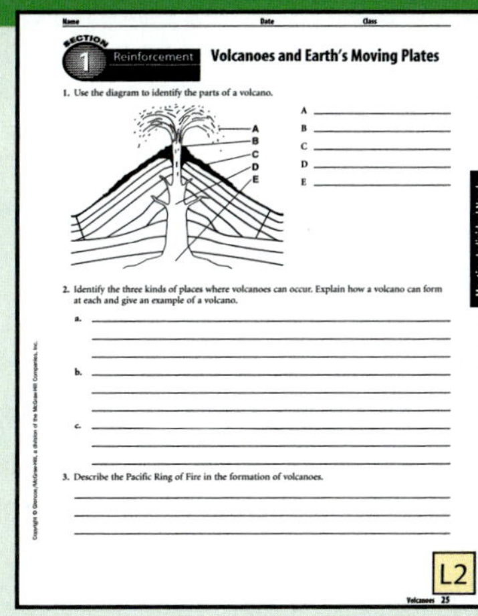

Enrichment

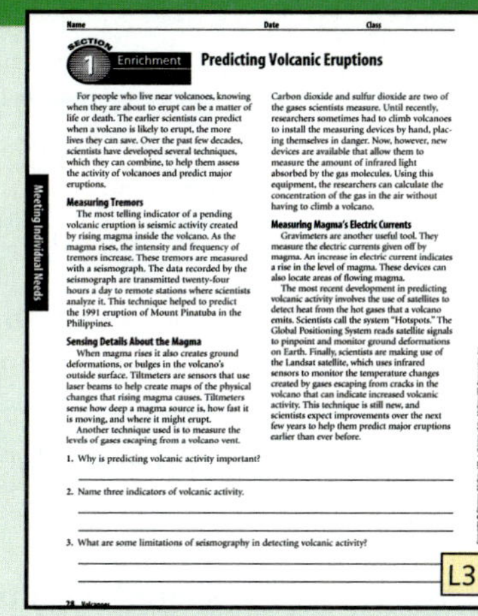

Directed Reading (English/Spanish)

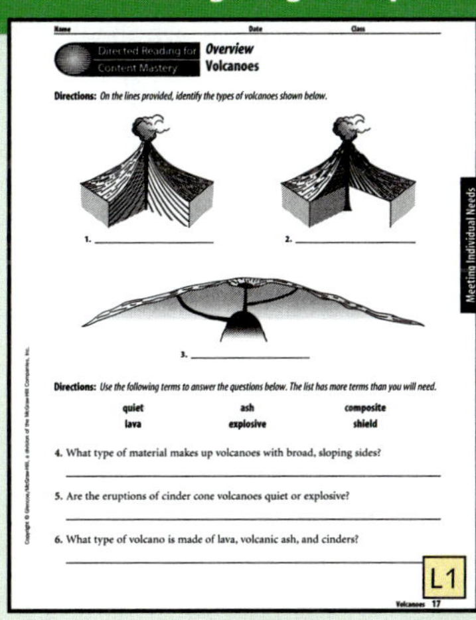

Study Guide

Reading Essentials

Assessment

Test Practice Workbook

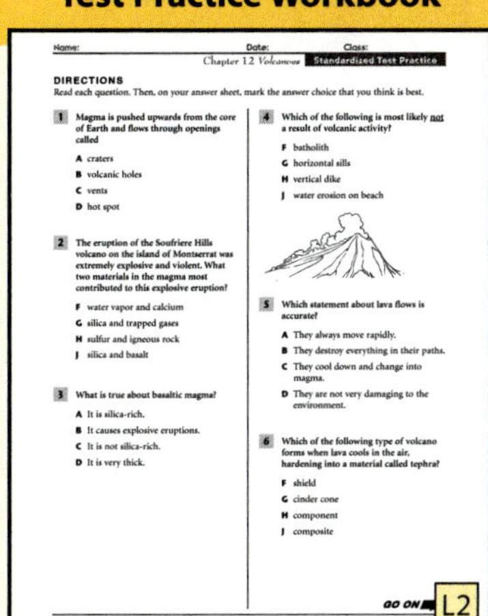

Chapter Review

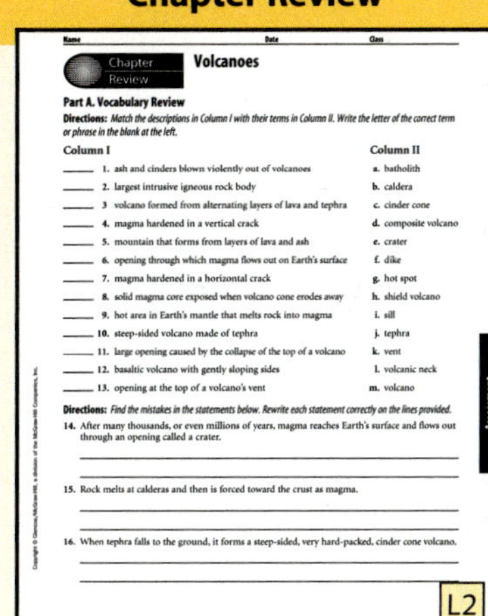

Chapter Tests

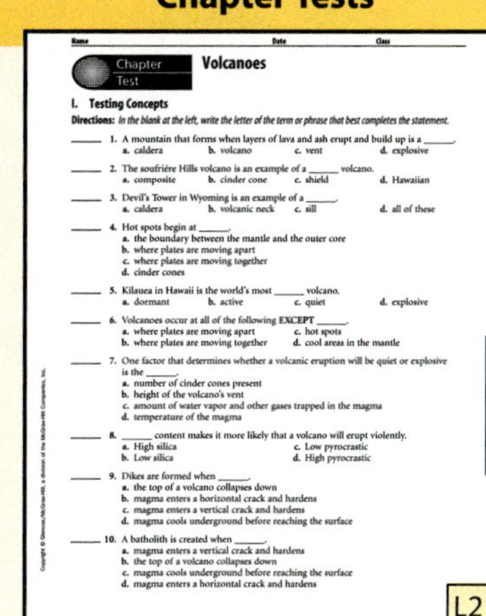

Science Content Background

section 1 Volcanoes and Earth's Moving Plates

How do volcanoes form?

Over 80 percent of Earth's surface, including the portions under the oceans, is of volcanic origin. We now know that most volcanoes form as a result of the movement of Earth's tectonic plates.

Despite the damage they cause, volcanoes have positive effects. Ancient volcanoes are associated with many of our major veins of minerals such as gold, silver, copper, lead, zinc, and diamonds. Volcanic rock weathers into some of the most fertile soil on Earth. Volcanic eruptions early in Earth's history provided the atmosphere and water that make life on Earth possible.

section 2 Types of Volcanoes

Trapped Gases

Most of the gas released from volcanoes is water vapor. Gases make up between one and five percent of the total weight of most magmas.

Composition of Magma

Volcanic eruptions that form at divergent plate boundaries and hot spots tend to expel basaltic lava. This lava flows readily, releases gas easily, and is associated with nonexplosive eruptions. Eruptions along convergent boundaries produce lava that is richer in silica. This lava traps gas more easily and is associated with more dangerous, explosive eruptions.

Teacher to Teacher
Steve Federman
Loveland Middle School
Loveland, Ohio

"Use a non-pudding cake mix and water to serve as an analogy for lava. Tape a grid with 10-cm spacing onto a cookie sheet and cover with plastic wrap. Pour the lava about 10-cm from the high end with the cookie sheet angled at 15°, then 25°. Students compare and record time flow at each 10-cm mark; measure the length, width, and depth of the flow, and observe distinct features such as levees and ridges."

Steve Federman

Forms of Volcanoes

Shield volcanoes form the largest cones. Mauna Loa measures 9 km from top to base and about 200 kilometers from side to side. Mount Rainer, a composite volcano, measures 3 km from top to base and about 20 km from side to side. Cinder cones form the smallest volcanoes. Sunset

Mark Lewis/Stone

Crater measures about 0.3 km from top to base and 1.8 km from side to side.

Materials that comprise tephra are referred to as cinders when they are the size of peas and lapilli when they are the size of walnuts. Materials larger than lapilli are called blocks, or bombs when ejected in a semi-molten state. Volcanic ash is tephra that is smaller than 2 mm.

Not all eruptions of lava come from volcanoes. The Columbia Plateau, located in the northwestern United States, formed from layer after layer of basaltic lava that was extruded from cracks, as fissure eruptions.

Viscous, or thick, lava does not flow easily. As this type of lava is extruded, it often produces a lava dome. Scientists study the growth of lava domes for signs of renewed eruptions. Changes in growth rate of lava domes can indicate changes inside the volcano.

Section 3 Igneous Rock Features

Plutons

Another name for intrusive igneous rock bodies is plutons. Plutons are described by position, shape, and/or size. Plutons that are flat and shaped similar to a tabletop such as dikes and sills are called tabular. Batholiths are large and massive. Sills are concordant. This means they form between layers of surrounding rock. Dikes are discordant; they cut across surrounding rock layers.

Crater Lake

Crater Lake, located in Oregon, formed about 7,000 years ago when a volcano spewed 40–50 cubic kilometers of volcanic material into the air. With the magma chamber then partly emptied, 1,500 meters of the original 3,600-meter cone collapsed. Rainwater filled the caldera to form a lake. Magma forced to the surface after the caldera formed built Wizard Island in the lake.

Darrell Gulin/DRK Photo

chapter content resources

Internet Resources

For additional content background, visit **bookf.msscience.com** to:

- access your book online
- find references to related articles in popular science magazines
- access Web links with related content background
- access current events with science journal topics

Print Resources

Geology Rocks! by Cindy Blobaum, Williamson Publishing, 1999

Phenomena by Henry and Melissa Billings, Jamestown Publishers, 1999

202 Oozing, Bubbling, Dripping & Bouncing Experiments by Janice VanCleave, John Wiley & Sons, Inc., 1996

Volcanoes by Janice VanCleave, John Wiley & Sons, Inc., 1994

chapter 6

Chapter Vocabulary

volcano, p. 156
vent, p. 158
crater, p. 158
hot spot, p. 160
shield volcano, p. 166
tephra, p. 166
cinder cone volcano, p. 166
composite volcano, p. 167
batholith, p. 172
dike, p. 173
sill, p. 173
volcanic neck, p. 173
caldera, p. 174

Science Journal Student answers will vary. Students may already know not all volcanoes begin with explosive, violent eruptions. Some volcanoes, such as Kilauea in Hawaii, erupt in a much less violent, quiet manner. Some volcanoes may begin to show activity by spewing ash or gas from their vents.

INTERACTIVE CHALKBOARD with Image Bank

PowerPoint® Presentations

This CD-ROM is an editable Microsoft® PowerPoint® presentation that includes:

- a pre-made presentation for every chapter
- interactive graphics
- animations
- audio clips
- image bank
- all new section and chapter questions
- Standardized Test Practice
- transparencies
- pre-lab questions for all labs
- Foldables directions
- links to bookf.msscience.com

chapter 6

Volcanoes

chapter preview

sections

Beautiful but Dangerous?

In late October, 2002, earthquakes triggered this vigorous eruption from Mt. Etna, a volcano on the Italian island of Sicily. Strombolian eruptions and lava fountains spewed gas, bombs, blocks, and liquid lava. Ash settled as far away as Libya.

Science Journal Do all volcanoes begin with violent, explosive eruptions? Write about your current beliefs, then do some research and write about your discoveries.

Theme Connection

Energy and Stability and Change The energy involved in the formation of magma, its movement upward, and its flow onto the surface as lava are focal points of the chapter. Stability and change is a secondary theme emphasized in the section on the eruption of volcanoes.

About the Photo

Mt. Etna One of the world's most active volcanoes, Mt. Etna, is a stratovolcano. It consists of many separate vents, cinder cones and domes on the volcano's flanks. Mt. Etna illustrates many volcanic characteristics as it rises above the ocean, creating andesite lava, slow-flowing basaltic magma, explosive lava fountains, Strombolian eruptions, and even smoke rings.

Start-Up Activities

Launch LAB

Map a Volcano

You've seen pictures of volcanoes from the ground, but what would a volcano look like on a map? Volcanoes can be represented on maps that show the elevation of the land, topographic maps.

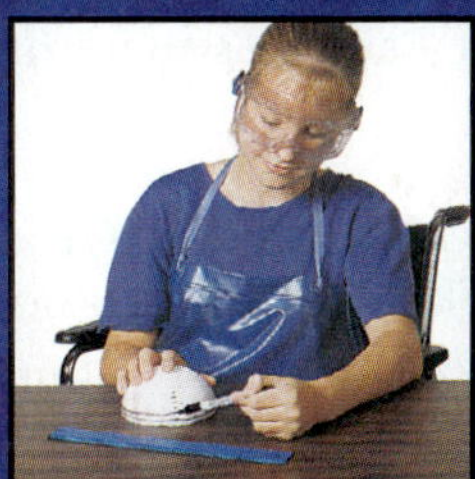

1. Obtain half of a foam ball from your teacher and place it on the top of a table with the flat side down.
2. Using a metric ruler and a permanent marker, mark 1-cm intervals on the foam ball. Start at the base of the ball and mark up at several places around the ball.
3. Connect the marks of equal elevation by drawing a line around the ball at the 1-cm mark, at the 2-cm mark, etc.
4. Look directly down on the top of the ball. Make a drawing of what you see in your Science Journal.
5. **Think Critically** In your Science Journal, write a paragraph that explains how your drawing shows the volcano's general shape.

FOLDABLES Study Organizer

Volcanoes Make the following Foldable to compare and contrast the characteristics of explosive and quiet volcanic eruptions.

STEP 1 **Fold** one sheet of paper lengthwise.

STEP 2 **Fold** into thirds.

STEP 3 **Unfold and draw** overlapping ovals. **Cut** the top sheet along the folds.

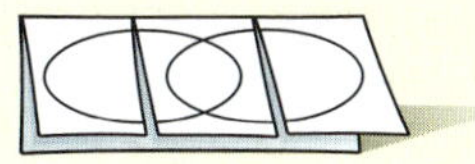

STEP 4 **Label** the ovals *Explosive Eruptions*, *Both*, and *Quiet Eruptions*, as shown.

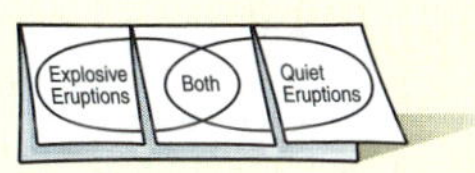

Construct a Venn Diagram As you read the chapter, list the characteristics unique to explosive eruptions under the left tab, those unique to quiet eruptions under the right tab, and those characteristics common to both under the middle tab.

Science Online — Preview this chapter's content and activities at bookf.msscience.com

Launch LAB

Purpose Students will create a topographic map of a model of a volcano. L2

Preparation Cut foam balls in half using a handheld jigsaw or sharp knife.

Materials half of a foam ball, metric ruler, permanent marker

Teaching Strategy As students perform the activity, remind them that real volcanoes have a variety of shapes and elevations. Ask them to think how their lines would look for different volcanoes.

Think Critically

Possible answer: The drawing shows how the volcano is wide at the base and narrow at the top. The lines connect similar elevations of the volcano.

Assessment

Oral Ask students to explain how their drawings differ from the topographic maps of real volcanoes. For real volcanoes, the lines would be more jagged because the elevations around the volcano would not be uniform. Use **Performance Assessment in the Science Classroom,** p. 89.

FOLDABLES Study Organizer — Dinah Zike Study Fold

Student preparation materials for this Foldable are available in the Chapter *FAST FILE* Resources.

1 Motivate

Bellringer

INTERACTIVE CHALKBOARD PowerPoint® Presentations

Section Focus Transparencies also are available on the Interactive Chalkboard CD-ROM. L2 ELL

SECTION 1 Section Focus Transparency **River Ablaze**

If you saw a black-and-white picture of this stream, it probably wouldn't impress you. But in living color, it's astonishing. The surface temperature is about 1,000°C (1,832°F).

1. What is this fiery, flowing substance?
2. Where did this glowing river originate?
3. What will become of this river of fire?

L2

Volcanoes

Tie to Prior Knowledge

Volcanoes Have students brainstorm words they associate with volcanoes. Record their ideas on the board. Possible answers: eruption, lava, ash, heat Discuss how the terms reveal what they think about volcanoes. Tell students that in this section they will learn why volcanoes form.

section 1

Volcanoes and Earth's Moving Plates

as you read

What You'll Learn

- **Describe** how volcanoes can affect people.
- **List** conditions that cause volcanoes to form.
- **Identify** the relationship between volcanoes and Earth's moving plates.

Why It's Important

Volcanoes can be dangerous to people and their communities.

Review Vocabulary
lava: molten rock material flowing from volcanoes onto Earth's surface

New Vocabulary
- volcano
- vent
- crater
- hot spot

What are volcanoes?

A **volcano** is an opening in Earth that erupts gases, ash, and lava. Volcanic mountains form when layers of lava, ash, and other material build up around these openings. Can you name any volcanoes? Did you know that Earth has more than 600 active volcanoes?

Most Active Volcanoes Kilauea (kee low AY ah), located in Hawaii, is the world's most active volcano. For centuries, this volcano has been erupting, but not explosively. In May of 1990, most of the town of Kalapana Gardens was destroyed, but no one was hurt because the lava moved slowly and people could escape. The most recent series of eruptions from Kilauea began in January 1983 and still continues.

The island country of Iceland is also famous for its active volcanoes. It sits on an area where Earth's plates move apart and is known as the land of fire and ice. The February 26, 2000, eruption of Hekla, in Iceland, is shown in **Figure 1.**

Figure 1 This photo of the February 26, 2000, eruption of Hekla shows why Iceland is known as the land of fire and ice.

Section 1 Resource Manager

Chapter *FAST FILE* Resources
- Transparency Activity, pp. 42, 45–46
- Directed Reading for Content Mastery, pp. 17, 18
- Note-taking Worksheets, pp. 31–33
- Enrichment, p. 25
- Reinforcement, p. 25
- MiniLAB, p. 3
- Lab Activity, pp. 9–10

Figure 2 This town on Montserrat was devastated by the eruption of Soufrière Hills volcano.

Effects of Eruptions

When volcanoes erupt, they often have direct, dramatic effects on the lives of people and their property. Lava flows destroy everything in their path. Falling volcanic ash can collapse buildings, block roads, and in some cases cause lung disease in people and animals. Sometimes, volcanic ash and debris rush down the side of the volcano. This is called a pyroclastic flow. The temperatures inside the flow can be high enough to ignite wood. When big eruptions occur, people often are forced to abandon their land and homes. People who live farther away from volcanoes are more likely to survive, but cities, towns, crops, and buildings in the area can be damaged by falling debris.

Human and Environmental Impacts The eruption of Soufrière (sew FREE er) Hills volcano in Montserrat, which began in July of 1995, was one of the largest recent volcanic eruptions near North America. Geologists knew it was about to erupt, and the people who lived near it were evacuated. On June 25, 1997, large pyroclastic flows swept down the volcano. As shown in **Figure 2,** they buried cities and towns that were in their path. The eruption killed 20 people who ignored the evacuation order.

INTEGRATE Earth Science When sulfurous gases from volcanoes mix with water vapor in the atmosphere, acid rain forms. The vegetation, lakes, and streams around Soufrière Hills volcano were impacted significantly by acid rain. As the vegetation died, shown in **Figure 3,** the organisms that lived in the forest were forced to leave or also died.

Figure 3 The vegetation near the volcano on Chances Peak, on the island of Montserrat in the West Indies, was destroyed by acid rain, heat, and ash.

2 Teach

Quick Demo

Toothpaste Volcano

Materials knife or other sharp object, small tube of toothpaste

Estimated Time 10 minutes

Procedure Cut a small slit in the side of a tube of toothpaste. With the cap on, squeeze the tube. The toothpaste oozing out of the slit and down the side of the tube models a quiet eruption of a shield volcano. L1 ELL LS **Visual-Spatial**

Discussion

Living Near a Volcano Most people know that volcanoes erupt, sometimes with deadly results. Yet many people still choose to live near volcanoes. Why do you think this is the case? Possible answers: There haven't been recent eruptions, so the people living there are not afraid; people think that if there is a problem, they will be able to escape in time; some people are unable to move because of economic or other reasons.

Teacher FYI

Pyroclastic Flows Pyroclastic flows of hot gases, ash, and rock can roll down the sides of a volcano at speeds of up to 200 km/h. Pyroclastic flows are responsible for most of the deaths and injuries caused by volcanic eruptions in the twentieth century. A pyroclastic flow caused almost all of the 29,000 deaths during Mount Pelée's eruption on the island of Martinique in 1902.

Active Reading

Problem/Solution Journal Have students label the left side of their paper *Problem* and the right side *Consequences.* Have students identify a problem, brainstorm alternatives, choose a solution, anticipate stumbling blocks, and propose arguments. Have students apply this strategy to dealing with the effects of a volcanic eruption. L2

Differentiated Instruction

Challenge Students may be familiar with human-made sources of acid rain. Therefore, the fact that acid rain does occur naturally may come as a surprise to them. Encourage them to research other natural causes of acid rain. They should then prepare a short newspaper article that addresses the fact that acid rain is natural. L3

INTEGRATE Career

Volcanologists The largest volcano in the solar system is Olympus Mons on Mars. This inactive volcano covers an area about the size of Ohio. Triton, a moon of Neptune, has volcanic-like eruptions of nitrogen and methane. Earth's Moon has ancient lava flows called maria. Evidence suggests that Venus has active volcanoes on its surface.

Discussion

Cooling Magma Why do you think magma cools quickly when it is forced out of a vent on Earth's surface? The temperature at Earth's surface is much lower than the temperature deep inside Earth, causing the rock to solidify as it is exposed to the air and water.

Reading Check

Answer It is less dense than the surrounding rock.

Visual Learning

Figure 4 Do people who live near volcanoes have to be aware of areas other than the vent when protecting themselves against lava flows? Yes, lava can also flow from the branch pipes that lead to cracks in the volcano's sides.

Fun Fact

Jupiter's moon Io is the most volcanically active object in the solar system.

INTEGRATE Career

Volcanology Volcanologists research many aspects of volcanoes, including space volcanoes. Io, a moon of Jupiter, has many active volcanoes. Become an amateur volcanologist; choose an aspect of volcanology that interests you, research your topic, and then create an exciting news broadcast to share the information with the class.

How do volcanoes form?

What happens inside Earth to create volcanoes? Why are some areas of Earth more likely to have volcanoes than others? Deep inside Earth, heat and pressure changes cause rock to melt, forming liquid rock or magma. Some deep rocks already are melted. Others are hot enough that a small rise in temperature or drop in pressure can cause them to melt and form magma. What makes magma come to the surface?

Magma Forced Upward Magma is less dense than the rock around it, so it is forced slowly toward Earth's surface. You can see this process if you turn a bottle of cold syrup upside down. Watch the dense syrup force the less dense air bubbles slowly toward the top.

Reading Check *Why is magma forced toward Earth's surface?*

After many thousands or even millions of years, magma reaches Earth's surface and flows out through an opening called a **vent.** As lava flows out, it cools quickly and becomes solid, forming layers of igneous rock around the vent. The steep-walled depression around a volcano's vent is the **crater. Figure 4** shows magma being forced out of a volcano.

Figure 4 This cutaway diagram shows how a volcano is formed and how magma from inside Earth is forced to the surface.

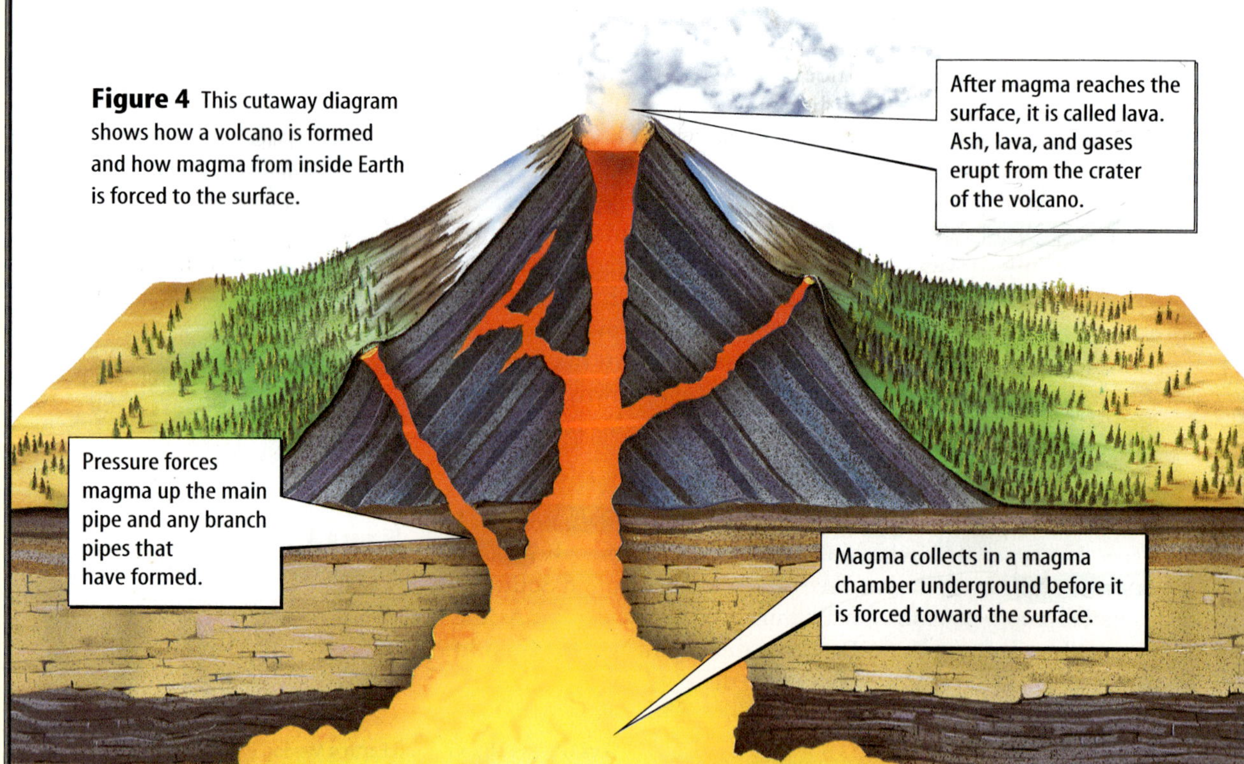

Cultural Diversity

Monitoring Philippine Volcanoes The Philippines is one of the most volcanically active places on Earth. Pinatubo and Mayon are two of the most active volcanoes on the islands. Direct students to research the work of the Philippines Institute of Volcanology and Seismology (PHIVOLS). What do the scientists there do, and why is their work important in a country such as the Philippines? PHIVOLS has nine observatories that transmit seismic phase readings, visual observations, wet tilt readings, and meteorological information twice a day to a central office. This data is interpreted to help warn people of possible volcanic eruptions. L2 LS **Linguistic**

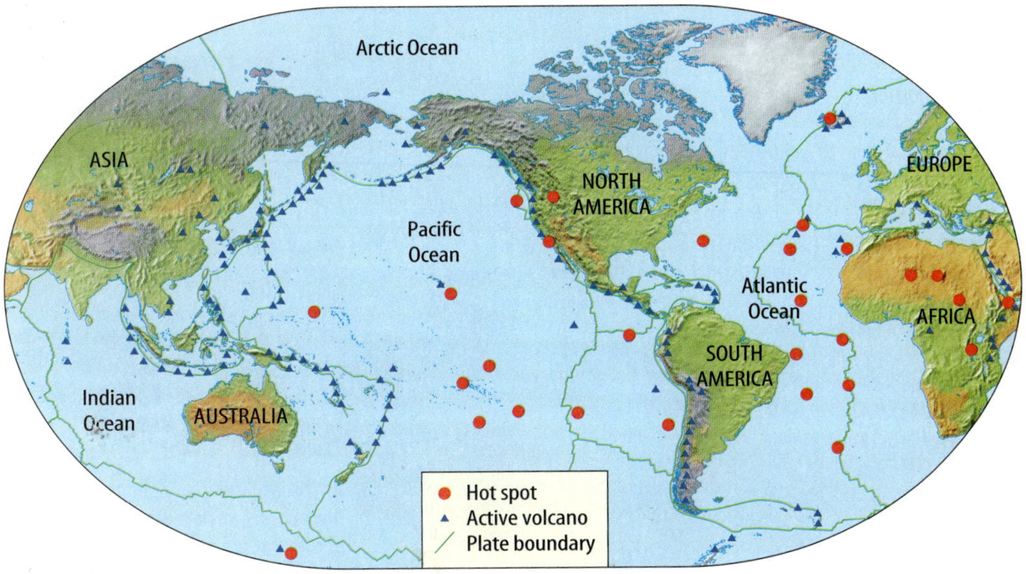

Figure 5 This map shows the locations of volcanoes, hot spots, and plate boundaries around the world. The Ring of Fire is a belt of active volcanoes that circles the Pacific Ocean.

Where do volcanoes occur?

Volcanoes often form in places where plates are moving apart, where plates are moving together, and at locations called hot spots. You can find locations of active volcanoes at plate boundaries and at hot spots on the map in **Figure 5.** Many examples can be found of volcanoes around the world that form at these three different kinds of areas. You'll explore volcanoes in Iceland, on the island of Montserrat, and in Hawaii.

Divergent Plate Boundaries Iceland is a large island in the North Atlantic Ocean. It is near the Arctic Circle and therefore has some glaciers. Iceland has volcanic activity because it is part of the Mid-Atlantic Ridge.

The Mid-Atlantic Ridge is a divergent plate boundary, which is an area where Earth's plates are moving apart. When plates separate, they form long, deep cracks called rifts. Lava flows from these rifts and is cooled quickly by seawater. **Figure 6** shows how magma rises at rifts to form new volcanic rock. As more lava flows and hardens, it builds up on the seafloor. Sometimes, the volcanoes and rift eruptions rise above sea level, forming islands such as Iceland. In 1963, the new island Surtsey was formed during a volcanic eruption.

Figure 6 This diagram shows how volcanic activity occurs where Earth's plates move apart.

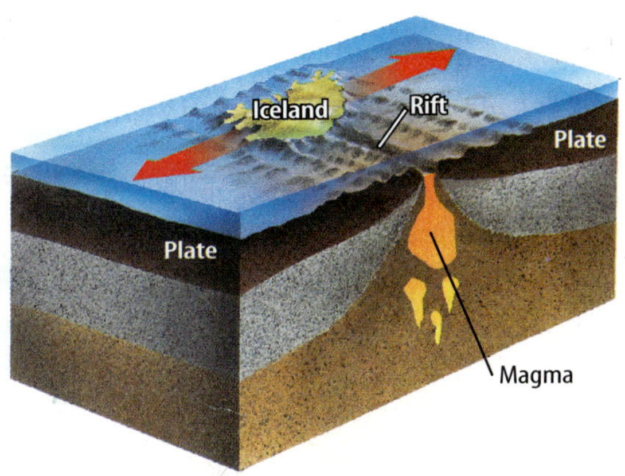

Visual Learning

Figure 5 Where in the United States are you most likely to find volcanoes? Along the Pacific coast and in Hawaii.

Quick Demo

Precursors to Eruptions

Materials hand held bike pump, large balloon, cloth

Estimated Time 10 minutes

Procedure Attach an air pump to a balloon neck and place the balloon under a large piece of cloth. As you inflate the balloon, explain that the bulge in the cloth represents magma that causes doming on a volcano. What is likely to happen if this build-up of magma continues? The volcano would eventually erupt. Inform students that bulging and earthquakes are often precursors to eruptions. L2 LS **Visual-Spatial**

IDENTIFYING Misconceptions

Active Volcanoes Some students may think that an active volcano is one that is now erupting or has erupted very recently. Volcanologists consider an active volcano one that has erupted within recorded history. As a result, volcanoes that have been "quiet" for many decades are still considered active and could pose a danger to those nearby.

Science Journal

Energy from Volcanoes Scientists have discovered communities that include giant clams and worms living near mid-ocean ridges. Have students find out how these communities obtain energy from volcanic activity, and write a summary of their findings in their Science Journals. Bacteria oxidize the hydrogen sulfide in plumes of hot water streaming from cracks along the ridge. L2 LS **Linguistic** P

Differentiated Instruction

English-Language Learners Working with a partner, have students study about a volcano from their native country or region. Ask students to prepare a creative buddy interview of questions and answers to include 15 vocabulary words and their meanings from the chapter. L2

Mini LAB

Purpose Students will model the movement of magma toward Earth's surface. L2

Materials two small transparent plastic cups, small amount of olive oil, water, dropper

Teaching Strategy Have students squeeze the oil into the water as slowly as possible to avoid stirring the water.

Analysis

1. The oil is forced upward to the surface of the water.
2. Oil is less dense than water, as magma is less dense than surrounding rock in Earth's crust.

Assessment

Oral Based on observations in this activity, ask students to explain why old, cool, oceanic plates sink beneath other plates at subduction zones. The old, cool, oceanic plates are more dense. Use **Performance Assessment in the Science Classroom,** p. 89.

Try at Home

Activity

Living Near a Volcano Many people choose to live in the shadow of a volcano. Give the students five minutes to quickly compile a list of disadvantages to living near a volcano. Next, have students brainstorm a list of advantages to living near a volcano. Remind students to keep these advantages in mind while they work in small groups to create a volcano evacuation plan. The plan may include a first aid kit with protective masks, a location for meeting family members, safety routes away from the volcano and lava flows, and toward a safe location for evacuation and rescue. Ask students how long is a reasonable time limit for evacuation? Time is dependent on the force of eruption and type of lava flow.

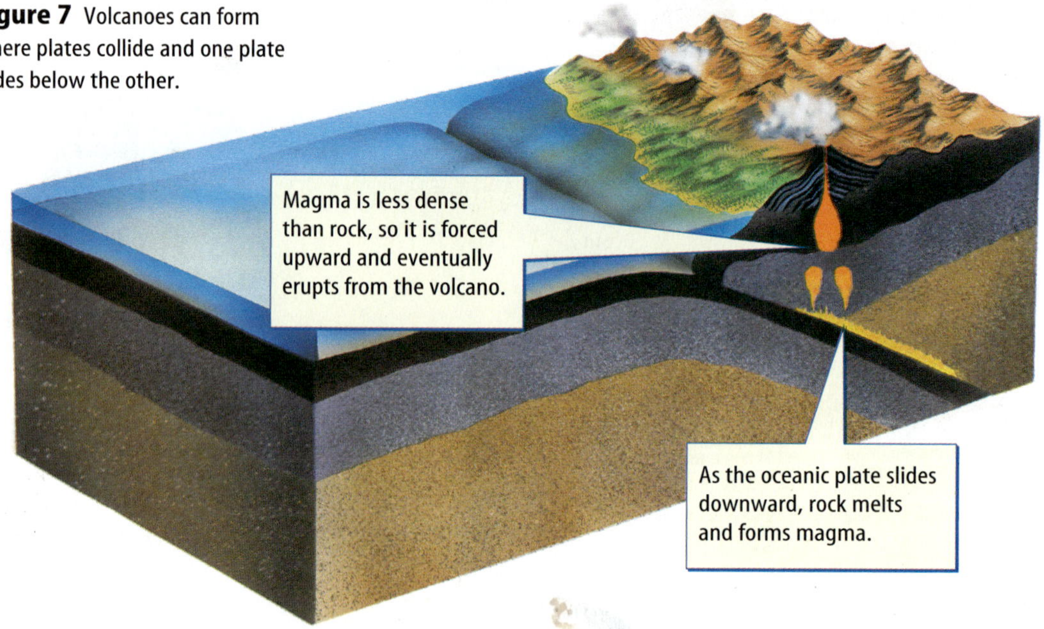

Figure 7 Volcanoes can form where plates collide and one plate slides below the other.

Mini LAB

Modeling Magma Movement

Procedure

1. Pour **water** into a **clear-plastic cup.**
2. Pour a small amount of **olive oil** into a separate plastic cup.
3. Extract a small amount of oil with a **dropper.**
4. Submerge the dropper tip into the water cup and slowly squeeze oil drops into the water.

Analysis

1. Describe what happened to the oil.
2. How do your observations compare with the movement of magma within Earth's crust?

Try at Home

Convergent Plate Boundaries Places where Earth's plates move together are called convergent plate boundaries. They include areas where an oceanic plate slides below a continental plate as in **Figure 7,** and where one oceanic plate slides below another oceanic plate. The Andes in South America began forming when an oceanic plate started sliding below a continental plate. Volcanoes that form on convergent plate boundaries tend to erupt more violently than other volcanoes do.

Magma forms when the plate sliding below another plate gets deep enough and hot enough to melt partially. The magma then is forced upward to the surface, forming volcanoes like Soufrière Hills on the island of Montserrat.

Hot Spots The Hawaiian Islands are forming as a result of volcanic activity. However, unlike Iceland, they haven't formed at a plate boundary. The Hawaiian Islands are in the middle of the Pacific Plate, far from its edges. What process could be forming them?

It is thought that some areas at the boundary between Earth's mantle and core are unusually hot. Hot rock at these areas is forced toward the crust where it melts partially to form a **hot spot.** The Hawaiian Islands sit on top of a hot spot under the Pacific Plate. Magma has broken through the crust to form several volcanoes. The volcanoes that rise above the water form the Hawaiian Islands, shown in **Figure 8.**

Curriculum Connection

Geography Display a map of the world. Have each student locate one active volcano on the map and place a pin with the name of the volcano at the proper location. When the map is complete, ask students to list the countries that have the greatest number of active volcanoes. Have them note whether these countries are around the Ring of Fire. L2 LS **Visual-Spatial**

Figure 8 This satellite photo shows five of the Hawaiian Islands, which actually are volcanoes. **Explain** *why they are in a relatively straight line.*

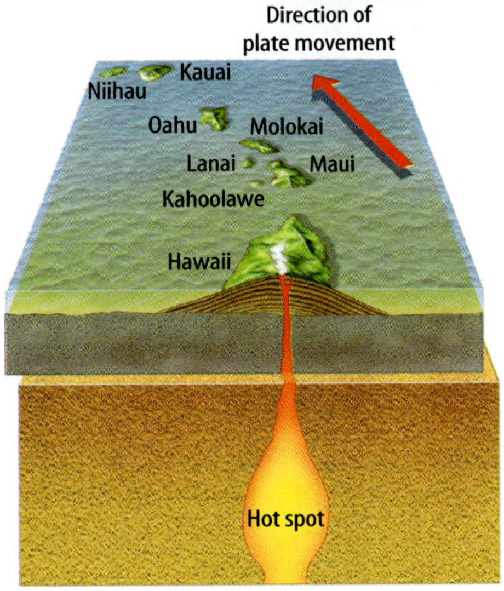

This illustration shows that the Hawaiian Islands were formed over a hot spot.

The Hawaiian Islands As you can see in **Figure 8,** the Hawaiian Islands are all in a line. This is because the Pacific Plate is moving over a stationary hot spot. Kauai, the oldest Hawaiian island, was once located where the big island, Hawaii, is situated today. As the plate moved, Kauai moved away from the hot spot and became dormant. As the Pacific Plate continued to move, the islands of Oahu, Molokai, Maui, and Hawaii were formed. The Hawaiian Islands formed over a period of about 5 million years.

section 1 review

Summary

What are volcanoes?

- A volcano is an opening in Earth's surface that erupts gases, ash, and lava.

Effects of Eruptions

- Direct effects of volcanic eruptions can be caused by lava flows, pyroclastic flows, and falling ash.
- Volcanic eruptions also produce indirect effects, such as acid rain.

How do volcanoes form?

- Volcanoes form when magma is forced up and flows onto Earth's surface as lava.
- A crater is a steep-walled depression around a volcano's vent.

Where do volcanoes occur?

- Volcanoes form where one plate sinks beneath another plate, where two plates are moving apart, and at hot spots.

Self Check

1. **Explain** why volcanoes are commonly found at the edges of Earth's moving plates.
2. **Describe** what effects pyroclastic flows have on people.
3. **Explain** why lava cools rapidly along a mid-ocean ridge. How might underwater lava differ from surface lava?
4. **Describe** the processes that cause Soufrière Hills volcano to erupt.
5. **Think Critically** If the Pacific Plate stopped moving, what might happen to the Big Island of Hawaii?

Applying Skills

6. **Concept Map** Make a concept map that shows how the Hawaiian Islands formed. Use the following phrases: *volcano forms, plate moves, volcano becomes dormant,* and *new volcano forms.* Draw and label an illustration of this process.

Science Online bookf.msscience.com/self_check_quiz

Caption Answer

Figure 8 The plate moves over the hot spot in one direction, leaving volcanoes in a straight line as it passes.

3 Assess

DAILY INTERVENTION

Check for Understanding

Kinesthetic In groups of two, have students demonstrate the movement of the Pacific Plate over a hot spot. One student should represent the stationary hot spot with their hands. The second student should use their hands to show the movement of the plate. They should be able to also explain what is occurring in their model. L2

Reteach

Subduction Have students draw a cross section of an area of Earth's crust where one plate is forced under another. Have them label the diagram to indicate where volcanoes form. L2 ELL IN **Visual-Spatial**

Assessment

Performance Have students draw concept maps that include the three main types of geological areas where volcanoes form. Convergent and divergent plate boundaries and hot spots Use **Performance Assessment in the Science Classroom,** p. 161.

section 1 review

1. Volcanoes occur most often at plate boundaries where magma flows upward toward the surface.
2. They can flow down the sides of volcanoes, burning most things and burying them under ash.
3. Lava flows from underwater cracks at mid-ocean ridges; the water cools lava quickly forming bubble-like lava, called pillow lava.
4. The North and South American plates are sinking under the converging Caribbean plate. The sinking plate melts, producing magma that is forced upward to form volcanoes such as Soufriére Hills.
5. The island would grow larger as volcanoes on it would continue to produce lava.
6. Volcano forms → plate moves → volcano becomes dormant → new volcano forms.

section 2 Types of Volcanoes

1 Motivate

Bellringer

Section Focus Transparencies also are available on the Interactive Chalkboard CD-ROM. L2 ELL

SECTION 2 Section Focus Transparency **An Island Escape**

Trapped gases under pressure sometimes are released by volcanic eruptions, but there are other ways that Earth vents gases, too. One example is Dominica's Boiling Lake. Located on an island with dormant volcanoes, the lake is heated by an underwater vent that releases hot gases.

1. Why is Boiling Lake shrouded in steam?
2. What might happen if the underwater vents closed?

L2

Volcanoes

Tie to Prior Knowledge

Volcanic Eruptions Ask students to recall volcanic eruptions they have seen or heard about. What did those eruptions do? Some students will recall violent, destructive eruptions such as Mount St. Helens, while others will recall lava flows such as those that ooze over the landscape near Kilauea. Use this information to reinforce the idea that eruptions differ. Tell students they will find out why they differ in this section.

as you read

What **You'll Learn**

- **Explain** how the explosiveness of a volcanic eruption is related to the silica and water vapor content of its magma.
- **List** three forms of volcanoes.

Why **It's Important**

If you know the type of volcano, you can predict how it will erupt.

Review Vocabulary
magma: hot, melted rock material beneath Earth's surface

New Vocabulary
- shield volcano
- tephra
- cinder cone volcano
- composite volcano

What controls eruptions?

Some volcanic eruptions are explosive, like those from Soufrière Hills volcano, Mount Pinatubo, and Mount St. Helens. In others, the lava quietly flows from a vent, as in the Kilauea eruptions. What causes these differences?

Two important factors control whether an eruption will be explosive or quiet. One factor is the amount of water vapor and other gases that are trapped in the magma. The second factor is how much silica is present in the magma. Silica is a compound composed of the elements silicon and oxygen.

Trapped Gases When you shake a soft-drink container and then quickly open it, the pressure from the gas in the drink is released suddenly, spraying the drink all over. In the same way, gases such as water vapor and carbon dioxide are trapped in magma by the pressure of the surrounding magma and rock. As magma nears the surface, it is under less pressure. This allows the gas to escape from the magma. Gas escapes easily from some magma during quiet eruptions. However, gas that builds up to high pressures eventually causes explosive eruptions such as the one shown in **Figure 9.**

Figure 9 Mount St. Helens erupted on May 18, 1980.

Section 2 Resource Manager

Chapter *FAST FILE* Resources
- Transparency Activity, p. 43
- Directed Reading for Content Mastery, p. 18
- Reinforcement, p. 26
- MiniLAB, p. 4
- Reinforcement, p. 29
- Lab Worksheets, pp. 5–6
- Lab Activity, pp. 11–14

Home and Community Involvement, p. 49

Water Vapor The magma at some convergent plate boundaries contains a lot of water vapor. This is because oceanic plate material and some of its water slide under other plate material at some convergent plate boundaries. The trapped water vapor in the magma can cause explosive eruptions.

Composition of Magma

The second major factor that affects the nature of the eruption is the composition of the magma. Magma can be divided into two major types—silica poor and silica rich.

Quiet Eruptions Magma that is relatively low in silica is called basaltic magma. It is fluid and produces quiet, nonexplosive eruptions such as those at Kilauea. This type of lava pours from volcanic vents and runs down the sides of a volcano. As this *pahoehoe* (pa-HOY-hoy) lava cools, it forms a ropelike structure. If the same lava flows at a lower temperature, a stiff, slowly moving *aa* (AH-ah) lava forms. In fact, you can walk right up to some aa lava flows on Kilauea.

Figure 10 shows some different types of lava. These quiet eruptions form volcanoes over hot spots such as the Hawaiian volcanoes. Basaltic magmas also flow from rift zones, which are long, deep cracks in Earth's surface. Many lava flows in Iceland are of this type. Because basaltic magma is fluid when it is forced upward in a vent, trapped gases can escape easily in a nonexplosive manner, sometimes forming lava fountains. Lavas that flow underwater form pillow lava formations. They consist of rock structures shaped like tubes, balloons, or pillows.

Science Online

Topic: Kilauea Volcano
Visit bookf.msscience.com for Web links to information about Kilauea volcano in Hawaii.

Activity On a map of the Hawaiian Islands, identify the oldest and most recent islands. Next, indicate where Kilauea volcano is located on the Big Island. Do you see a directional pattern? Indicate on your Hawaiian map where you believe the next Hawaiian island will form.

42 seconds later

53 seconds later

2 Teach

Visual Learning

Figure 9 Ask students to describe how Mount St. Helens changed as the volcano erupted. Accept any answer supported by the photos. In less than one minute a part of the top of the mountain was removed.

Inquiry Lab

Viscosity

Purpose To explore the properties of different types of magma, students will manipulate and observe a variety of cooking ingredients. L2

Possible Materials paper plates, candy thermometer, tub of ice, tub of very warm water, paper cups containing molasses, honey, hot fudge sauce, water, corn syrup

Estimated Time 40 minutes

Teaching Strategies

- Students can model different magma compositions using a variety of substances of different viscosity. They should observe the behavior of each substance as it moves around a paper plate. Ask students design a chart or scale to help compare and contrast what they observed with each substance.
- Students can conduct a series of experiments that changes the angle at which the plate is placed, representing the steepness of the sides of the volcano.
- They can chill the fluids in the tub of ice, and warm the fluids in the warm tub of water to observe how the viscosity changes with temperature.
- Allow students to explore other questions about magma and viscosity as they arise.
- Students should be able to conclude which magma is more likely to cause explosive eruptions, and which would cause basaltic magma based on viscosity.

For additional inquiry activities, see *Science Inquiry Labs*.

Curriculum Connection

History The eruption of Mount St. Helens in 1980 was a major event that made headlines all over the world. Assign groups of students to look for newspaper and magazine articles that discuss different aspects of the volcano either before, during, or in the aftermath of the eruption. Have groups work together to make a class report on the eruption. L2 COOP LEARN IS **Interpersonal and Naturalist**

NATIONAL GEOGRAPHIC

Visualizing Lava

Have students examine the pictures and read the captions. Then ask the following questions.

Which type of lava of the three shown probably cools most rapidly? Explain your answer. Pillow lava cools most rapidly because water conducts heat away more efficiently than does air.

Viscosity is a property of fluids. Viscous fluids resist flowing, while less viscous fluids flow easily. For example, a milkshake is more viscous than milk.

Which type of lava flow, an aa flow or a pahoehoe flow, is more viscous? an aa flow

Which kind of lava flow, an aa or a pahoehoe flow, do you think would present more of a danger to slow-moving animals? Why? Possible answer: A pahoehoe flow might present more of a danger because it is less viscous and so moves faster.

Activity

Internal Pressure Place a 16-ounce clear, plastic soda bottle in the center of a cookie sheet. Pour 1/2 cup of tap water into the bottle. Break two antacid tablets into pieces and place them in the bottle. Quickly secure a small balloon onto the mouth of the bottle. Observe the contents of the bottle, the balloon, and the reaction. Have students predict what might happen when the balloon is removed, more tablets are added, the water is heated, or the bottle is shaken. How is this model like a volcano eruption? Gas is produced when the water and antacid tablets are mixed. Covering the bottle with a balloon represents the pressure build up of gases in a volcano. Heating or shaking the bottle may increase the reaction time, similar to plate tectonic movement. Adding tablets represents different magma composition that would effect types of eruption.

NATIONAL GEOGRAPHIC VISUALIZING LAVA

Figure 10

Lava rarely travels faster than a few kilometers an hour. Therefore, it poses little danger to people. However, homes and property can be damaged. On land, there are two main types of lava flows—aa and pahoehoe. When lava comes out of cracks in the ocean floor, it is called pillow lava. The lava cooling here came from a volcanic eruption on the island of Hawaii.

Aa flows, like this one on Mount Etna in Italy, carry sharp angular chunks of rock called scoria. Aa flows move slowly and are intensely hot.

Pillow lava occurs where lava oozes out of cracks in the ocean floor. It forms pillow-shaped lumps as it cools. Pillow lava is the most common type of lava on Earth.

Pahoehoe flows, like this one near Kilauea's Mauna Ulu Crater in Hawaii, are more fluid than aa flows. They develop a smooth skin and form ropelike patterns when they cool.

Differentiated Instruction

Challenge Encourage students to research the impact the eruption of Krakatoa had on the global climate. They might then add the impact of more andesitic volcanoes (such as Mt. St. Helens) to their research. Ask them to create a television weather report about the unusual weather caused by such eruptions. L3

Figure 11 Magmas that are rich in silica produce violent eruptions.

Violent eruptions, such as this one in Alaska, often produce a lot of volcanic ash.

Magnification: 450×

This color enhanced view of volcanic ash, from a 10 million year old volcano in Nebraska, shows the glass particles that make up ash.

X 11/17/08

Explosive Magma Silica-rich, or granitic, magma produces explosive eruptions such as those at Soufrière Hills volcano. This magma sometimes forms where Earth's plates are moving together and one plate slides under another. As the plate that is sliding under the other goes deeper, some rock is melted. The magma is forced upward by denser surrounding rock, comes in contact with the crust, and becomes enriched in silica. Silica-rich granitic magma is thick, and gas gets trapped inside, causing pressure to build up. When an explosive eruption occurs, as shown in **Figure 11,** the gases expand rapidly, often carrying pieces of lava in the explosion.

Reading Check *What type of magmas produce violent eruptions?*

Some magmas have an andesitic composition. Andesitic magma is more silica rich than basaltic magma is, but it is less silica rich than granitic magma. It often forms at convergent plate boundaries where one plate slides under the other. Because of their higher silica content, they also erupt more violently than basaltic magmas. One of the biggest eruptions in recorded history, Krakatau, was primarily andesitic in composition. The word *andesitic* comes from the Andes, which are mountains located along the western edge of South America, where andesite rock is common. Many of the volcanoes encircling the Pacific Ocean also are made of andesite.

INTEGRATE Health

Volcanic Ash When volcanoes erupt, ash often is spread over a great distance. People who live near volcanoes must be careful not to inhale too much of the ash particles because the particles can cause respiratory problems. In your Science Journal, describe what people can do to prevent exposure to volcanic ash.

Teacher FYI

Soufriére Hills Volcano When Soufriére Hills volcano erupted on Montserrat, many of the inhabitants were moved to the far side of the island. Others were removed from the island entirely. Many cities, including the capital, Plymouth, were covered by volcanic ash.

Reading Check

Answer silica-rich, or granitic, magma

INTEGRATE Health

Volcanic Ash People can wear masks over their noses and mouths made from material that will not allow the ash to penetrate. People also can wear long sleeves and long pants and stay indoors.

Research Have students research specific respiratory problems that might arise from prolonged exposure to volcanic ash. They should also include a list of possible preexisting conditions that could be aggravated by the ash in the atmosphere. L2

Science Journal

Volcanoes and You Encourage students to use maps and other resources to determine whether or not volcanic activity might affect your geographic location. Have students write in their Science Journals what they discovered, and whether or not they were surprised by the information. L2 LS **Intrapersonal**

Curriculum Connection

Literature Have each student read a book or magazine article about a famous volcano, such as Mt. St. Helens, Soufriére Hills, Vesuvius, or Kilauea. Have them examine how the eruptions affect nearby people. Have students write short stories describing what it would have been like to be near the volcano when it erupted. L2 LS P **Intrapersonal**

Mini LAB

Purpose Students will make models of cinder cone and shield volcanoes to learn about their shapes. L2

Materials sugar or sand, plaster of paris, paper plates, protractor

Teaching Strategies

- Mix the plaster of paris so that it flows but isn't runny.
- Students should pour the sand or sugar slowly.
- Review how to measure angles with a protractor.

Analysis

The model with steeper sides represents a cinder cone volcano.

Assessment

Performance Have students draw a cross-sectional diagram of their models in their Science Journals and identify the type of volcano each represents. The sand or sugar forms steeper sides and represents a cinder cone volcano. The plaster of paris model has gently sloping sides and represents a shield volcano. Use **Performance Assessment in the Science Classroom,** p. 127.

Visual Learning

Figure 12 Why is this type of volcano called a "shield" volcano? Its shape is broad and flat, like that of a shield.

Mini LAB

Modeling Volcanic Cones

Procedure

1. Pour **dry sand** or **sugar** onto one spot on a **paper plate. WARNING:** *Do not taste, eat, or drink any materials used in the lab.*
2. Mix a batch of **plaster of paris** and pour it onto one spot on another paper plate.
3. Allow the plaster of paris to dry. Use a **protractor** to measure the slope angles of the sides of the volcanoes.

Analysis

What form of volcano is represented by the model with steeper sides?

Forms of Volcanoes

A volcano's form depends on whether it is the result of a quiet or an explosive eruption and the type of lava it is made of—basaltic, granitic, or andesitic (intermediate). The three basic types of volcanoes are shield volcanoes, cinder cone volcanoes, and composite volcanoes.

Shield Volcano Quiet eruptions of basaltic lava spread out in flat layers. The buildup of these layers forms a broad volcano with gently sloping sides called a **shield volcano,** as seen in **Figure 12.** The Hawaiian Islands are examples of shield volcanoes. Basaltic lava also can flow onto Earth's surface through large cracks called fissures. This type of eruption forms flood basalts, not volcanoes, and accounts for the greatest volume of erupted volcanic material. The basaltic lava flows over Earth's surface, covering large areas with thick deposits of basaltic igneous rock when it cools. The Columbia Plateau located in the northwestern United States was formed in this way. Much of the new seafloor that originates at mid-ocean ridges forms as underwater flood basalts.

Cinder Cone Volcano Explosive eruptions throw lava and rock high into the air. Bits of rock or solidified lava dropped from the air are called **tephra** (TEH fruh). Tephra varies in size from volcanic ash, to cinders, to larger rocks called bombs and blocks. When tephra falls to the ground, it forms a steep-sided, loosely packed **cinder cone volcano,** as seen in **Figure 13.**

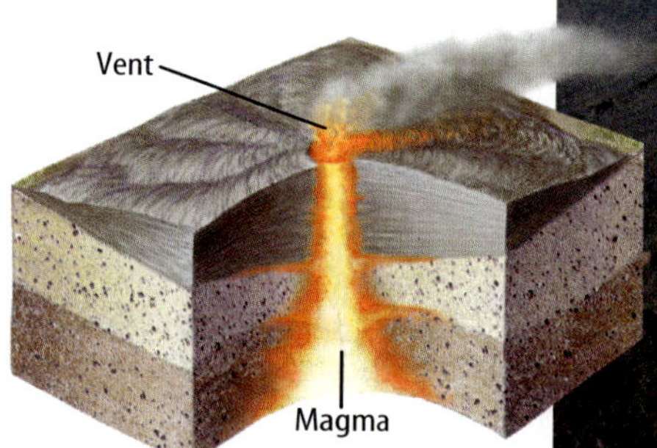

Figure 12 A shield volcano like Mauna Loa, shown here, is formed when lava flows from one or more vents without erupting violently.

LAB DEMONSTRATION

Purpose to observe properties of rocks formed from granitic and basaltic magma

Materials scoria, pumice, basalt, aquarium filled with water

Procedure Show students a light-colored volcanic rock (pumice) and dark-colored rocks (scoria and basalt). Point out the holes in pumice and scoria and explain that both form from lava with gas trapped inside. Place all three rocks in the aquarium as students observe.

Expected Outcome Pumice floats; scoria sinks.

Assessment

Why does the basalt and scoria sink and the pumice float? Basalt and scoria are iron- and magnesium-rich. This makes the rocks dense. Pumice is silica-rich and thus is less dense. Air trapped in pumice makes it buoyant. Note that not all pumice floats.

Paricutín On February 20, 1943, a Mexican farmer learned about cinder cones when he went to his cornfield. He noticed that a hole in his cornfield that had been there for as long as he could remember was giving off smoke. Throughout the night, hot glowing cinders were thrown high into the air. In just a few days, a cinder cone several hundred meters high covered his cornfield. This is the volcano named Paricutín.

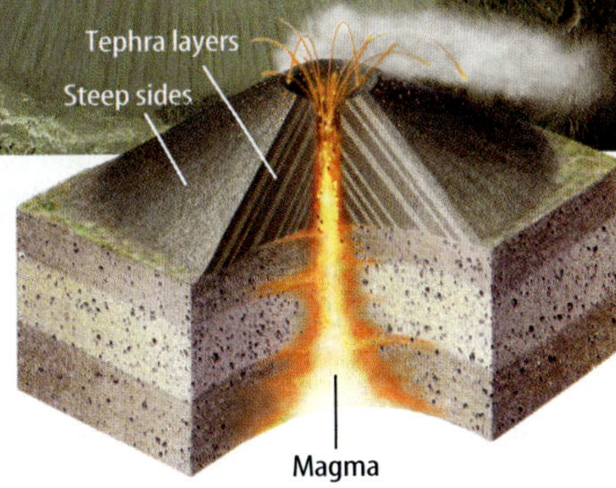

Figure 13 Paricutín is a large, cinder cone volcano located in Mexico.

Composite Volcano Some volcanic eruptions can vary between quiet and violent, depending on the amount of trapped gases and how rich in silica the magma is. An explosive period can release gas and ash, forming a tephra layer. Then, the eruption can switch to a quieter period, erupting lava over the top of the tephra layer. When this cycle of lava and tephra is repeated over and over in alternating layers, a **composite volcano** is formed. Composite volcanoes, shown in **Figure 14,** are found mostly where Earth's plates come together and one plate slides below the other. Soufrière Hills volcano is an example. As you can see in **Table 1** on the next page, many things affect eruptions and the form of a volcano.

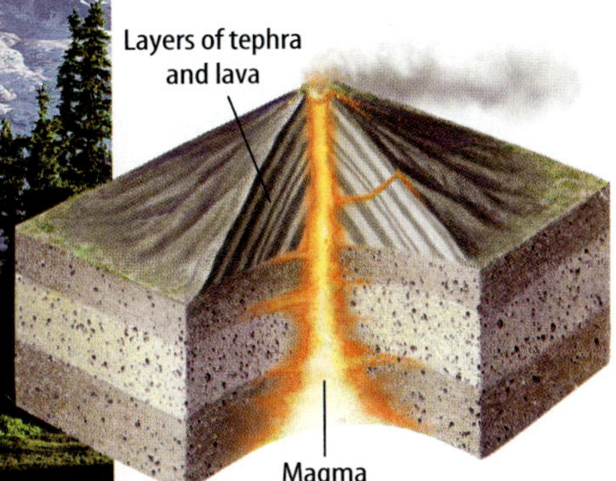

Figure 14 Mount Rainier in the state of Washington is an example of a composite volcano.

Fun Fact

Mauna Loa in Hawaii is the largest shield volcano in the Hawaiian chain. It rises more than 9 km above the ocean floor.

Make a Model

Clay Volcanoes Have students work in small groups to make clay models of shield volcanoes, cinder cone volcanoes, and composite volcanoes. Suggest that some groups make cross-section models. The shapes should reflect the broad, low angle of shield volcanoes, the steep sides of cinder cone volcanoes, and the structured layers of composite volcanoes. L1 COOP LEARN

ELL LS **Kinesthetic and Interpersonal**

Discussion

Dealing with an Eruption Would it be more difficult for a community to deal with an eruption from a nearby composite, shield, or cinder cone volcano? Explain. Most students will suggest that a shield volcano would be easiest to deal with, because its eruptions are less violent. Cinder cone eruptions are always violent, and the threat of tephra falling on a community would be difficult to prepare for or deal with. Eruptions of composite volcanoes vary. Some would be quiet, and thus easier to deal with, than the explosive eruptions.

Teacher FYI

Mt. Rainier Although Mt. Rainier has not erupted for many years, it is still considered an active volcano by volcanologists. An eruption of Mt. Rainier would affect nearby populous areas, including Seattle.

Differentiated Instruction

Learning Disabled Have students list the three different types of volcanoes in their Science Journals. Next to each name, encourage them to write as many of the characteristics of each form that they can remember or find in their textbooks. Encourage students to use **Figures 12, 13,** and **14** as references for drawing diagrams of each form of volcano. L1 ELL LS **Visual-Spatial**

Visual Learning

Table 1 What are the characteristics of the most explosive volcanoes? They are cinder cones or composite volcanoes that are mostly high in silica and water content; their magma has a low ability to flow and they produce ash along with cinders, gas, rock, and lava. Describe each eruption that occurred in the United States. Kilauea, 1989: shield volcano that erupted with "low" force releasing lava; Mount St. Helens, 1980: composite volcano that erupted explosively with gas and ash; Mount Katmai, 1912: composite volcano that erupted explosively with lava, ash, and gas

Quick Demo

A Sticky Volcano

Materials paper plate, colored flour and water paste, pastry bag with tip

Estimated Time 10 minutes

Procedure Cut a small hole in the middle of a paper plate. Fill the pastry bag with flour and water paste. Explain to students that the paste represents basaltic magma. Place the pastry bag tip through the underside of the plate, so that the tip is pointing up. Slowly squeeze the paste onto the plate, several drops at a time. Ask students to observe the behavior of the paste and to infer what type of volcano would form from basaltic lava. a shield volcano

Virtual Labs

Magma and Eruptions

How does magma's composition affect a volcano's eruption?

Violent Eruptions Soufrière Hills volcano formed as ocean floor of the North American Plate and the South American Plate slid beneath the Caribbean Plate, causing magma to form. Successive eruptions of lava and tephra produced the majestic composite volcanoes that tower above the surrounding landscape on Montserrat and other islands in the Lesser Antilles. Before the 1995 eruption, silica-rich magma rose and was trapped beneath the surface. As the magma was forced toward Earth's surface, the pressure on the underlying magma was released. This started a series of eruptions that were still continuing in the year 2003.

Table 1 Thirteen Selected Eruptions

Volcano and Location	Year	Type	Eruptive Force	Magma Content		Ability of Magma to Flow	Products of Eruption
				Silica	H_2O		
Mount Etna, Sicily	1669	composite	moderate	high	low	medium	lava, ash
Tambora, Indonesia	1815	cinder cone	high	high	high	low	cinders, ash
Krakatau, Indonesia	1883	composite	high	high	high	low	cinders, ash
Mount Pelée, Martinique	1902	cinder cone	high	high	high	low	gas, ash
Vesuvius, Italy	1906	composite	moderate	high	low	medium	lava, ash
Mount Katmai, Alaska	1912	composite	high	high	high	low	lava, ash, gas
Paricutín, Mexico	1943	cinder cone	moderate	high	low	medium	ash, cinders
Surtsey, Iceland	1963	shield	moderate	low	low	high	lava, ash
Mount St. Helens, Washington	1980	composite	high	high	high	low	gas, ash
Kilauea, Hawaii	1983	shield	low	low	low	high	lava
Mount Pinatubo, Philippines	1991	composite	high	high	high	low	gas, ash
Soufrière Hills, Montserrat	1995	composite	high	high	high	low	gas, ash, rocks
Popocatépetl, Mexico	2000	composite	moderate	high	low	medium	gas, ash

Teacher FYI

Explosive Eruption The granitic magma inside Mount St. Helens is rich in silica and tends to trap gases. When an earthquake caused the north slope of Mount St. Helens to avalanche, pressure was released and the gases escaped in a sudden explosive eruption. Soufriére Hills volcano on Montserrat erupted explosively for similar reasons.

168 ◆ F CHAPTER 6 Volcanoes

Figure 15 Not much was left after Krakatau erupted in 1883.

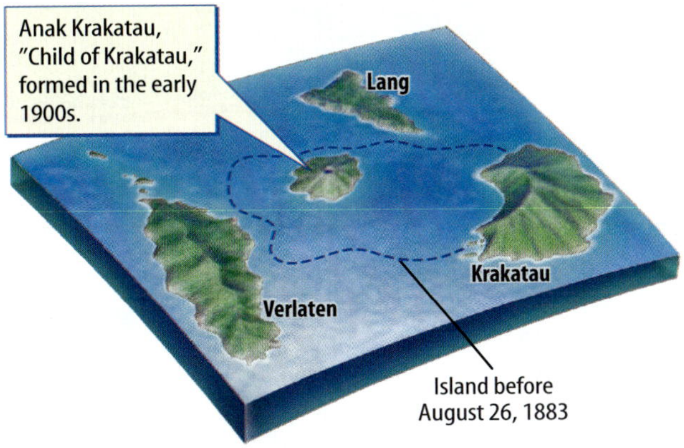

Krakatau One of the most violent eruptions in recent times occurred on an island in the Sunda Straits near Indonesia in August of 1883. Krakatau, a volcano on the island, erupted with such force that the island disappeared, as shown in **Figure 15A.** Most of the island collapsed into the emptied magma chamber. The noise of the eruption was so loud that it woke people in Australia and was heard as far away as 4,653 km from the island. Ash from the eruption fell in Singapore, which is 840 km to the north, and the area around the volcano was in complete darkness for 24 h. More than 36,000 people were killed, most by the giant tsunami waves created by the eruption. Global temperatures were lowered as much as 1.2°C by particles blown into the atmosphere and didn't return to normal until 1888.

section 2 review

Summary

What controls eruptions?

- The amount of water vapor and other gases control the type of eruption and the amount of silica present in the magma.

Composition of Magma

- Magma can be divided into two major types—silica rich and silica poor.

Forms of Volcanoes

- A shield volcano is a broad, gently sloping volcano formed by quiet eruptions of basaltic lava.
- A cinder cone volcano is a steep-sided, loosely packed volcano formed from tephra.
- Composite volcanoes are formed by alternating explosive and quiet eruptions that produce layers of tephra and lava.

Self Check

1. **Define** the term *tephra,* and where it can be found.
2. **Describe** the differences between basaltic and granitic magma.
3. **Identify** the specific water vapor and silica conditions that cause differences in eruptions.
4. **Describe** how the Hawaiian Islands formed.
5. **Think Critically** In 1883, Krakatau in Indonesia erupted. Infer which kind of lava Krakatau erupted—lava rich in silica or lava low in silica. Support your inference using data in **Table 1.**

Applying Skills

6. **Compare and contrast** Kilauea and Mount Pinatubo using information from **Table 1.**

Science Online bookf.msscience.com/self_check_quiz

3 Assess

DAILY INTERVENTION

Check for Understanding

Logical-Mathematical Have students, in groups of two or three, create a table from memory of the differences between basaltic and granitic magma. They should include eruption type, behavior of lava or magma, and specific examples in their table. L2

Reteach

Volcanoes Have students make their own tables listing the three types of volcanoes, with two examples of an eruption of each type. Columns in the table should show the date the volcano erupted, its location, and the type of material ejected. Students should not use the volcanoes listed in the text in **Table 1.** L2

ELL IS **Visual-Spatial**

Assessment

Oral Ask each student to use the text as a guide to draft a question and answer about volcanoes. Collect the questions and answers and use them to compose an oral quiz for the class. Organize the class into two teams, and act as a moderator and ask the questions. Make sure you do not ask students to answer the question they wrote. L2

section 2 review

1. bits of rock or solidified lava dropped from the air during an explosive volcanic eruption; found on the ground, buildings, plants; or inhaled.
2. Magma that is low in silica, is fluid, and produces quiet eruptions is basaltic. Magma that is high in silica and produces explosive eruptions is granitic.
3. low silica and low water vapor usually make for non-explosive, basaltic lava eruptions; higher silica and higher water vapor content usually produce explosive eruptions.
4. The islands formed as the Pacific Plate moved over a hot spot.
5. lava rich in silica, because of the explosive nature of the eruption
6. Kilauea is a shield volcano that erupts flowing lava with low silica and water contents. It generally erupts in a non-explosive manner. Pinatubo is a composite volcano. Its lava has higher amounts of silica and water vapor, and it erupts explosively.

BENCH TESTED

Real-World Question

Purpose Relate the physical and chemical properties of magma to the nature of the volcanic eruptions and the volcanic form. L2 COOP LEARN LS **Interpersonal**

Process Skills classify, infer, communicate, interpret data, hypothesize

Time Required 40 minutes

Procedure

Teaching Strategy Enlarge the figure to be used in this activity, and distribute copies onto which students are to plot the data given.

Conclude and Apply

1. The greater the ability of the magma to flow, the lower the eruptive force of the volcano, and vice versa.
2. magma that flows easily
3. Volcanoes with magma that is low in silica and water tend to produce lavas that are more fluid. Volcanoes with magma that is high in silica and water tend to produce cinders, ash, and superheated gases.
4. The ability of magma to flow is greater when the silica content is low.
5. silica content
6. Shield volcanoes result from magma with relatively low amounts of silica and water. The lavas from cinder cone volcanoes have relatively high amounts of silica and water. Composite volcanoes erupt lava that is high in silica with variable water content.

LAB

Identifying Types of Volcanoes

You have learned that certain properties of magma are related to the type of eruption and the form of the volcano that will develop. Do this lab to see how to make and use a table that relates the properties of magma to the form of volcano that develops.

Real-World Question

Are the silica and water content of a magma related to the form of volcano that develops?

Goals

- **Determine** any relationship between the ability of magma to flow and eruptive force.
- **Determine** any relationship between magma composition and eruptive force.

Materials

Table 1 (thirteen selected eruptions)
paper
pencil

Procedure

1. Copy the graph shown above.
2. Using the information from **Table 1,** plot the magma content for each of the volcanoes listed by writing the name of the basic type of volcano in the correct spot on the graph.

Conclude and Apply

1. What relationship appears to exist between the ability of the magma to flow and the eruptive force of the volcano?
2. Which would be more liquidlike: magma that flows easily or magma that flows with difficulty?

Types of Volcanoes

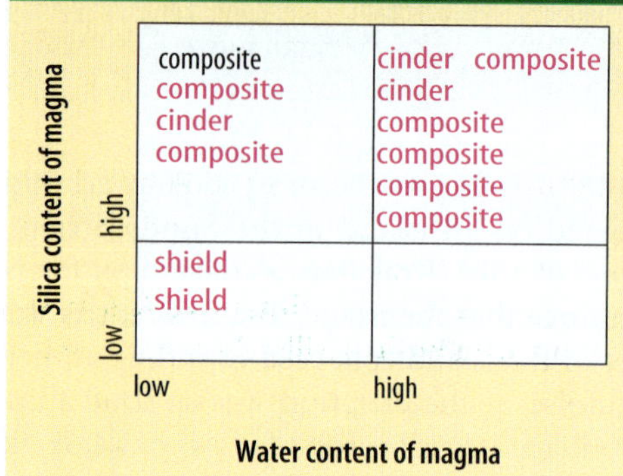

3. What relationship appears to exist between the silica or water content of the magma and the nature of the material ejected from the volcano?
4. How is the ability of a magma to flow related to its silica content?
5. **Infer** which of the two variables, silica or water content, appears to have the greater effect on the eruptive force of the volcano.
6. **Describe** the relationship that appears to exist between the silica and water content of the magma and the type of volcano that is produced.

Communicating Your Data

Create a flowchart that shows the relationship between magma composition and the type of volcano formed. **For more help, refer to the** Science Skill Handbook.

Assessment

Content To further assess students' understanding of volcano type identification, have them research the 1991 eruption of Mount Unzen in Japan and the 1982 eruption of El Chichón in Mexico and plot them on their tables. Use **Performance Assessment in the Science Classroom,** p. 89. L2

Communicating Your Data

Students can use computer graphics software to prepare their flowcharts showing the relationship between magma composition and the type of volcano formed.

section 3

Igneous Rock Features

Intrusive Features

You can observe volcanic eruptions because they occur at Earth's surface. However, far more activity occurs underground. In fact, most magma never reaches Earth's surface to form volcanoes or to flow as flood basalts. This magma cools slowly underground and produces underground rock bodies that could become exposed later at Earth's surface by erosion. These rock bodies are called intrusive igneous rock features. There are several different types of intrusive features. Some of the most common are batholiths, sills, dikes, and volcanic necks. What do intrusive igneous rock bodies look like? You can see illustrations of these features in **Figure 16.**

as you read

What You'll Learn

- **Describe** intrusive igneous rock features and how they form.
- **Explain** how a volcanic neck and a caldera form.

Why It's Important

Many features formed underground by igneous activity are exposed at Earth's surface by erosion.

Review Vocabulary

intrude: to enter by force; cut in
extrude: to force or push out

New Vocabulary

- batholith
- dike
- sill
- volcanic neck
- caldera

Figure 16 This diagram shows intrusive and other features associated with volcanic activity.
Identify *which features shown are formed above ground. Which are formed by intrusive activities?*

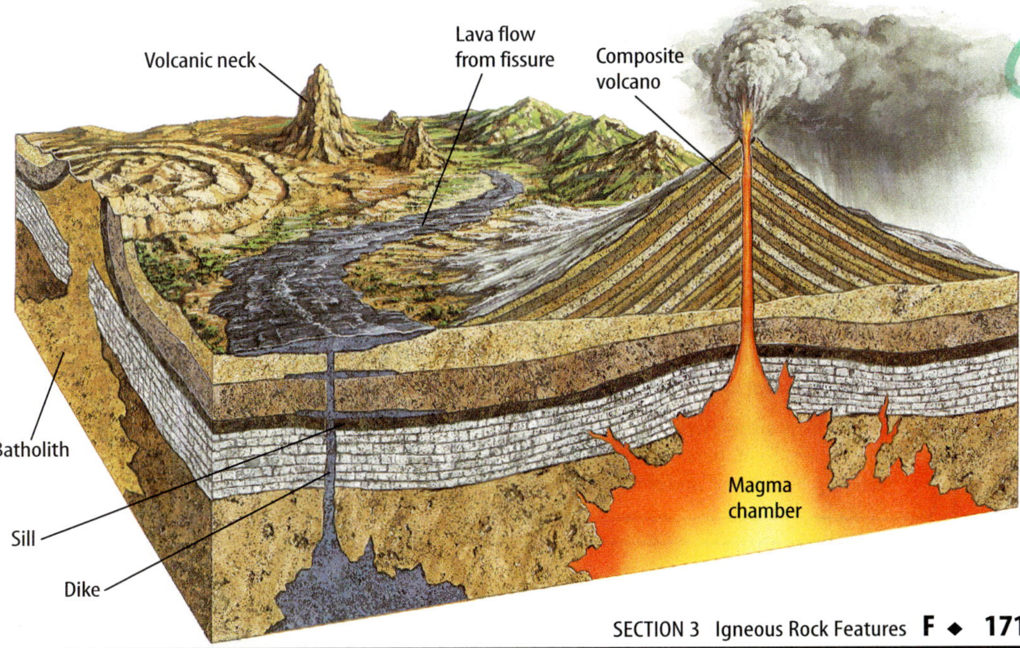

1 Motivate

Bellringer

INTERACTIVE CHALKBOARD PowerPoint® Presentations

Section Focus Transparencies also are available on the Interactive Chalkboard CD-ROM. L2 ELL

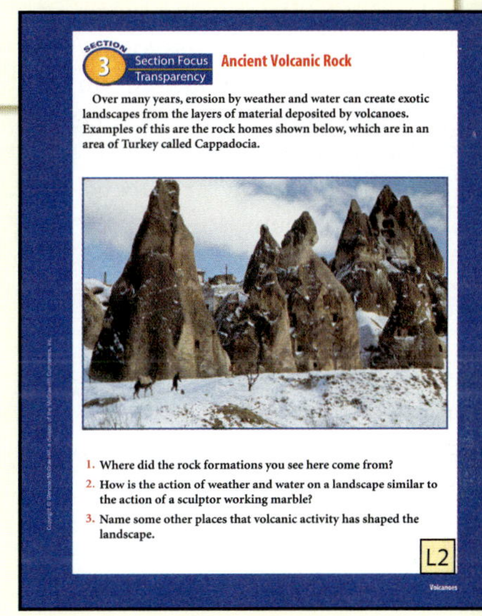
Section 3 Section Focus Transparency — Ancient Volcanic Rock

Over many years, erosion by weather and water can create exotic landscapes from the layers of material deposited by volcanoes. Examples of this are the rock homes shown below, which are in an area of Turkey called Cappadocia.

1. Where did the rock formations you see here come from?
2. How is the action of weather and water on a landscape similar to the action of a sculptor working marble?
3. Name some other places that volcanic activity has shaped the landscape.

L2

Tie to Prior Knowledge

Volcano Cross Section Have students recall how a volcano looks in cross section. Explain that the magma in the central chamber leading to the vent can eventually cool and become rock. Show students a photo of Shiprock in New Mexico (available in **Figure 17** of the chapter). Shiprock is the rocky central neck of a volcano that has been exposed by erosion. Students will learn in this section about many other volcanic features.

Caption Answer

Figure 16 above: lava flows; below: batholith, sill, dike, volcanic neck

Section 3 Resource Manager

Chapter *FAST FILE* Resources

- Transparency Activity, p. 44
- Directed Reading for Content Mastery, pp. 19, 20
- Enrichment, p. 30
- Reinforcement, p. 27
- Lab Worksheets, pp. 7–8

2 Teach

Visual Learning

Figure 16 Ask students to hypothesize whether the igneous intrusions shown are younger or older than the rocks they cut through. They are younger. The rocks that they cut through had to be there first. LS **Logical-Mathematical**

Discussion

Batholiths Remind students that batholiths are exposed on the surface by erosion. Is the rock that forms these batholiths relatively hard or relatively soft rock? Explain. It is relatively hard rock, or it would have been eroded in the same way as the material was that once surrounded it.

Activity

Identifying Rocks Have students collect several rocks in their neighborhood or on the way to school. Or, obtain rocks from a garden supply store or scientific supply company. Have groups of four students work together to identify any igneous rocks among the samples. Make available several rock and mineral guides that include color photos and descriptions. L2 COOP LEARN LS **Interpersonal and Naturalist**

Applying Math

National Math Standards

Correlation to Mathematics Objectives

1, 2, 9

Answers to Practice Problems

1. dark
2. about 35 to 45 kg

Batholiths The largest intrusive igneous rock bodies are **batholiths.** They can be many hundreds of kilometers in width and length and several kilometers thick. Batholiths form when magma bodies that are being forced upward from inside Earth cool slowly and solidify before reaching the surface. However, not all of them remain hidden inside Earth. Some batholiths have been exposed at Earth's surface by many millions of years of erosion. The granite domes of Yosemite National Park are the remains of a huge batholith that stretches across much of the length of California.

Applying Math — Calculate Percent

CLASSIFYING IGNEOUS ROCKS Igneous rocks are classified into three types depending on the amount of silica they contain. Basaltic rocks contain approximately 45 percent to 52 percent silica. Andesitic, or intermediate, rocks contain about 52 percent to 66 percent silica, and granitic rocks have more than 66 percent silica. The lighter the color is, the higher the silica content is. A 900-kg block of igneous rock contains 630 kg of silica. Calculate the percent of silica in the rock to classify it.

Solution

1. *This is what you know:*
   - rock = 900 kg
   - silica = 630 kg
2. *This is what you need to find:* The percentage of silica: x
3. *This is the equation you need to use:* Mass of silica / mass of rock = x / 100
4. *Solve the equation for* x*:*
   - $x = (630 \text{ kg}/900 \text{ kg}) \times 100$
   - $x = 70$ percent, therefore, the rock is granitic.

Check your answer by dividing it by 100, then multiplying by 900. Did you get the given amount of silica?

Practice Problems

1. A 250-kg boulder of basalt contains 125 kg of silica. Use the classification system to determine whether basalt is light or dark.
2. Andesite is an intermediate, medium-colored rock with a silica content ranging from 52 percent to 66 percent. About how many kilograms of silica would you predict to be in a 68-kg boulder of andesite?

Science Online — For more practice, visit bookf.msscience.com/math_practice

Teacher FYI

Plutons Intrusive structures are classified as plutons. The name comes from Pluto, the Roman god of the underworld. Some of the largest plutons are thought by geologists to be the remnants of magma reservoirs that fed the eruptions of volcanoes of the distant past.

172 ◆ F CHAPTER 6 Volcanoes

Dikes and Sills Magma sometimes squeezes into cracks in rock below the surface. This is like squeezing toothpaste into the spaces between your teeth. Magma that is forced into a crack that cuts across rock layers and hardens is called a **dike.** Magma that is forced into a crack parallel to rock layers and hardens is called a **sill.** These features are shown in **Figure 17.** Most dikes and sills run from a few meters to hundreds of meters long.

Other Features

When a volcano stops erupting, the magma hardens inside the vent. Erosion, usually by water and wind, begins to wear away the volcano. The cone is much softer than the solid igneous rock in the vent. Thus, the cone erodes first, leaving behind the solid igneous core as a **volcanic neck.** Ship Rock in New Mexico, shown in **Figure 17,** is a good example of a volcanic neck.

Science Online

Topic: Igneous Rock Features

Visit bookf.msscience.com for Web links to information about igneous rock features.

Activity Create a collage for artistic competition by using a variety of pictures of igneous rock features. For extra challenge, research Devils Tower, Wyoming. Develop your own hypothesis for its formation, and present your ideas as a panel discussion with other classmates.

Figure 17 Igneous features can form in many different sizes and shapes.

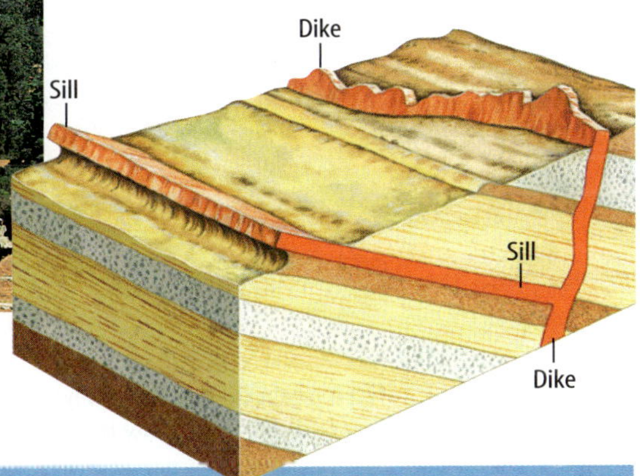

Use an Analogy

Sills Use this analogy to help students remember the orientation of a sill. Point to the window sills. Have students note that these sills are horizontal in orientation, in the same way volcanic sills are. The threshold of a door is also called a sill. L2 LS **Visual-Spatial**

Use Science Words

Word Origin Have students find the origin of *-lith.* Greek lithos, meaning "stone" Have them find other words that contain *-lith* and explain their meanings. Possible answers: lithosphere—Earth's solid surface or rocky crust and upper mantle; lithography—printing on a flat surface, such as a rock; lithophyte—a plant that grows on rock L2 LS **Linguistic**

Activity

Edible Igneous Features Prepare a small hole in the bottom of a small, clear, plastic cup. Insert the tip of a pastry bag filled with frosting into the hole. Have students fill the clear cup 1/3 full with finely crushed chocolate cookies, representing soil. Have students record their observations as they carefully squeeze the frosting, representing lava, up into the cup. What igneous features are represented in this model? batholith, basaltic magma, vent Using a clean clear, plastic cup with a frosting bag attached, have students fill the cup with pieces of wafers, chocolate chips, and cookie crumbs. Have students use a second cup to gently press down on the ingredients as they squeeze the frosting up into the layers. Have students observe the flow of the frosting through the layers creating igneous features such as dikes, sills, laccoliths, batholith and a basaltic plateau.

Fun Fact

One well-known example of a sill is a huge, 80-km-long sheer cliff on the west bank of the Hudson River facing New York. The cliff is known as the Palisades.

SECTION 3 Igneous Rock Features F ◆ 173

Make a Model

Caldera Show students photographs and topographic maps of Crater Lake in Oregon. Have students use modeling clay to make a model of the caldera. Explain that Wizard Island formed after the caldera formed and should also be included in their model. L1 LS **Visual-Spatial**

Discussion

Types of Volcanoes Have students think back to the differences in the types of volcanoes they learned about in Section 2. Is Mount Mazama, the volcano that exploded to form the Crater Lake caldera, more likely to have been a shield volcano or a composite volcano? Explain. A composite volcano, because these volcanoes are most likely to erupt violently, as Mount Mazama did; shield volcanoes have fairly low eruptive force.

Quick Demo

Tiltmeter

Materials 2 paper cups, drinking straw, clay, serving tray, water

Estimated Time 10 minutes

Procedure Make a small hole through the side of each paper cup, near the bottom. Insert the end of the straw into the paper cups, so they are now connected, and seal with some clay. Place the cups in the center of the tray, one down hill from the other. Fill the cups half full with water, mark the water level on the cup. Lift one end of the tray to a small incline (5° incline). Have students observe the water level in the cups. Increase the angle of incline and have students observe the water level in the cups. Ask students how this model could be used by volcanologists to help predict when a volcano might erupt. A sudden or large uprising of the ground near a volcano might tell scientists that an eruption is likely to happen soon.

Calderas Sometimes after an eruption, the top of a volcano can collapse, as seen in **Figure 18.** This produces a large depression called a **caldera.** Crater Lake in Oregon, shown in **Figure 19,** is a caldera that filled with water and is now a lake. Crater Lake formed after the violent eruption and destruction of Mount Mazama about 7,000 years ago.

Figure 18 Calderas form when the top of a volcano collapses.

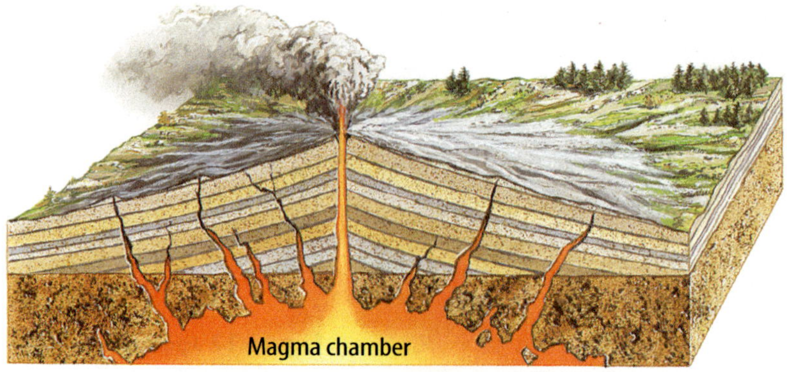

Magma is forced upward, causing volcanic activity to occur.

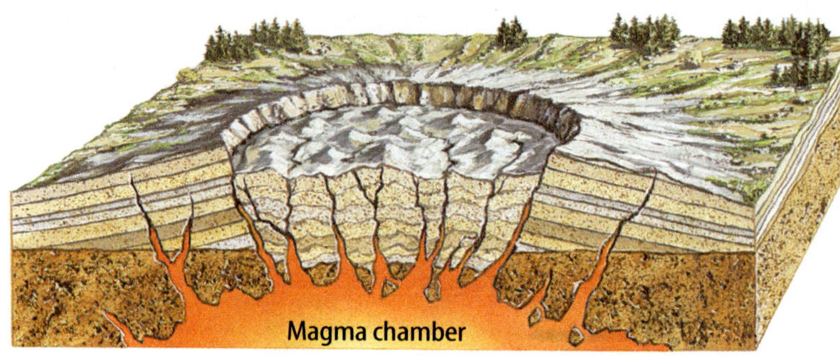

The magma chamber partially empties, causing rock to collapse into the emptied chamber below the surface. This forms a circular-shaped caldera.

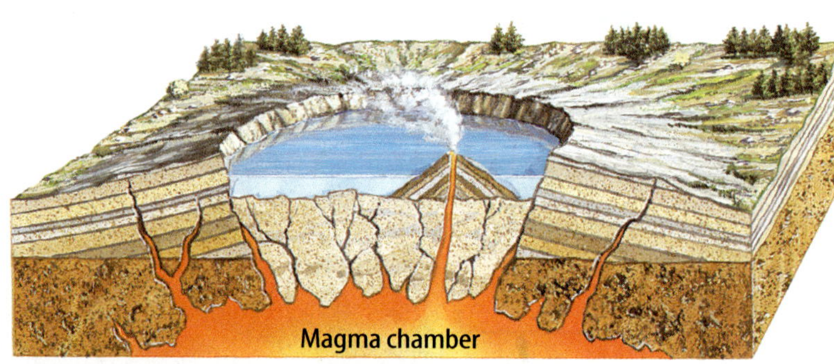

Crater Lake in Oregon formed when water collected in the circular space left when surface material collapsed.

Differentiated Instruction

Visually Impaired Construct a hollow, clay model of a volcano. Allow students to become familiar with the model's shape. Cut away the upper portion of the volcano, then drop it down into the hollow, lower section of the mountain. Again, allow your students to study the model and discover caldera formation. L1 LS **Kinesthetic**

Challenge Some students might want to videotape the formation of a model caldera during the lab later in this chapter. Students can play the tapes, displaying what happens to the volcano for their classmates, providing news commentary. L2 P

Figure 19 Wizard Island in Crater Lake is a cinder cone volcano that erupted after the formation of the caldera.
Explain *what causes a caldera to form.*

Igneous Features Exposed You have learned in this chapter that Earth's surface is built up and worn down continually. The surface of Earth is built up by volcanoes. Also, igneous rock is formed when magma hardens below ground. Eventually, the processes of weathering and erosion wear down rock at the surface, exposing features like batholiths, dikes, and sills.

section 3 review

Summary

Intrusive Features

- Intrusive igneous rock features are formed from magma that is forced upward toward Earth's crust, then slowly cools and solidifies underground before reaching the surface.
- Batholiths, dikes, and sills are a few examples of intrusive igneous rock features.

Other Features

- A volcanic neck is the solid igneous core of a volcano left behind after the softer cone has been eroded.
- A caldera is a large, circular-shaped depression that forms when the top of a volcano collapses.

Self Check

1. **Compare and contrast** a caldera and a crater.
2. **Illustrate** how a sill forms. How is it different from a dike?
3. **Describe** a batholith and explain how it forms.
4. **Think Critically** Why are the large, granite dome features of Yosemite National Park in California considered to be intrusive volcanic features when they are exposed at the surface?

Applying Math

5. **Calculate** Basaltic rocks contain approximately 45 percent to 52 percent silica. About how many kilograms of silica would you predict to be in a 68-kg boulder of basalt?

Science Online bookf.msscience.com/self_check_quiz

3 Assess

DAILY INTERVENTION

Check for Understanding

Visual-Spatial Give each student a copy of an outline map of the United States. Using their text, have them identify the location of at least three igneous features found in this section. Ask them to label the feature on the map. L1

Reteach

Igneous Rocks Have students make and use flash cards of igneous rock features. They should draw each feature on one side of a card and place its name and a short description on the other. L1 ELL LS **Linguistic and Visual-Spatial**

Assessment

Process Give each student an unlabeled diagram (such as **Figure 16**) that shows all of the igneous rock features covered in this section. Place arrows on the diagram pointing to each feature to be labeled. Have students label the features. L2

Caption Answer

Figure 19 An explosion caldera results from a volcanic explosion, and is relatively rare and small compared to collapse calderas. A collapse caldera is produced by the collapse of the roof of a magma chamber when the magma is removed by eruption or by subterranean withdrawal.

section 3 review

1. A caldera forms when the top of a volcano collapses. A crater is the steep-walled depression found around a volcano's vent.
2. A sill forms when magma squeezes between parallel rock layers. Dikes form when magma squeezes into cracks that cut across rock layers.
3. A very large intrusive igneous rock body that is exposed at Earth's surface by erosion. It forms when magma bodies that are forced upward from inside Earth cool slowly and solidify before reaching the surface.
4. The features formed when magma cooled underground. Erosion later exposed them.
5. between 30–35 kg of silica

LAB Design Your Own

How do calderas form?

Goals

- **Design** a volcano setup that will demonstrate how a caldera could form.
- **Observe** what happens during trials with your volcano setup.
- **Describe** what you observe.

Possible Materials

small box
small balloon
paper
newspaper
flour
plastic tubing
clamp for tubing
tape
scissors

Safety Precautions

Real-World Question

A caldera is a depression that forms when the top of a volcano collapses after an eruption. What might cause the top of a volcano to collapse?

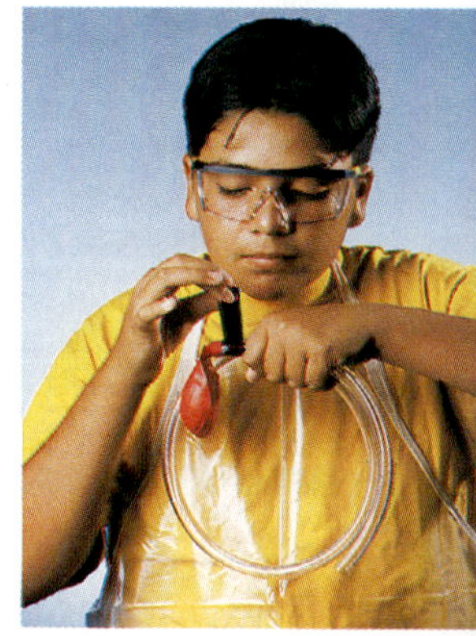

Form a Hypothesis

Based on your reading about volcanoes, state a hypothesis about what would happen if the magma inside the magma chamber of a volcano were suddenly removed.

Test Your Hypothesis

Make a Plan

1. As a group, agree upon the hypothesis and identify which results will support the hypothesis.
2. **Design** a volcano that allows you to test your hypothesis. What materials will you use to build your volcano?
3. What will you remove from inside your volcano to represent the loss of magma? How will you remove it?
4. Where will you place your volcano? What will you do to minimize messes?
5. **Identify** all constants, variables, and controls of the experiment.

Follow Your Plan

1. Make sure your teacher approves your plan before you start.
2. **Construct** your volcano with any features that will be required to test your hypothesis.
3. **Conduct** one or more appropriate trials to test your hypothesis. Record any observations that you make and any other data that are appropriate to test your hypothesis.

BENCH TESTED

Real-World Question

Purpose Students experiment to discover how the sudden removal of magma would affect a volcano. L2 COOP LEARN LS **Logical-Mathematical**

Process Skills experiment, make a model, record, analyze, communicate, hypothesize, observe

Time Required 30 minutes to design the experiment; 40 minutes to carry out the experiment, analyze data, and draw conclusions

Materials Students might help supply the materials by contributing common items such as newspapers and balloons from home.

Safety Precautions Caution students to use care with scissors.

Form a Hypothesis

Possible Hypothesis Hypotheses should include the idea that suddenly removing magma would cause the volcano to collapse in on itself.

Test Your Hypothesis

Possible Procedure

1. Tape the end of a balloon around the plastic tubing.
2. Use the tubing to blow up the balloon, but not all the way, and then clamp the tubing to prevent the air from escaping.
3. Push the tubing through a pre-cut hole in the bottom of a small box until the balloon rests on the bottom.
4. Pour flour over the balloon to form a volcano.
5. Release the clamp on the tubing and let the air out of the balloon.
6. The result: The balloon deflates and the surface collapses to form a caldera.

Alternative Inquiry Lab

Igneous Rock Features Encourage students to use what they have learned about the formation of calderas to predict how other igneous features would form. Ask them to create a plan to model the formation of one of the other types of igneous features mentioned in the text. Once this plan has been approved, provide students with appropriate materials so that they may create this model. L3

Using Scientific Methods

Analyze Your Data

1. **Describe** in words or with a drawing what your volcano looked like before you began.
2. **Observe** what happened to your volcano during the experiment that you conducted? Did its appearance change?
3. **Describe** in words or with a drawing what your volcano looked like after the trial.
4. **Observe** What other observations did you make?
5. **Describe** any other data that you recorded.

Conclude and Apply

1. **Draw Conclusions** Did your observations support your hypothesis? Explain.
2. **Explain** how your demonstration was similar to what might happen to a real volcano. How was it different?

Communicating Your Data

Make a 4-sequence time-lapse diagram with labels and descriptions of how a caldera forms. Use your visual aid to describe caldera formation to students in another class.

Teaching Strategy Encourage students to think in terms of something that will allow the volcano to hold its shape, but from which material can be removed quickly to make the volcano collapse.

Expected Outcome Students will make a model volcano, then quickly withdraw material from its interior, causing the top of the volcano to collapse.

Analyze Your Data

Answers to Questions

1. Answers will vary but students should note shape and material used to make and fill the volcano.
2. Students should describe the process for quickly removing material from the interior and the collapse of the cone.
3. Students should describe a collapsed volcano.
4. Students should report any other observations.
5. Students should report additional recorded data.

Error Analysis Errors may have occurred if air escaped from the balloon too soon, or was uncontrolled.

Conclude and Apply

1. Observations should support hypotheses. Possible answer: Yes; quickly emptying the magma chamber produces a caldera.
2. In a real volcano, the eruption removes material quickly from the volcano's interior, causing the top to collapse. In the experiment, removing material from the interior of the model volcano causes the model to collapse in on itself.

Assessment

Performance Have pairs of students write short critiques of each other's designs and procedures for carrying them out. Use **PASC**, p. 89. L2

Communicating Your Data

Encourage students to use graphic art and word-processing programs to prepare their 4-sequence time-lapse diagrams.

Content Background

Vesuvius is a 1,281-m-high, composite volcano. It is the only active volcano on the mainland of Europe.

Before the 79 A.D. eruption that buried Herculaneum, Pompeii, and Stabiae, Vesuvius was thought to be extinct. Since 1631, Vesuvius has been continuously active and its activity is now closely monitored.

Despite the destruction that volcanoes cause, they also bring benefits. The fertile volcanic soil below the volcano supports many crops—vegetable gardens and orchards near the base and vineyards on the lower slopes.

Discussion

Living Near a Volcano Explain to students that today more than 2 million people live near the base of Vesuvius. Why would people still live so close to an active volcano? Possible answers: Volcanic soil is very fertile so it is a good location for growing crops. Also, people whose families have lived there for generations may not want to leave the area.

Analyze the Event

Discovery Ask students to speculate how they might have felt if they had been the farmer digging the well. Possible answer: curious or excited Ask what they might do if they were digging on their property and came across some trace of a buried city. Possible answers: Contact a local historical society, an archaeologist, or scientists at a nearby university to help them find out more.

Oops! Accidents in SCIENCE

SOMETIMES GREAT DISCOVERIES HAPPEN BY ACCIDENT!

Buried in Ash

A long-forgotten city is accidentally found after 2,000 years

In the heat of the Italian Sun, a farmer digs a new well for water. He thrusts his shovel into the ground one more time. But instead of hitting water, the shovel strikes something hard; a slab of smooth white marble.

Under the ground lay the ancient city of Herculaneum (her kew LAY nee um). The city, and its neighbor Pompeii (pom PAY) had been buried for more than 1,600 years. On August 24, 79 A.D., Mount Vesuvius, a nearby volcano, erupted and buried both cities with pumice, rocks, mud, and ash.

Back in Time

The Sun shone over the peaceful town of Herculaneum on that August morning almost 2,000 years ago. But at about 1 P.M., that peace was shattered forever.

With massive force, the peak of Vesuvius exploded, sending six cubic kilometers of ash and pumice into the sky. Hours later, a fiery surge made its way from the volcano to the city. These pyroclastic flows continued as more buildings were crushed and buried by falling ash and pumice. Within six hours, much of the city was totally buried under the flows. After six surges from Vesuvius, the deadly eruption ceased. But the city had disappeared under approximately 21 m of ash, rock, and mud.

Excavated ruins with Mount Vesuvius in the background.

A City Vanishes

More than 3,600 people were killed in the natural disaster. Scientists believe that many died trying to protect their faces from the pyroclastic surges that filled the air with hot ash. Those able to escape returned to find no trace of their city. Over hundreds of years, grass and fields covered Herculaneum, erasing it from human memory.

Archaeologists have unearthed perfectly preserved mosaics and a library with ancient scrolls in excellent condition. Archaeologists found skeletons and voids that were filled with plaster to form casts of people who died the day Vesuvius erupted. Visitors to the site can see a Roman woman, a teen-aged girl, and a soldier with his sword still in his hand.

Much of Herculaneum still lies buried beneath thick layers of volcanic ash, and archaeologists still are digging to expose more of the ruins. Their work is helping scientists better understand everyday life in an ancient Italian town. But, if it weren't for a farmer's search for water, Herculaneum might not have been discovered at all!

Research the history of your town. Ask your local librarian to help "unearth" maps, drawings, or photos that let you travel back in time! Design a two-layer map that shows the past and the present.

Science Online

For more information, visit bookf.msscience.com/oops

Research What students find in their research will vary depending on your location. Many areas of the United States have artifacts and other evidence of cultures that inhabited the areas thousands of years ago. Suggest that students also ask local librarians to help them locate information on the prehistory of your area.

Resources for Teachers and Students

In Search of Pompeii: Uncovering a Buried Roman City, by Giovanni Caselli, P. Bedrick Books, 1999

Mount Vesuvius: Europe's Mighty Volcano of Smoke and Ash, by Kathy Furgang, Powerkids Press, 2001

Reviewing Main Ideas

Section 1 Volcanoes and Earth's Moving Plates

1. Volcanoes can be dangerous to people because they can cause deaths and destroy property.
2. Rocks in the crust and mantle melt to form magma, which is forced toward Earth's surface. When the magma flows through vents, it's called lava and forms volcanoes.
3. Volcanoes can form over hot spots or when Earth's plates pull apart or come together.

Section 2 Types of Volcanoes

1. The three types of volcanoes are composite volcanoes, cinder cone volcanoes, and shield volcanoes.
2. Shield volcanoes produce quiet eruptions. Cinder cone and composite volcanoes can produce explosive eruptions.
3. Some lavas are thin and flow easily, producing quiet eruptions. Other lavas are thick and stiff, producing violent eruptions.

Section 3 Igneous Rock Features

1. Intrusive igneous rock bodies such as batholiths, dikes, and sills form when magma solidifies underground.
2. Batholiths are the most massive igneous rock bodies. Dikes and sills form when magma squeezes into cracks.
3. A caldera forms when the top of a volcano collapses, forming a large depression.

Visualizing Main Ideas

Copy and complete the following concept map on types of volcanic eruptions.

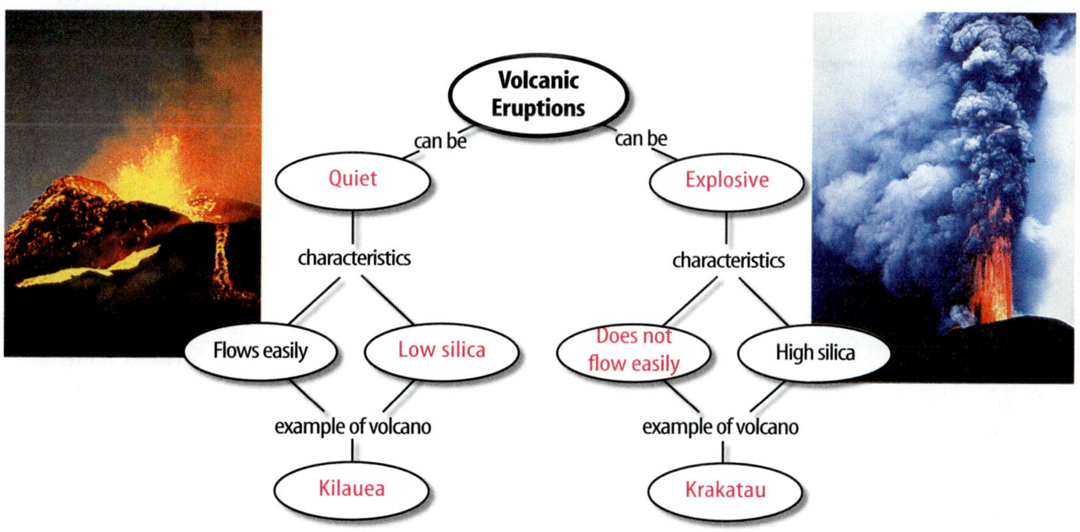

Reviewing Main Ideas

Summary statements can be used by students to review the major concepts of the chapter.

Visualizing Main Ideas

See student page.

Science Online

Visit **bookf.msscience.com**
- /self_check_quiz
- /interactive_tutor
- /vocabulary_puzzlemaker
- /chapter_review
- /standardized_test

Assessment Transparency

For additional assessment questions, use the *Assessment Transparency* located in the transparency book.

Assessment

Assessment Transparency **Volcanoes**

Directions: *Carefully review the table and answer the following questions.*

Volcanic Eruptions at Mount St. Helens				
Date	Explosive activity	Broken flows	Lava flows (dome)	Volume (million cubic yds)
6/12/80	x	x	x	54.5
7/22/80	x	x		13.1
10/16/80	x	x	x	3.6
12/27/80			x	2.1
6/18/81			x	5.4
10/30/81			x	4.7
3/19/82	x		x	4.4
8/18/82			x	6.0
2/7/83	x		x	29.3
3/29/84			x	1.4

1. According to the table, which year did Mount St. Helens experience the most volcanic activity?
   A 1980 B 1981 C 1982 D 1983
2. According to the table, there was no explosive activity between ___.
   F 06/12/80–10/16/80 H 02/01/83–03/29/84
   G 12/27/80–03/20/82 J 10/17/80–03/18/82
3. All of the following days had lava flows from the dome EXCEPT ___.
   A December 27, 1980 C June 12, 1980
   B July 22, 1980 D August 18, 1982

L2

Volcanoes

chapter 6 Review

Using Vocabulary

1. shield volcano
2. tephra
3. sill
4. crater
5. dike

Checking Concepts

6. C
7. B
8. A
9. C
10. C
11. A
12. D
13. D

chapter 6 Review

Using Vocabulary

batholith p. 162
caldera p. 174
cinder cone volcano p. 166
composite volcano p. 167
crater p. 158
dike p. 173
hot spot p. 160
shield volcano p. 166
sill p. 173
tephra p. 166
vent p. 158
volcanic neck p. 173
volcano p. 156

Fill in the blanks with the correct vocabulary word or words.

1. A broad volcano with gently sloping sides is called a(n) ______.
2. Bits of rock or solidified lava dropped from the air after a volcanic eruption are ______.
3. Magma squeezed into a horizontal crack between rock layers is a(n) ______.
4. The steep-walled depression around a volcano's vent is called a(n) ______.
5. Magma squeezed into a vertical crack across rock layers is called a(n) ______.

Checking Concepts

Choose the word or phrase that best answers the question.

6. What type of boundary is associated with composite volcanoes?
   A) plates moving apart
   B) plates sticking and slipping
   C) plates moving together
   D) plates sliding past each other
7. Why is Hawaii made of volcanoes?
   A) Plates are moving apart.
   B) A hot spot exists.
   C) Plates are moving together.
   D) Rift zones exist.
8. What kind of magmas produce violent volcanic eruptions?
   A) those rich in silica
   B) those that are fluid
   C) those forming shield volcanoes
   D) those rich in iron
9. Magma that is low in silica generally produces what kind of eruptions?
   A) thick **C)** quiet
   B) caldera **D)** explosive

Use the photo below to answer question 10.

10. Which type of volcano, shown above, is made entirely of tephra?
    A) shield **C)** cinder cone
    B) caldera **D)** composite
11. What kind of volcano is Kilauea?
    A) shield **C)** cinder cone
    B) composite **D)** caldera cone
12. What is the largest intrusive igneous rock body?
    A) dike **C)** sill
    B) volcanic neck **D)** batholith
13. What is the process that formed Soufrière Hills volcano on Montserrat?
    A) plates sticking and slipping
    B) caldera formation
    C) plates sliding sideways
    D) plates moving together

Science Online bookf.msscience.com/vocabulary_puzzlemaker

Use the ExamView® Pro Testmaker CD-ROM to:

- create multiple versions of tests
- create modified tests with one mouse click for inclusion students
- edit existing questions and add your own questions
- build tests aligned with state standards using built-in State Curriculum Tags
- change English tests to Spanish with one mouse click and vice versa

Thinking Critically

14. Explain how glaciers and volcanoes can exist on Iceland.

15. Describe what kind of eruption is produced when basaltic lava that is low in silica flows from a volcano.

16. Explain how volcanoes are related to earthquakes.

17. Infer Misti is a volcano in Peru. Peru is on the western edge of South America. How might this volcano have formed?

18. Describe the layers of a composite volcano. Which layers represent violent eruptions?

19. Classify the volcano Fuji, which has steep sides and is made of layers of silica-rich lava and ash.

Use the map below to answer question 20.

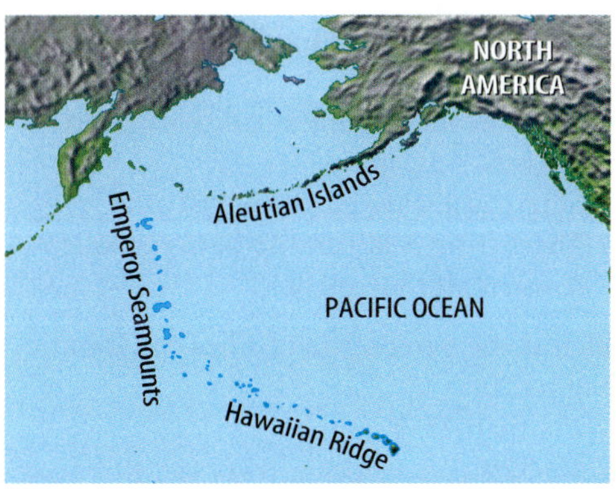

20. Interpret Scientific Illustrations Look at the map above. The Hawaiian Islands and Emperor Seamounts were formed when the Pacific Plate moved over a fixed hot spot. If the Emperor chain trends in a direction different from the Hawaiian Islands, what can you infer about the Pacific Plate?

21. Concept Map Make a network-tree concept map about where volcanoes can occur. Include the following words and phrases: *hot spots, divergent plate boundaries, convergent plate boundaries, volcanoes, can occur, examples, Iceland, Soufrière Hills,* and *Hawaiian Islands.*

Performance Activities

22. Poster Make a Venn diagram of the three basic types of volcanoes. Label them and indicate what cone formation, lava composition, eruption, and geologic location are expected of each type of volcano.

Applying Math

23. Sea Level The base of the volcano Mauna Loa is about 5,000 m below sea level. The total height of the volcano is 9,170 m. What percentage of the volcano is above sea level? Below sea level?

Use the table below to answer questions 24 and 25.

Volcano	Year of Eruption	Amount of Material Ejected
Tambora	1815	131 km^3
Katmai	1912	30 km^3
Novarupta	1912	15 km^3
Mt. St. Helens	1980	1.3 km^3
Pinatubo	1991	5.5 km^3

24. Ejected Material How many times greater was the volume of ejected material from Tambora, as compared to Mt. St. Helens?

25. Graph Design a bar graph to show the amount of ejected material from the volcanoes. Present the information from least to greatest volume.

chapter 6 Review

Thinking Critically

14. Glaciers form because of Iceland's cold climate. Volcanoes also exist there because Iceland is located on the Mid-Atlantic Ridge, a divergent plate boundary.

15. The lava is very fluid. Because it flows easily, a relatively quiet eruption is produced.

16. They both occur at the boundaries between plates and are related to geological movements in Earth's crust.

17. Students should infer that the volcano formed as a result of one plate sliding under another.

18. A composite volcano is made up of alternating layers of tephra and lava. Tephra is deposited by violent, explosive eruptions. Lava is deposited by less-violent eruptions.

19. Fuji is a composite volcano.

20. The motion of the Pacific Plate over the hot spot changed from nearly north to nearly northwest.

21. Check students' work.

Performance Activities

22. Venn diagrams should show each type of volcano in cross section and labels with the type of eruption it produces. Use **PASC**, p. 145.

Applying Math

National Math Standards

1, 2, 5, 9

23. 54.5 % below, 45.5 % above; $(5{,}000 \text{ m} / 9{,}170 \text{ m}) \times 100 = 54.5\,\%$

24. approximately 100 times greater

25. Student graphs may vary, but should contain labels and a title, and indicate amount of ejected material from least to greatest volume, Mt. St. Helens, Pinatubo, Novarupta, Katmai, and Tambora.

Assessment Resources

Reproducible Masters

Chapter *Fast File* Resources
- Chapter Review, pp. 35–36
- Chapter Tests, pp. 37–40
- Assessment Transparency Activity, p. 47

Glencoe Science Web site
- Chapter Review Test
- Standardized Test Practice

Glencoe Technology

- Assessment Transparency
- Exam*View*® Pro Testmaker
- MindJogger Videoquiz
- Interactive Chalkboard

FAST FILE

Answer Sheet A practice answer sheet can be found at bookf.msscience.com/answer_sheet.

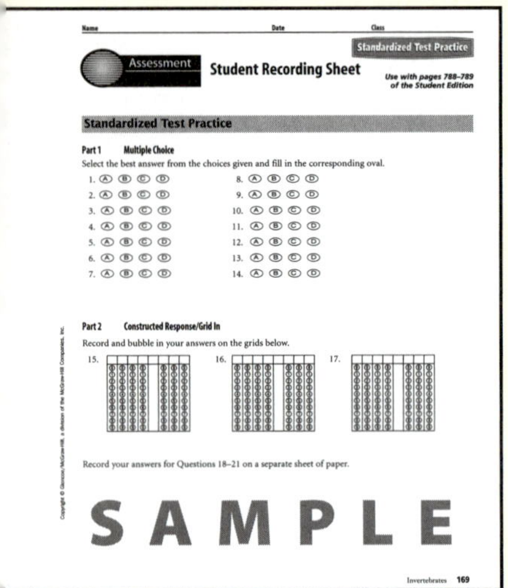
Name Date Class

Standardized Test Practice

Assessment Student Recording Sheet

Use with pages 788–789 of the Student Edition

Standardized Test Practice

Part 1 Multiple Choice

Select the best answer from the choices given and fill in the corresponding oval.

1. Ⓐ Ⓑ Ⓒ Ⓓ 8. Ⓐ Ⓑ Ⓒ Ⓓ
2. Ⓐ Ⓑ Ⓒ Ⓓ 9. Ⓐ Ⓑ Ⓒ Ⓓ
3. Ⓐ Ⓑ Ⓒ Ⓓ 10. Ⓐ Ⓑ Ⓒ Ⓓ
4. Ⓐ Ⓑ Ⓒ Ⓓ 11. Ⓐ Ⓑ Ⓒ Ⓓ
5. Ⓐ Ⓑ Ⓒ Ⓓ 12. Ⓐ Ⓑ Ⓒ Ⓓ
6. Ⓐ Ⓑ Ⓒ Ⓓ 13. Ⓐ Ⓑ Ⓒ Ⓓ
7. Ⓐ Ⓑ Ⓒ Ⓓ 14. Ⓐ Ⓑ Ⓒ Ⓓ

Part 2 Constructed Response/Grid In

Record and bubble in your answers on the grids below.

15. 16. 17.

Record your answers for Questions 18–21 on a separate sheet of paper.

SAMPLE

Invertebrates 169

Part 1 Multiple Choice

1. B
2. C
3. D
4. C
5. C
6. A
7. D
8. A

Part 1 Multiple Choice

Record your answers on the answer sheet provided by your teacher or on a sheet of paper.

Use the photo below to answer question 1.

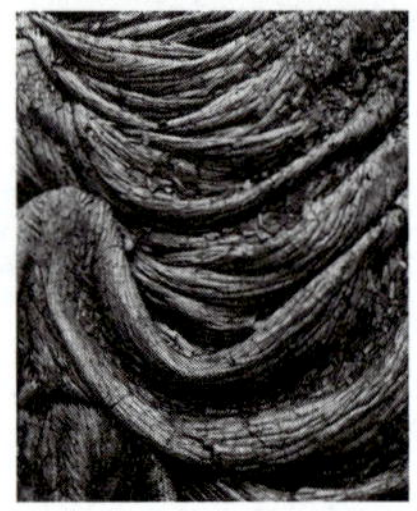

1. Which of the following terms best describes the rock in the photo above?
   - **A.** aa
   - **B.** pahoehoe
   - **C.** pillow lava
   - **D.** ash
2. Which of the following is made of layers of ash and cooled lava flows?
   - **A.** shield volcano
   - **B.** plateau basalts
   - **C.** composite volcano
   - **D.** cinder cone volcano
3. Which of the following volcanoes is located in the United States?
   - **A.** Hekla
   - **B.** Paricutin
   - **C.** Mount Vesuvius
   - **D.** Mount St. Helens
4. Which of the following igneous features is parallel to the rock layers that it intrudes?
   - **A.** batholith
   - **B.** volcanic neck
   - **C.** sill
   - **D.** dike
5. Which of the following forms when the top of a volcano collapses into a partially emptied magma chamber?
   - **A.** fissure
   - **B.** crater
   - **C.** caldera
   - **D.** volcanic neck

Test-Taking Tip

Relax Stay calm during the test. If you feel yourself getting nervous, close your eyes and take five slow, deep breaths.

Use the graph below to answer questions 6 and 7.

Percentage of Silica v. Viscosity

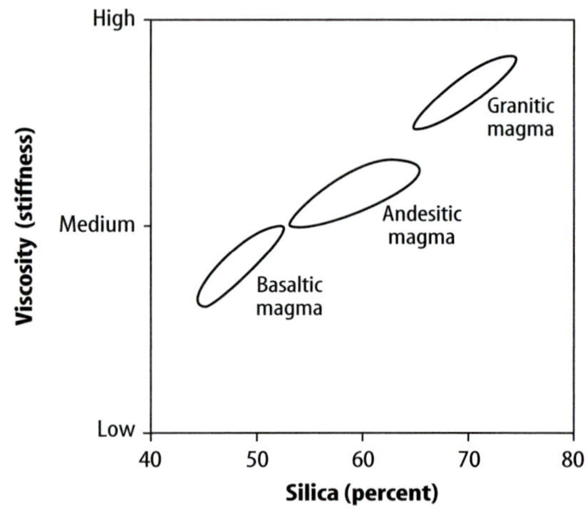

6. What relationship can be inferred from the graph?
   - **A.** Magmas that have more silica are more viscous.
   - **B.** Magmas that have less silica are more viscous.
   - **C.** Magmas always have low viscosity.
   - **D.** There is no relationship between silica content and viscosity.
7. What is the percentage of silica in Granitic magma?
   - **A.** less than 45%
   - **B.** 45–52%
   - **C.** 53–65%
   - **D.** greater than 65%
8. Which of the following is the finest type of tephra?
   - **A.** volcanic ash
   - **B.** volcanic bombs
   - **C.** volcanic cinders
   - **D.** volcanic blocks

Part 2 Short Response/Grid In

9. A hot spot is a volcanic region on the surface of a rising mantle plume. As the plume rises, the pressure decreases and melting occurs.
10. The Hawaiian Islands are in a line because the Pacific Plate moved in a relatively consistent direction over a hot spot as they formed.
11. Both dikes and sills are tabular intrusive bodies. Dikes cut across rock layers, whereas sills are parallel to rock layers.
12. mean–59.8%, median–59.6%, range–58.7–61.4, 2.7%
13. 100% − 59.8% = 40.2%
14. When magma consists of high percentages of water vapor, eruptions tend to be explosive lava fountains.
15. Magma is molten rock that is beneath Earth's surface. Lava is molten rock that is on Earth's surface.
16. Igneous rock forms when lava cools and minerals crystallize from it.

Part 2 Short Response/Grid In

Record your answers on the answer sheet provided by your teacher or on a sheet of paper.

9. What is a hot spot? Why do volcanoes often form at hot spots?

10. Why are the Hawaiian Islands in a line?

11. How is a dike different from a sill? Support your answer with a Venn diagram.

Use the table below to answer questions 12–14.

Eruption	Volume Percent Water Vapor
1	58.7
2	60.1
3	61.4
4	59.3
5	59.6

12. Calculate the mean, median, and range of the water vapor data in the table? Describe how this information would be helpful to a volcanologist.

13. Using the mean value that you calculated in question 12, what percentage of the volcanic gas consists of gases other than water vapor?

14. The water vapor content of Kilauea is above average when compared to other volcanoes. How might these data help to explain why lava fountains often occur on Kilauea?

15. What is the difference between magma and lava?

16. Explain how igneous rock forms from lava.

Part 3 Open Ended

Record your answers on a sheet of paper.

17. Explain how volcanic necks, such as Ship Rock, form. Support your answer with a labeled diagram.

18. How does tephra form?

19. Why do some volcanoes occur where one plate sinks beneath another plate? Support your answer with a labeled diagram.

20. How can pillow-shaped bodies form from lava?

Use the map below to answer questions 21 and 22.

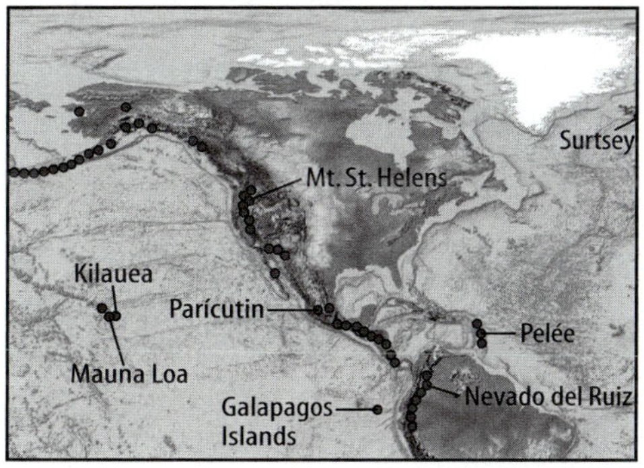

21. What kind of magma was required to create the Columbia Plateau as compared to Mt. St. Helen's, only 200 miles away? What would you predict would be the percentage of water vapor in the 2 types of magma?

22. Where do volcanoes occur in the United States?

23. How do volcanoes affect people and their property? List four safety precautions for people living in volcanic areas.

21. The Columbia Plateau formed when large fissure eruptions of basaltic lava, low in water vapor erupted. The eruption of Mt. St. Helens would contain a higher percentage of water vapor.

22. Volcanoes occur along the northwest coast, in Alaska, in Hawaii, and in some of the other western states.

23. Property may be destroyed or burned from pyroclastic flows, lahars, landslides, or tsunami waves, inhaling ash or toxic gas can be a health risk; but volcanoes produce fertile soil for farming, and geothermal energy. Student safety tips will vary; but may include: heed volcanologists' warnings, have an evacuation plan, protective masks, build towns away from active volcanoes, or be aware of earthquakes and rumblings.

Rubrics

For more help evaluating open-ended assessment questions, see the rubric on p. 10T.

Part 3 Open Ended

17. Volcanic necks are made of igneous rock that formed in the volcanic conduit of a volcano. Because the igneous rock is more resistant to erosion than the softer volcanic cone, it survives after the cone is eroded away. Check student diagrams.

18. Tephra forms from pieces of lava and rock thrown into the air during a volcanic eruption.

19. The sinking plate and the mantle wedge rub together and melt. This melting is encouraged by water that is liberated from water-containing minerals in the ocean crust and upper mantle rock. This water lowers the melting temperature of the surrounding rock, thereby causing melting to occur. Check student diagrams.

20. Pillow-lava forms when bubbles of lava are erupted onto the seafloor.

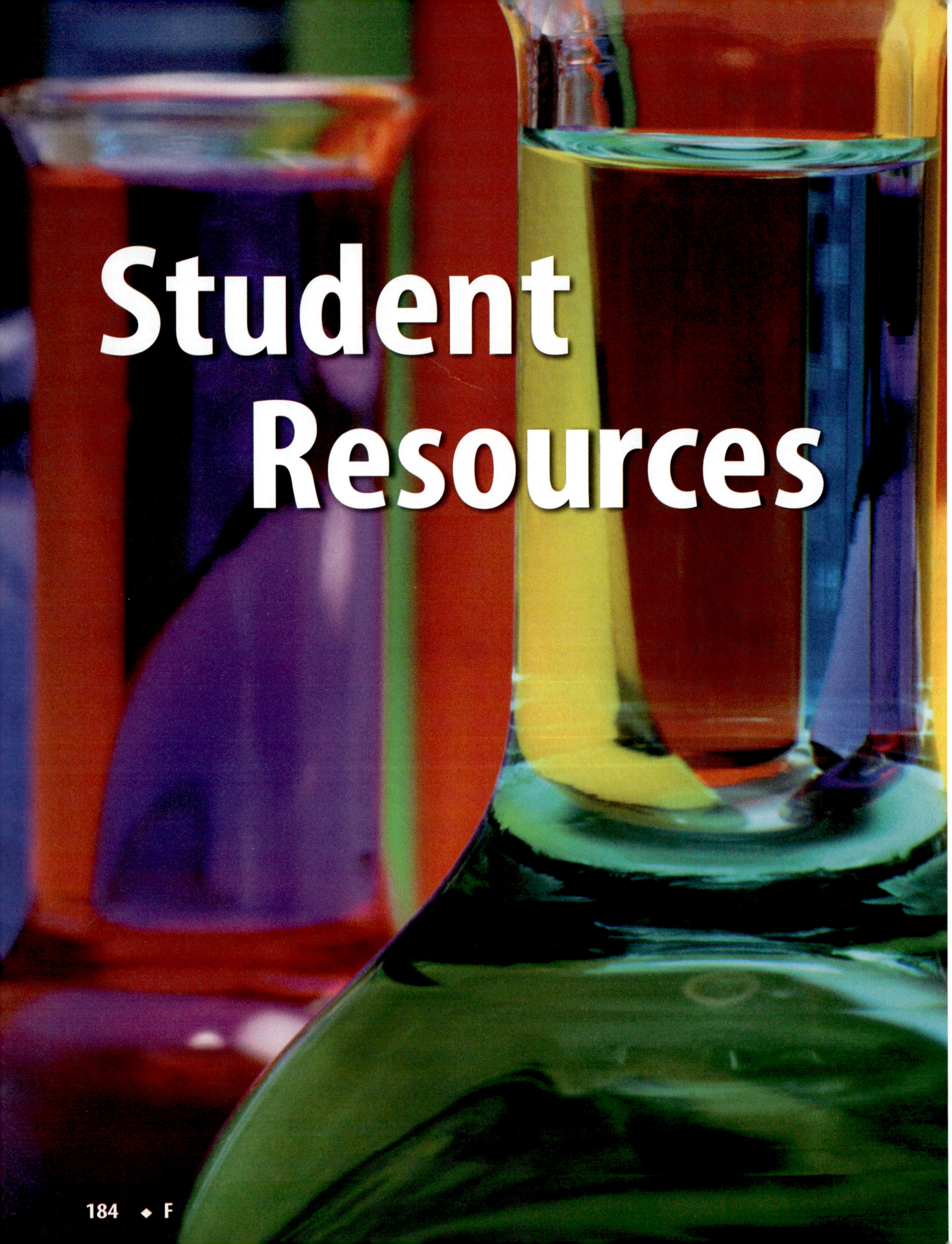

Student Resources

CONTENTS

Scientific Methods

Scientists use an orderly approach called scientific methods to solve problems. These include organizing and recording data so others can understand them. Scientists use many variations in these methods when they solve problems.

Identify a Question

The first step in a scientific investigation or experiment is to identify a question to be answered or a problem to be solved. For example, you might ask which gasoline is the most efficient.

Gather and Organize Information

After you have identified your question, begin gathering and organizing information. There are many ways to gather information, such as researching in a library, interviewing those knowledgeable about the subject, testing and working in the laboratory and field. Fieldwork is investigations and observations done outside of a laboratory.

Researching Information Before moving in a new direction, it is important to gather the information that already is known about the subject. Start by asking yourself questions to determine exactly what you need to know. Then you will look for the information in various reference sources, like the student is doing in **Figure 1.** Some sources may include textbooks, encyclopedias, government documents, professional journals, science magazines, and the Internet. Always list the sources of your information.

Figure 1 The Internet can be a valuable research tool.

Evaluate Sources of Information Not all sources of information are reliable. You should evaluate all of your sources of information, and use only those you know to be dependable. For example, if you are researching ways to make homes more energy efficient, a site written by the U.S. Department of Energy would be more reliable than a site written by a company that is trying to sell a new type of weatherproofing material. Also, remember that research always is changing. Consult the most current resources available to you. For example, a 1985 resource about saving energy would not reflect the most recent findings.

Sometimes scientists use data that they did not collect themselves, or conclusions drawn by other researchers. This data must be evaluated carefully. Ask questions about how the data were obtained, if the investigation was carried out properly, and if it has been duplicated exactly with the same results. Would you reach the same conclusion from the data? Only when you have confidence in the data can you believe it is true and feel comfortable using it.

Interpret Scientific Illustrations As you research a topic in science, you will see drawings, diagrams, and photographs to help you understand what you read. Some illustrations are included to help you understand an idea that you can't see easily by yourself, like the tiny particles in an atom in **Figure 2.** A drawing helps many people to remember details more easily and provides examples that clarify difficult concepts or give additional information about the topic you are studying. Most illustrations have labels or a caption to identify or to provide more information.

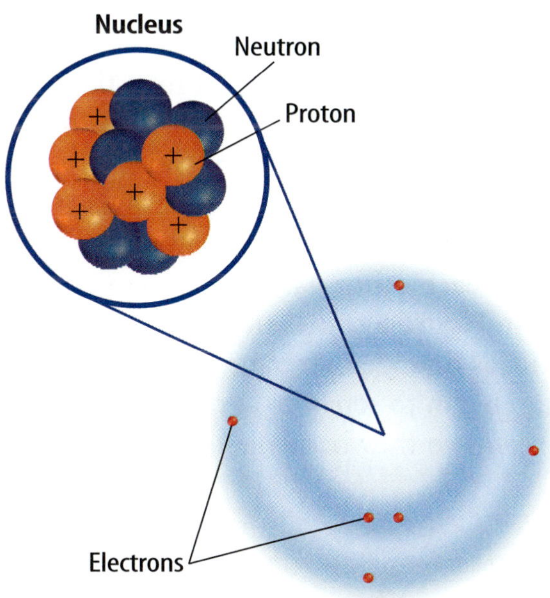

Figure 2 This drawing shows an atom of carbon with its six protons, six neutrons, and six electrons.

Concept Maps One way to organize data is to draw a diagram that shows relationships among ideas (or concepts). A concept map can help make the meanings of ideas and terms more clear, and help you understand and remember what you are studying. Concept maps are useful for breaking large concepts down into smaller parts, making learning easier.

Network Tree A type of concept map that not only shows a relationship, but how the concepts are related is a network tree, shown in **Figure 3.** In a network tree, the words are written in the ovals, while the description of the type of relationship is written across the connecting lines.

When constructing a network tree, write down the topic and all major topics on separate pieces of paper or notecards. Then arrange them in order from general to specific. Branch the related concepts from the major concept and describe the relationship on the connecting line. Continue to more specific concepts until finished.

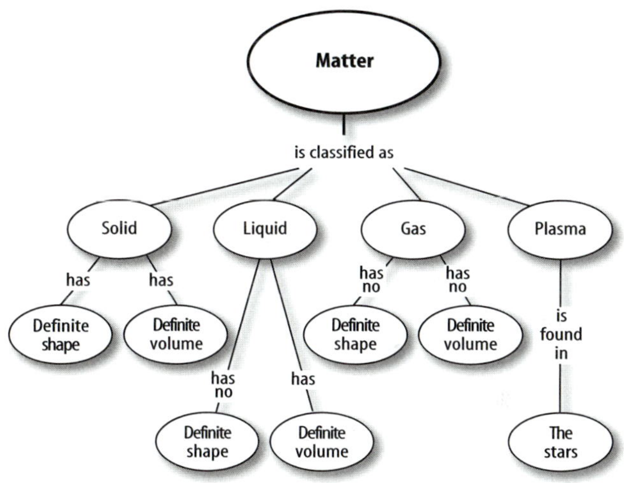

Figure 3 A network tree shows how concepts or objects are related.

Events Chain Another type of concept map is an events chain. Sometimes called a flow chart, it models the order or sequence of items. An events chain can be used to describe a sequence of events, the steps in a procedure, or the stages of a process.

When making an events chain, first find the one event that starts the chain. This event is called the initiating event. Then, find the next event and continue until the outcome is reached, as shown in **Figure 4.**

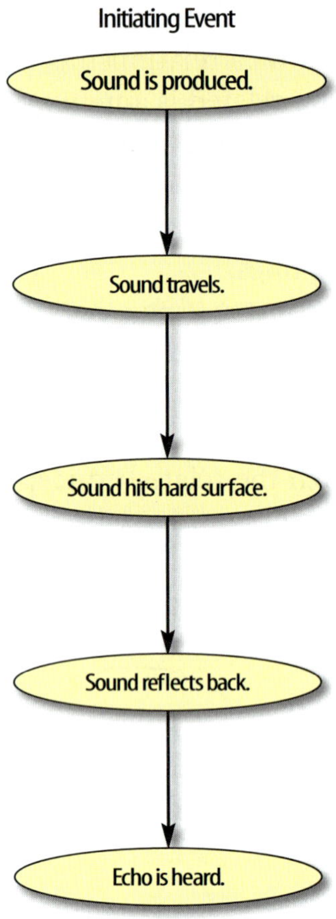

Figure 4 Events-chain concept maps show the order of steps in a process or event. This concept map shows how a sound makes an echo.

Cycle Map A specific type of events chain is a cycle map. It is used when the series of events do not produce a final outcome, but instead relate back to the beginning event, such as in **Figure 5.** Therefore, the cycle repeats itself.

To make a cycle map, first decide what event is the beginning event. This is also called the initiating event. Then list the next events in the order that they occur, with the last event relating back to the initiating event. Words can be written between the events that describe what happens from one event to the next. The number of events in a cycle map can vary, but usually contain three or more events.

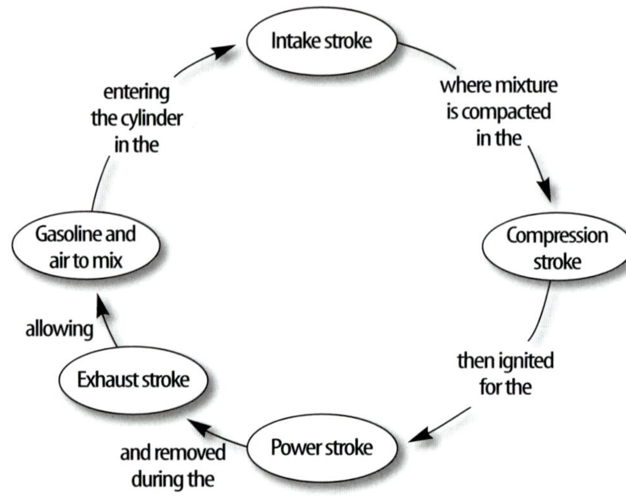

Figure 5 A cycle map shows events that occur in a cycle.

Spider Map A type of concept map that you can use for brainstorming is the spider map. When you have a central idea, you might find that you have a jumble of ideas that relate to it but are not necessarily clearly related to each other. The spider map on sound in **Figure 6** shows that if you write these ideas outside the main concept, then you can begin to separate and group unrelated terms so they become more useful.

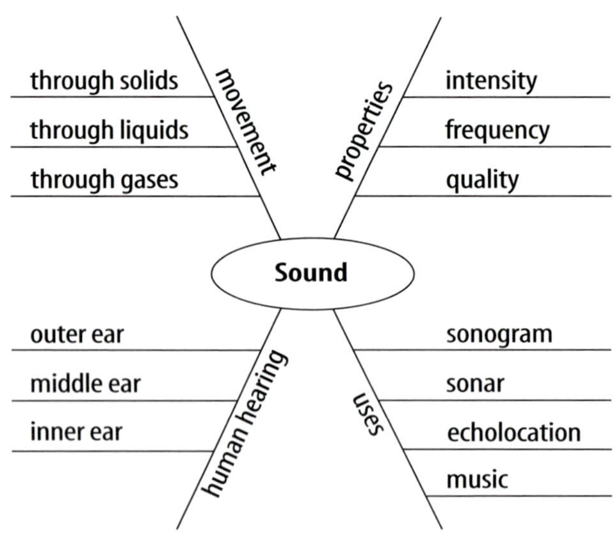

Figure 6 A spider map allows you to list ideas that relate to a central topic but not necessarily to one another.

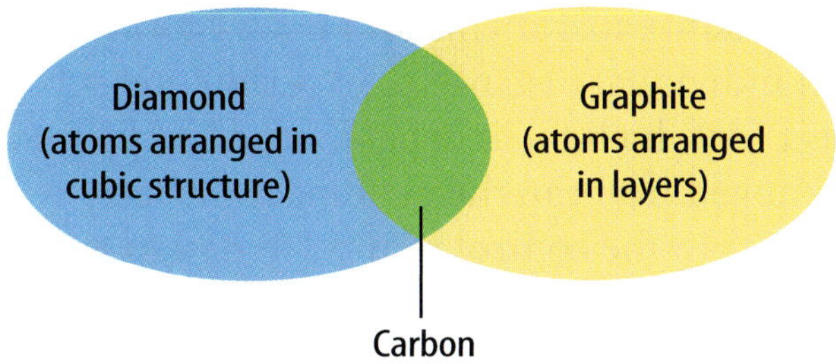

Figure 7 This Venn diagram compares and contrasts two substances made from carbon.

Venn Diagram To illustrate how two subjects compare and contrast you can use a Venn diagram. You can see the characteristics that the subjects have in common and those that they do not, shown in **Figure 7.**

To create a Venn diagram, draw two overlapping ovals that that are big enough to write in. List the characteristics unique to one subject in one oval, and the characteristics of the other subject in the other oval. The characteristics in common are listed in the overlapping section.

Make and Use Tables One way to organize information so it is easier to understand is to use a table. Tables can contain numbers, words, or both.

To make a table, list the items to be compared in the first column and the characteristics to be compared in the first row. The title should clearly indicate the content of the table, and the column or row heads should be clear. Notice that in **Table 1** the units are included.

Table 1 Recyclables Collected During Week

Day of Week	Paper (kg)	Aluminum (kg)	Glass (kg)
Monday	5.0	4.0	12.0
Wednesday	4.0	1.0	10.0
Friday	2.5	2.0	10.0

Make a Model One way to help you better understand the parts of a structure, the way a process works, or to show things too large or small for viewing is to make a model. For example, an atomic model made of a plastic-ball nucleus and pipe-cleaner electron shells can help you visualize how the parts of an atom relate to each other. Other types of models can by devised on a computer or represented by equations.

Form a Hypothesis

A possible explanation based on previous knowledge and observations is called a hypothesis. After researching gasoline types and recalling previous experiences in your family's car you form a hypothesis—our car runs more efficiently because we use premium gasoline. To be valid, a hypothesis has to be something you can test by using an investigation.

Predict When you apply a hypothesis to a specific situation, you predict something about that situation. A prediction makes a statement in advance, based on prior observation, experience, or scientific reasoning. People use predictions to make everyday decisions. Scientists test predictions by performing investigations. Based on previous observations and experiences, you might form a prediction that cars are more efficient with premium gasoline. The prediction can be tested in an investigation.

Design an Experiment A scientist needs to make many decisions before beginning an investigation. Some of these include: how to carry out the investigation, what steps to follow, how to record the data, and how the investigation will answer the question. It also is important to address any safety concerns.

Test the Hypothesis

Now that you have formed your hypothesis, you need to test it. Using an investigation, you will make observations and collect data, or information. This data might either support or not support your hypothesis. Scientists collect and organize data as numbers and descriptions.

Follow a Procedure In order to know what materials to use, as well as how and in what order to use them, you must follow a procedure. **Figure 8** shows a procedure you might follow to test your hypothesis.

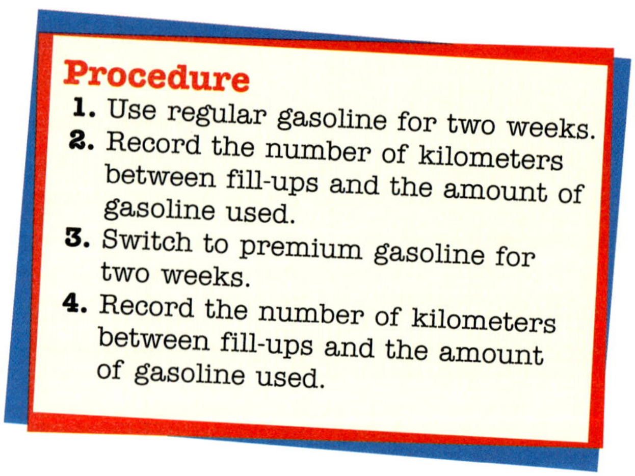

Figure 8 A procedure tells you what to do step by step.

Identify and Manipulate Variables and Controls In any experiment, it is important to keep everything the same except for the item you are testing. The one factor you change is called the independent variable. The change that results is the dependent variable. Make sure you have only one independent variable, to assure yourself of the cause of the changes you observe in the dependent variable. For example, in your gasoline experiment the type of fuel is the independent variable. The dependent variable is the efficiency.

Many experiments also have a control—an individual instance or experimental subject for which the independent variable is not changed. You can then compare the test results to the control results. To design a control you can have two cars of the same type. The control car uses regular gasoline for four weeks. After you are done with the test, you can compare the experimental results to the control results.

Collect Data

Whether you are carrying out an investigation or a short observational experiment, you will collect data, as shown in **Figure 9.** Scientists collect data as numbers and descriptions and organize it in specific ways.

Observe Scientists observe items and events, then record what they see. When they use only words to describe an observation, it is called qualitative data. Scientists' observations also can describe how much there is of something. These observations use numbers, as well as words, in the description and are called quantitative data. For example, if a sample of the element gold is described as being "shiny and very dense" the data are qualitative. Quantitative data on this sample of gold might include "a mass of 30 g and a density of 19.3 g/cm^3."

Figure 9 Collecting data is one way to gather information directly.

Figure 10 Record data neatly and clearly so it is easy to understand.

When you make observations you should examine the entire object or situation first, and then look carefully for details. It is important to record observations accurately and completely. Always record your notes immediately as you make them, so you do not miss details or make a mistake when recording results from memory. Never put unidentified observations on scraps of paper. Instead they should be recorded in a notebook, like the one in **Figure 10.** Write your data neatly so you can easily read it later. At each point in the experiment, record your observations and label them. That way, you will not have to determine what the figures mean when you look at your notes later. Set up any tables that you will need to use ahead of time, so you can record any observations right away. Remember to avoid bias when collecting data by not including personal thoughts when you record observations. Record only what you observe.

Estimate Scientific work also involves estimating. To estimate is to make a judgment about the size or the number of something without measuring or counting. This is important when the number or size of an object or population is too large or too difficult to accurately count or measure.

Sample Scientists may use a sample or a portion of the total number as a type of estimation. To sample is to take a small, representative portion of the objects or organisms of a population for research. By making careful observations or manipulating variables within that portion of the group, information is discovered and conclusions are drawn that might apply to the whole population. A poorly chosen sample can be unrepresentative of the whole. If you were trying to determine the rainfall in an area, it would not be best to take a rainfall sample from under a tree.

Measure You use measurements everyday. Scientists also take measurements when collecting data. When taking measurements, it is important to know how to use measuring tools properly. Accuracy also is important.

Length To measure length, the distance between two points, scientists use meters. Smaller measurements might be measured in centimeters or millimeters.

Length is measured using a metric ruler or meter stick. When using a metric ruler, line up the 0-cm mark with the end of the object being measured and read the number of the unit where the object ends. Look at the metric ruler shown in **Figure 11.** The centimeter lines are the long, numbered lines, and the shorter lines are millimeter lines. In this instance, the length would be 4.50 cm.

Figure 11 This metric ruler has centimeter and millimeter divisions.

Mass The SI unit for mass is the kilogram (kg). Scientists can measure mass using units formed by adding metric prefixes to the unit gram (g), such as milligram (mg). To measure mass, you might use a triple-beam balance similar to the one shown in **Figure 12.** The balance has a pan on one side and a set of beams on the other side. Each beam has a rider that slides on the beam.

When using a triple-beam balance, place an object on the pan. Slide the largest rider along its beam until the pointer drops below zero. Then move it back one notch. Repeat the process for each rider proceeding from the larger to smaller until the pointer swings an equal distance above and below the zero point. Sum the masses on each beam to find the mass of the object. Move all riders back to zero when finished.

Instead of putting materials directly on the balance, scientists often take a tare of a container. A tare is the mass of a container into which objects or substances are placed for measuring their masses. To mass objects or substances, find the mass of a clean container. Remove the container from the pan, and place the object or substances in the container. Find the mass of the container with the materials in it. Subtract the mass of the empty container from the mass of the filled container to find the mass of the materials you are using.

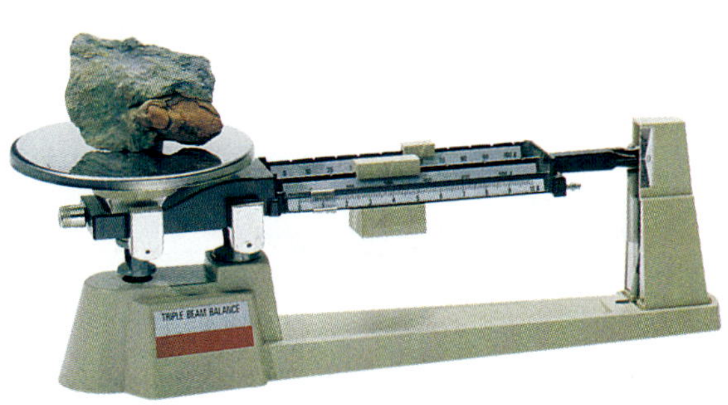

Figure 12 A triple-beam balance is used to determine the mass of an object.

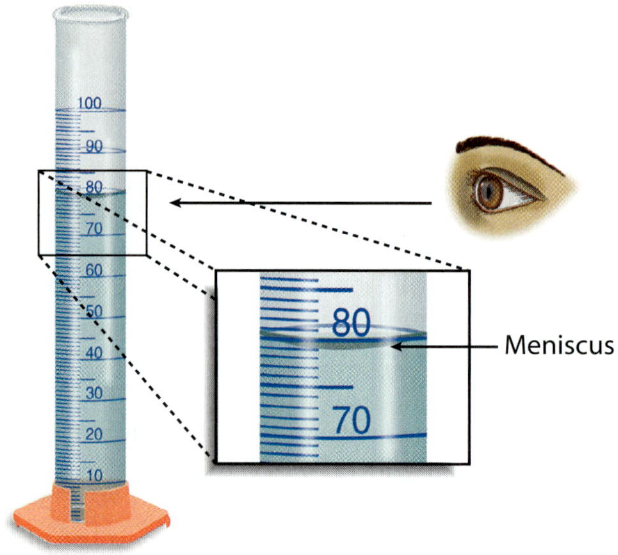

Figure 13 Graduated cylinders measure liquid volume.

Liquid Volume To measure liquids, the unit used is the liter. When a smaller unit is needed, scientists might use a milliliter. Because a milliliter takes up the volume of a cube measuring 1 cm on each side it also can be called a cubic centimeter ($cm^3 = cm \times cm \times cm$).

You can use beakers and graduated cylinders to measure liquid volume. A graduated cylinder, shown in **Figure 13,** is marked from bottom to top in milliliters. In lab, you might use a 10-mL graduated cylinder or a 100-mL graduated cylinder. When measuring liquids, notice that the liquid has a curved surface. Look at the surface at eye level, and measure the bottom of the curve. This is called the meniscus. The graduated cylinder in **Figure 13** contains 79.0 mL, or 79.0 cm^3, of a liquid.

Temperature Scientists often measure temperature using the Celsius scale. Pure water has a freezing point of 0°C and boiling point of 100°C. The unit of measurement is degrees Celsius. Two other scales often used are the Fahrenheit and Kelvin scales.

Figure 14 A thermometer measures the temperature of an object.

Scientists use a thermometer to measure temperature. Most thermometers in a laboratory are glass tubes with a bulb at the bottom end containing a liquid such as colored alcohol. The liquid rises or falls with a change in temperature. To read a glass thermometer like the thermometer in **Figure 14,** rotate it slowly until a red line appears. Read the temperature where the red line ends.

Form Operational Definitions An operational definition defines an object by how it functions, works, or behaves. For example, when you are playing hide and seek and a tree is home base, you have created an operational definition for a tree.

Objects can have more than one operational definition. For example, a ruler can be defined as a tool that measures the length of an object (how it is used). It can also be a tool with a series of marks used as a standard when measuring (how it works).

Analyze the Data

To determine the meaning of your observations and investigation results, you will need to look for patterns in the data. Then you must think critically to determine what the data mean. Scientists use several approaches when they analyze the data they have collected and recorded. Each approach is useful for identifying specific patterns.

Interpret Data The word *interpret* means "to explain the meaning of something." When analyzing data from an experiement, try to find out what the data show. Identify the control group and the test group to see whether or not changes in the independent variable have had an effect. Look for differences in the dependent variable between the control and test groups.

Classify Sorting objects or events into groups based on common features is called classifying. When classifying, first observe the objects or events to be classified. Then select one feature that is shared by some members in the group, but not by all. Place those members that share that feature in a subgroup. You can classify members into smaller and smaller subgroups based on characteristics. Remember that when you classify, you are grouping objects or events for a purpose. Keep your purpose in mind as you select the features to form groups and subgroups.

Compare and Contrast Observations can be analyzed by noting the similarities and differences between two more objects or events that you observe. When you look at objects or events to see how they are similar, you are comparing them. Contrasting is looking for differences in objects or events.

Recognize Cause and Effect A cause is a reason for an action or condition. The effect is that action or condition. When two events happen together, it is not necessarily true that one event caused the other. Scientists must design a controlled investigation to recognize the exact cause and effect.

Draw Conclusions

When scientists have analyzed the data they collected, they proceed to draw conclusions about the data. These conclusions are sometimes stated in words similar to the hypothesis that you formed earlier. They may confirm a hypothesis, or lead you to a new hypothesis.

Infer Scientists often make inferences based on their observations. An inference is an attempt to explain observations or to indicate a cause. An inference is not a fact, but a logical conclusion that needs further investigation. For example, you may infer that a fire has caused smoke. Until you investigate, however, you do not know for sure.

Apply When you draw a conclusion, you must apply those conclusions to determine whether the data supports the hypothesis. If your data do not support your hypothesis, it does not mean that the hypothesis is wrong. It means only that the result of the investigation did not support the hypothesis. Maybe the experiment needs to be redesigned, or some of the initial observations on which the hypothesis was based were incomplete or biased. Perhaps more observation or research is needed to refine your hypothesis. A successful investigation does not always come out the way you originally predicted.

Avoid Bias Sometimes a scientific investigation involves making judgments. When you make a judgment, you form an opinion. It is important to be honest and not to allow any expectations of results to bias your judgments. This is important throughout the entire investigation, from researching to collecting data to drawing conclusions.

Communicate

The communication of ideas is an important part of the work of scientists. A discovery that is not reported will not advance the scientific community's understanding or knowledge. Communication among scientists also is important as a way of improving their investigations.

Scientists communicate in many ways, from writing articles in journals and magazines that explain their investigations and experiments, to announcing important discoveries on television and radio. Scientists also share ideas with colleagues on the Internet or present them as lectures, like the student is doing in **Figure 15.**

Figure 15 A student communicates to his peers about his investigation.

SAFETY SYMBOLS	HAZARD	EXAMPLES	PRECAUTION	REMEDY
DISPOSAL	Special disposal procedures need to be followed.	certain chemicals, living organisms	Do not dispose of these materials in the sink or trash can.	Dispose of wastes as directed by your teacher.
BIOLOGICAL	Organisms or other biological materials that might be harmful to humans	bacteria, fungi, blood, unpreserved tissues, plant materials	Avoid skin contact with these materials. Wear mask or gloves.	Notify your teacher if you suspect contact with material. Wash hands thoroughly.
EXTREME TEMPERATURE	Objects that can burn skin by being too cold or too hot	boiling liquids, hot plates, dry ice, liquid nitrogen	Use proper protection when handling.	Go to your teacher for first aid.
SHARP OBJECT	Use of tools or glassware that can easily puncture or slice skin	razor blades, pins, scalpels, pointed tools, dissecting probes, broken glass	Practice common-sense behavior and follow guidelines for use of the tool.	Go to your teacher for first aid.
FUME	Possible danger to respiratory tract from fumes	ammonia, acetone, nail polish remover, heated sulfur, moth balls	Make sure there is good ventilation. Never smell fumes directly. Wear a mask.	Leave foul area and notify your teacher immediately.
ELECTRICAL	Possible danger from electrical shock or burn	improper grounding, liquid spills, short circuits, exposed wires	Double-check setup with teacher. Check condition of wires and apparatus.	Do not attempt to fix electrical problems. Notify your teacher immediately.
IRRITANT	Substances that can irritate the skin or mucous membranes of the respiratory tract	pollen, moth balls, steel wool, fiberglass, potassium permanganate	Wear dust mask and gloves. Practice extra care when handling these materials.	Go to your teacher for first aid.
CHEMICAL	Chemicals can react with and destroy tissue and other materials	bleaches such as hydrogen peroxide; acids such as sulfuric acid, hydrochloric acid; bases such as ammonia, sodium hydroxide	Wear goggles, gloves, and an apron.	Immediately flush the affected area with water and notify your teacher.
TOXIC	Substance may be poisonous if touched, inhaled, or swallowed.	mercury, many metal compounds, iodine, poinsettia plant parts	Follow your teacher's instructions.	Always wash hands thoroughly after use. Go to your teacher for first aid.
FLAMMABLE	Flammable chemicals may be ignited by open flame, spark, or exposed heat.	alcohol, kerosene, potassium permanganate	Avoid open flames and heat when using flammable chemicals.	Notify your teacher immediately. Use fire safety equipment if applicable.
OPEN FLAME	Open flame in use, may cause fire.	hair, clothing, paper, synthetic materials	Tie back hair and loose clothing. Follow teacher's instruction on lighting and extinguishing flames.	Notify your teacher immediately. Use fire safety equipment if applicable.

Eye Safety
Proper eye protection should be worn at all times by anyone performing or observing science activities.

Clothing Protection
This symbol appears when substances could stain or burn clothing.

Animal Safety
This symbol appears when safety of animals and students must be ensured.

Handwashing
After the lab, wash hands with soap and water before removing goggles.

Safety in the Science Laboratory

The science laboratory is a safe place to work if you follow standard safety procedures. Being responsible for your own safety helps to make the entire laboratory a safer place for everyone. When performing any lab, read and apply the caution statements and safety symbol listed at the beginning of the lab.

General Safety Rules

1. Obtain your teacher's permission to begin all investigations and use laboratory equipment.
2. Study the procedure. Ask your teacher any questions. Be sure you understand safety symbols shown on the page.
3. Notify your teacher about allergies or other health conditions which can affect your participation in a lab.
4. Learn and follow use and safety procedures for your equipment. If unsure, ask your teacher.
5. Never eat, drink, chew gum, apply cosmetics, or do any personal grooming in the lab. Never use lab glassware as food or drink containers. Keep your hands away from your face and mouth.
6. Know the location and proper use of the safety shower, eye wash, fire blanket, and fire alarm.

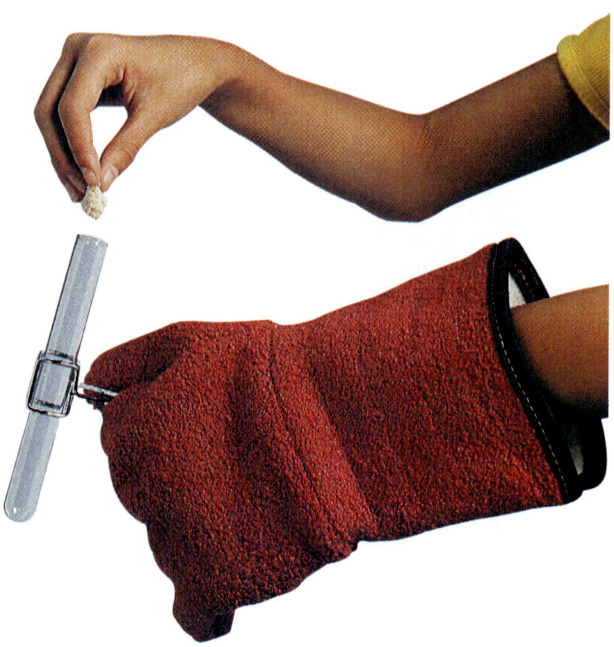

Prevent Accidents

1. Use the safety equipment provided to you. Goggles and a safety apron should be worn during investigations.
2. Do NOT use hair spray, mousse, or other flammable hair products. Tie back long hair and tie down loose clothing.
3. Do NOT wear sandals or other open-toed shoes in the lab.
4. Remove jewelry on hands and wrists. Loose jewelry, such as chains and long necklaces, should be removed to prevent them from getting caught in equipment.
5. Do not taste any substances or draw any material into a tube with your mouth.
6. Proper behavior is expected in the lab. Practical jokes and fooling around can lead to accidents and injury.
7. Keep your work area uncluttered.

Laboratory Work

1. Collect and carry all equipment and materials to your work area before beginning a lab.
2. Remain in your own work area unless given permission by your teacher to leave it.

3. Always slant test tubes away from yourself and others when heating them, adding substances to them, or rinsing them.
4. If instructed to smell a substance in a container, hold the container a short distance away and fan vapors towards your nose.
5. Do NOT substitute other chemicals/substances for those in the materials list unless instructed to do so by your teacher.
6. Do NOT take any materials or chemicals outside of the laboratory.
7. Stay out of storage areas unless instructed to be there and supervised by your teacher.

Laboratory Cleanup

1. Turn off all burners, water, and gas, and disconnect all electrical devices.
2. Clean all pieces of equipment and return all materials to their proper places.
3. Dispose of chemicals and other materials as directed by your teacher. Place broken glass and solid substances in the proper containers. Never discard materials in the sink.
4. Clean your work area.
5. Wash your hands with soap and water thoroughly BEFORE removing your goggles.

Emergencies

1. Report any fire, electrical shock, glassware breakage, spill, or injury, no matter how small, to your teacher immediately. Follow his or her instructions.
2. If your clothing should catch fire, STOP, DROP, and ROLL. If possible, smother it with the fire blanket or get under a safety shower. NEVER RUN.
3. If a fire should occur, turn off all gas and leave the room according to established procedures.
4. In most instances, your teacher will clean up spills. Do NOT attempt to clean up spills unless you are given permission and instructions to do so.
5. If chemicals come into contact with your eyes or skin, notify your teacher immediately. Use the eyewash or flush your skin or eyes with large quantities of water.
6. The fire extinguisher and first-aid kit should only be used by your teacher unless it is an extreme emergency and you have been given permission.
7. If someone is injured or becomes ill, only a professional medical provider or someone certified in first aid should perform first-aid procedures.

1 Panning Minerals

Time Required one period to pan the sand; an additional one to two periods to examine the black sand

Materials
- The sand should be mostly free of silt, clay, and organic matter.
- Clean sand can be collected at many locations. Clean, natural sand also can be obtained from some hardware suppliers, concrete plants, and aggregate pits.

Safety Precaution To ensure safety, obtain the sand for the students in advance.

Teaching Strategies
- You might want to practice the panning technique in advance.
- Inform students that the black sand appears black because it contains a high percentage of dark minerals, such as magnetite, ilmenite, pyroxene, and hornblende.

Conclude and Apply
1. The sand grains that were left are denser than the sand grains that washed over the pan's rim.
2. Students should see many well-formed crystals with a variety of shapes, sizes, and colors. If very lucky, someone might find a fleck of gold.
3. Grains of magnetite were attracted to the magnet.

EXTRA Try at Home Labs

From Your Kitchen, Junk Drawer, or Yard

1 Panning Minerals

Real-World Question

How can minerals be separated from sand?

Possible Materials
- large, aluminum pie pan
- gallon jug filled with water
- empty gallon jug
- clean sand
- funnel
- coffee filter
- squirt bottle of water
- magnifying lens
- white paper
- hand magnet

Procedure
1. Conduct this lab outdoors.
2. Line the funnel with a coffee filter. Insert the funnel stem into an empty gallon jug.
3. Add a small amount of sand to the pie pan. Add some water and swirl the pan.
4. Continue to shake and swirl the pan until only black sand is left in the pan.
5. Use the squirt bottle to wash the black sand into the coffee filter. Repeat steps 3–5 until you have a good sample of black sand.
6. Let the black sand dry. Then observe it with a magnifying lens. Test the sand with a magnet.

Conclude and Apply
1. Why was black sand left in the gold pan after swirling it?
2. Describe how the sand looked under the lens. Did you see any well-shaped crystals?
3. What happened when you tested the sand with a magnet? Explain.

2 Changing Rocks

Real-World Question

How can the change of metamorphic rock be modeled?

Possible Materials
- soil
- water
- measuring cup
- bowl
- spoon
- shale sample
- slate sample
- schist sample
- gneiss sample

Procedure
1. Mix equal parts of soil and water in a measuring cup or bowl. Stir the mixture until you make mud.
2. Place the bowl of mud on the table near the top edge.
3. Lay a sample of shale below the mud, a sample of slate below the shale, a sample of schist below the slate, a sample of gneiss below the schist.
4. Observe the different stages of sedimentary and metamorphic rocks that are formed by heat and pressure over long periods of time.

Conclude and Apply
1. Identify which rock sample(s) are sedimentary rock and which sample(s) are metamorphic rock.
2. Infer which type of rock is found at the greatest depth beneath the surface of Earth.

Adult supervision required for all labs.

These labs are available at bookf.msscience.com.

2 Changing Rocks

Time required 30 minutes

Materials Save class time by preparing the mud beforehand.

Safety Precaution Students should wear aprons.

Teaching Strategy Have students hypothesize the approximate depth beneath Earth's surface where each type of rock is found. shale—5 km below; slate—10 km below; schist—15 km below; and gneiss—20 km below

Conclude and Apply
1. Shale is a sedimentary rock; all other samples are metamorphic rocks.
2. gneiss

Extra Try at Home Labs

3 Why recycle?

Real-World Question
What are the effects of throwing out aluminum cans instead of recycling them?

Possible Materials
- calculator
- an aluminum can

Procedure
1. An aluminum can has a mass of about 13 g.

2. Convert the can's mass from grams to kilograms by dividing the mass by 1,000.
3. Find the volume of the can (in milliliters) on the label.
4. Convert the volume from milliliters to liters by dividing it by 1,000.

Conclude and Apply
1. Calculate the mass of aluminum cans thrown out by Americans each year by multiplying the mass of the can in kilograms times 50,000,000,000.
2. Calculate the amount of fuel needed to remake the cans thrown out by Americans each year by multiplying the volume of the can in liters times 50,000,000,000 and dividing your total by 2.
3. Infer the environmental effects of throwing out aluminum cans instead of recycling them.

4 Measuring Movement

Real-World Question
How can we model continental drift?

Possible Materials
- flashlight, nail, rubber band or tape, thick circle of paper
- protractor
- mirror
- stick-on notepad paper
- marker
- metric ruler
- calculator

Procedure
1. Cut a circle of paper to fit around the lens of the flashlight. Use a nail to make a hole in the paper. Fasten the paper with the rubber band or tape. You should now have a flashlight that shines a focused beam of light.
2. Direct the light beam of the flashlight on a protractor held horizontally so that the beam lines up to the 90° mark.
3. Darken a room and aim the light beam at a mirror from an angle. Measure the angle. Observe where the reflected beam hits the wall.
4. Have a partner place a stick-on note on the wall and mark the location of the beam on the paper with a marker.
5. Move the flashlight to a 100° angle and mark the beam's location on the wall with a second note.
6. Measure the distance between the two points on the wall and divide by ten to determine the distance per degree.

Conclude and Apply
1. What was the distance per degree of your measurements?
2. Calculate what the distance would be between the first spot and a third spot marking the location of the flashlight at a 40° angle. Test your calculations.
3. Explain how this lab models measuring continental drift.

Adult supervision required for all labs.

3 Why Recycle?

Time Required 15 minutes

Teaching Strategy
Americans discard 50 billion aluminum cans each year instead of recycling them. To make a can from raw materials instead of recycled materials requires the energy equivalent of 178 mL of gasoline.

Conclude and Apply
1. 650,000,000 kg
2. 4,128,000,000 L
3. Aluminum metal is wasted, fossil fuels are wasted, and there is an increase in air and water pollution.

4 Measuring Movement

Time Required 15 minutes

Materials A laser pointer can be used with caution in place of the flashlight.

Safety Precaution Caution students to never shine a laser beam or other light into someone's eyes or their own eyes.

Teaching Strategy Students should hold the laser pointer or flashlight beam in the same location for each measurement.

Conclude and Apply
1. Answers will vary.
2. Students should multiply the "distance per degree" by 45.
3. The mirror acts like a satellite in space reflecting a laser beam to an exact location back on earth for a precise measurement.

5 Making Waves

Time Required ten minutes

Materials
- Plastic coiled springs sold at toy stores can also be used for this lab.
- Tape the ends of any cut rope to prevent fraying.

Safety Precautions
- Students should be careful not to wave the rope or spring near other people.
- Caution both partners to hold on to the spring or rope firmly.

Teaching Strategies
- Clear away floor space in your classroom or go outside before starting this lab.
- Choose the pair of students who make the best waves to model the three types of seismic waves for the class while you and the class discuss the characteristics of each type of wave.

Conclude and Apply
1. primary (P) waves
2. secondary (S) waves
3. surface waves

Extra Try at Home Labs

Extra Try at Home Labs

5 Making Waves

Real-World Question

What do earthquake waves look like?

Possible Materials
- rope (3-m length)
- coiled spring
- garden hose

Procedure
1. With a partner, stretch a coiled spring out on the floor. Firmly push your side of the spring in and out toward your partner and observe the waves you created.
2. With a partner, stretch the rope out on the floor. Quickly wave your end side to side and observe the waves you created.
3. Stand with a partner, stretch the rope out, and hold it waist high. Quickly move your hand up and down and observe the waves you created.

Conclude and Apply
1. Infer the type of seismic wave you modeled with the coiled spring.
2. Infer the type of seismic waves you modeled with the rope on the floor.
3. Infer the type of seismic waves you modeled with the rope in the air.

6 Mini Eruptions

Real-World Question

How can we model the eruptions of shield and cinder cone volcanoes?

Possible Materials
- tube of toothpaste
- pin
- unopened bottle of carbonated soda
- newspaper
- paper towels

Procedure
1. Lay down newspaper or paper towels.
2. Press down on the back end of a full tube of toothpaste to move all the paste to the front of the tube.
3. Have a partner press a long pin into the center of the tube. Observe what happens to the toothpaste.
4. Go outside and vigorously shake a bottle of carbonated soda for 1 min.
5. Point the bottle away from other people and quickly remove the cap. Observe what happens to the soda.

Conclude and Apply
1. Describe what happened to the toothpaste and soda.
2. Infer how you modeled a shield volcano eruption.
3. Infer how you modeled a cinder cone volcano eruption.

Adult supervision required for all labs.

These labs are available at bookf.msscience.com.

6 Mini Eruptions

Time Required 20 minutes

Materials
- Cans of carbonated soda can also be used but with less dramatic results.

Safety Precautions
- Students should use pins with care.
- Caution students not to point their bottles at anyone during the eruptions.

Conclude and Apply
1. The toothpaste oozed out of the tube, but the soda sprayed out of the bottle up to 2 m.
2. The toothpaste oozed out of the pinhole and spread out without a violent eruption like the lava spreading out of a shield cone.
3. Soda sprayed out of the bottle with great force like the explosive eruption of a cinder cone volcano.

Computer Skills

People who study science rely on computers, like the one in **Figure 16,** to record and store data and to analyze results from investigations. Whether you work in a laboratory or just need to write a lab report with tables, good computer skills are a necessity.

Using the computer comes with responsibility. Issues of ownership, security, and privacy can arise. Remember, if you did not author the information you are using, you must provide a source for your information. Also, anything on a computer can be accessed by others. Do not put anything on the computer that you would not want everyone to know. To add more security to your work, use a password.

Use a Word Processing Program

A computer program that allows you to type your information, change it as many times as you need to, and then print it out is called a word processing program. Word processing programs also can be used to make tables.

Figure 16 A computer will make reports neater and more professional looking.

Learn the Skill To start your word processing program, a blank document, sometimes called "Document 1," appears on the screen. To begin, start typing. To create a new document, click the *New* button on the standard tool bar. These tips will help you format the document.

- The program will automatically move to the next line; press *Enter* if you wish to start a new paragraph.
- Symbols, called non-printing characters, can be hidden by clicking the *Show/Hide* button on your toolbar.
- To insert text, move the cursor to the point where you want the insertion to go, click on the mouse once, and type the text.
- To move several lines of text, select the text and click the *Cut* button on your toolbar. Then position your cursor in the location that you want to move the cut text and click *Paste.* If you move to the wrong place, click *Undo.*
- The spell check feature does not catch words that are misspelled to look like other words, like "cold" instead of "gold." Always reread your document to catch all spelling mistakes.
- To learn about other word processing methods, read the user's manual or click on the *Help* button.
- You can integrate databases, graphics, and spreadsheets into documents by copying from another program and pasting it into your document, or by using desktop publishing (DTP). DTP software allows you to put text and graphics together to finish your document with a professional look. This software varies in how it is used and its capabilities.

Use a Database

A collection of facts stored in a computer and sorted into different fields is called a database. A database can be reorganized in any way that suits your needs.

Learn the Skill A computer program that allows you to create your own database is a database management system (DBMS). It allows you to add, delete, or change information. Take time to get to know the features of your database software.

- Determine what facts you would like to include and research to collect your information.
- Determine how you want to organize the information.
- Follow the instructions for your particular DBMS to set up fields. Then enter each item of data in the appropriate field.
- Follow the instructions to sort the information in order of importance.
- Evaluate the information in your database, and add, delete, or change as necessary.

Use the Internet

The Internet is a global network of computers where information is stored and shared. To use the Internet, like the students in **Figure 17,** you need a modem to connect your computer to a phone line and an Internet Service Provider account.

Learn the Skill To access internet sites and information, use a "Web browser," which lets you view and explore pages on the World Wide Web. Each page is its own site, and each site has its own address, called a URL. Once you have found a Web browser, follow these steps for a search (this also is how you search a database).

Figure 17 The Internet allows you to search a global network for a variety of information.

- Be as specific as possible. If you know you want to research "gold," don't type in "elements." Keep narrowing your search until you find what you want.
- Web sites that end in *.com* are commercial Web sites; *.org*, *.edu*, and *.gov* are nonprofit, educational, or government Web sites.
- Electronic encyclopedias, almanacs, indexes, and catalogs will help locate and select relevant information.
- Develop a "home page" with relative ease. When developing a Web site, NEVER post pictures or disclose personal information such as location, names, or phone numbers. Your school or community usually can host your Web site. A basic understanding of HTML (hypertext mark-up language), the language of Web sites, is necessary. Software that creates HTML code is called authoring software, and can be downloaded free from many Web sites. This software allows text and pictures to be arranged as the software is writing the HTML code.

Use a Spreadsheet

A spreadsheet, shown in **Figure 18,** can perform mathematical functions with any data arranged in columns and rows. By entering a simple equation into a cell, the program can perform operations in specific cells, rows, or columns.

Learn the Skill Each column (vertical) is assigned a letter, and each row (horizontal) is assigned a number. Each point where a row and column intersect is called a cell, and is labeled according to where it is located—Column A, Row 1 (A1).

- Decide how to organize the data, and enter it in the correct row or column.
- Spreadsheets can use standard formulas or formulas can be customized to calculate cells.
- To make a change, click on a cell to make it activate, and enter the edited data or formula.
- Spreadsheets also can display your results in graphs. Choose the style of graph that best represents the data.

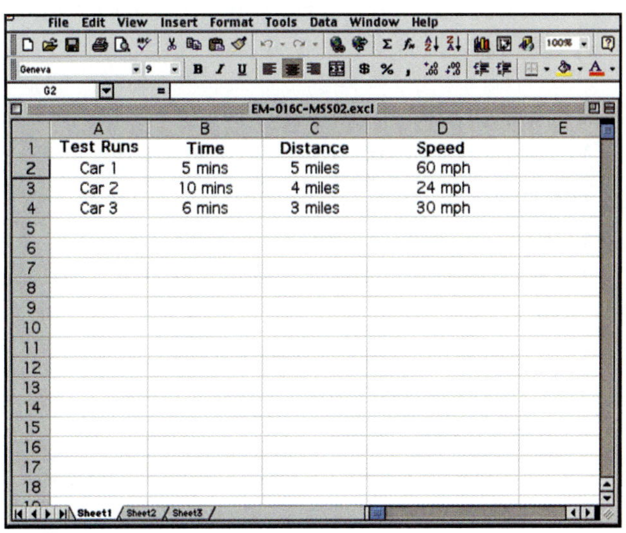

Figure 18 A spreadsheet allows you to perform mathematical operations on your data.

Use Graphics Software

Adding pictures, called graphics, to your documents is one way to make your documents more meaningful and exciting. This software adds, edits, and even constructs graphics. There is a variety of graphics software programs. The tools used for drawing can be a mouse, keyboard, or other specialized devices. Some graphics programs are simple. Others are complicated, called computer-aided design (CAD) software.

Learn the Skill It is important to have an understanding of the graphics software being used before starting. The better the software is understood, the better the results. The graphics can be placed in word-processing document.

- Clip art can be found on a variety of internet sites, and on CDs. These images can be copied and pasted into your document.
- When beginning, try editing existing drawings, then work up to creating drawings.
- The images are made of tiny rectangles of color called pixels. Each pixel can be altered.
- Digital photography is another way to add images. The photographs in the memory of a digital camera can be downloaded into a computer, then edited and added to the document.
- Graphics software also can allow animation. The software allows drawings to have the appearance of movement by connecting basic drawings automatically. This is called in-betweening, or tweening.
- Remember to save often.

Presentation Skills

Develop Multimedia Presentations

Most presentations are more dynamic if they include diagrams, photographs, videos, or sound recordings, like the one shown in **Figure 19.** A multimedia presentation involves using stereos, overhead projectors, televisions, computers, and more.

Learn the Skill Decide the main points of your presentation, and what types of media would best illustrate those points.

- Make sure you know how to use the equipment you are working with.
- Practice the presentation using the equipment several times.
- Enlist the help of a classmate to push play or turn lights out for you. Be sure to practice your presentation with him or her.
- If possible, set up all of the equipment ahead of time, and make sure everything is working properly.

Figure 19 These students are engaging the audience using a variety of tools.

Computer Presentations

There are many different interactive computer programs that you can use to enhance your presentation. Most computers have a compact disc (CD) drive that can play both CDs and digital video discs (DVDs). Also, there is hardware to connect a regular CD, DVD, or VCR. These tools will enhance your presentation.

Another method of using the computer to aid in your presentation is to develop a slide show using a computer program. This can allow movement of visuals at the presenter's pace, and can allow for visuals to build on one another.

Learn the Skill In order to create multimedia presentations on a computer, you need to have certain tools. These may include traditional graphic tools and drawing programs, animation programs, and authoring systems that tie everything together. Your computer will tell you which tools it supports. The most important step is to learn about the tools that you will be using.

- Often, color and strong images will convey a point better than words alone. Use the best methods available to convey your point.
- As with other presentations, practice many times.
- Practice your presentation with the tools you and any assistants will be using.
- Maintain eye contact with the audience. The purpose of using the computer is not to prompt the presenter, but to help the audience understand the points of the presentation.

Math Review

Use Fractions

A fraction compares a part to a whole. In the fraction $\frac{2}{3}$, the 2 represents the part and is the numerator. The 3 represents the whole and is the denominator.

Reduce Fractions To reduce a fraction, you must find the largest factor that is common to both the numerator and the denominator, the greatest common factor (GCF). Divide both numbers by the GCF. The fraction has then been reduced, or it is in its simplest form.

Example Twelve of the 20 chemicals in the science lab are in powder form. What fraction of the chemicals used in the lab are in powder form?

Step 1 Write the fraction.

$$\frac{\text{part}}{\text{whole}} = \frac{12}{20}$$

Step 2 To find the GCF of the numerator and denominator, list all of the factors of each number.

Factors of 12: 1, 2, 3, 4, 6, 12 (the numbers that divide evenly into 12)

Factors of 20: 1, 2, 4, 5, 10, 20 (the numbers that divide evenly into 20)

Step 3 List the common factors.

1, 2, 4.

Step 4 Choose the greatest factor in the list.

The GCF of 12 and 20 is 4.

Step 5 Divide the numerator and denominator by the GCF.

$$\frac{12 \div 4}{20 \div 4} = \frac{3}{5}$$

In the lab, $\frac{3}{5}$ of the chemicals are in powder form.

Practice Problem At an amusement park, 66 of 90 rides have a height restriction. What fraction of the rides, in its simplest form, has a height restriction?

Add and Subtract Fractions To add or subtract fractions with the same denominator, add or subtract the numerators and write the sum or difference over the denominator. After finding the sum or difference, find the simplest form for your fraction.

Example 1 In the forest outside your house, $\frac{1}{8}$ of the animals are rabbits, $\frac{3}{8}$ are squirrels, and the remainder are birds and insects. How many are mammals?

Step 1 Add the numerators.

$$\frac{1}{8} + \frac{3}{8} = \frac{(1 + 3)}{8} = \frac{4}{8}$$

Step 2 Find the GCF.

$$\frac{4}{8} \text{ (GCF, 4)}$$

Step 3 Divide the numerator and denominator by the GCF.

$$\frac{4}{4} = 1, \ \frac{8}{4} = 2$$

$\frac{1}{2}$ of the animals are mammals.

Example 2 If $\frac{7}{16}$ of the Earth is covered by freshwater, and $\frac{1}{16}$ of that is in glaciers, how much freshwater is not frozen?

Step 1 Subtract the numerators.

$$\frac{7}{16} - \frac{1}{16} = \frac{(7 - 1)}{16} = \frac{6}{16}$$

Step 2 Find the GCF.

$$\frac{6}{16} \text{ (GCF, 2)}$$

Step 3 Divide the numerator and denominator by the GCF.

$$\frac{6}{2} = 3, \ \frac{16}{2} = 8$$

$\frac{3}{8}$ of the freshwater is not frozen.

Practice Problem A bicycle rider is going 15 km/h for $\frac{4}{9}$ of his ride, 10 km/h for $\frac{2}{9}$ of his ride, and 8 km/h for the remainder of the ride. How much of his ride is he going over 8 km/h?

Reduce Fractions

$$\frac{66 \div 6}{90 \div 6} = \frac{11}{15}$$

Add and Subtract Fractions

$$\frac{4}{9} + \frac{2}{9} = \frac{6}{9}$$

$$\frac{6 \div 3}{9 \div 3} = \frac{2}{3}$$

Math Skill Handbook

MATH SKILL HANDBOOK F ◆ 205

Unlike Denominators

Problem 1

$1 \times 5 = 5, 8 \times 5 = 40$

$1 \times 4 = 4, 10 \times 4 = 40$

$\frac{5}{40} + \frac{4}{40} = \frac{9}{40}$

Problem 2

If $\frac{7}{10}$ are involuntary, the remainder are voluntary.

$\frac{10}{10} - \frac{7}{10} = \frac{3}{10}$

Unlike Denominators To add or subtract fractions with unlike denominators, first find the least common denominator (LCD). This is the smallest number that is a common multiple of both denominators. Rename each fraction with the LCD, and then add or subtract. Find the simplest form if necessary.

Example 1 A chemist makes a paste that is $\frac{1}{2}$ table salt (NaCl), $\frac{1}{3}$ sugar ($C_6H_{12}O_6$), and the rest water (H_2O). How much of the paste is a solid?

Step 1 Find the LCD of the fractions.

$\frac{1}{2} + \frac{1}{3}$ (LCD, 6)

Step 2 Rename each numerator and each denominator with the LCD.

$1 \times 3 = 3, \quad 2 \times 3 = 6$

$1 \times 2 = 2, \quad 3 \times 2 = 6$

Step 3 Add the numerators.

$\frac{3}{6} + \frac{2}{6} = \frac{(3 + 2)}{6} = \frac{5}{6}$

$\frac{5}{6}$ of the paste is a solid.

Example 2 The average precipitation in Grand Junction, CO, is $\frac{7}{10}$ inch in November, and $\frac{3}{5}$ inch in December. What is the total average precipitation?

Step 1 Find the LCD of the fractions.

$\frac{7}{10} + \frac{3}{5}$ (LCD, 10)

Step 2 Rename each numerator and each denominator with the LCD.

$7 \times 1 = 7, \quad 10 \times 1 = 10$

$3 \times 2 = 6, \quad 5 \times 2 = 10$

Step 3 Add the numerators.

$\frac{7}{10} + \frac{6}{10} = \frac{(7 + 6)}{10} = \frac{13}{10}$

$\frac{13}{10}$ inches total precipitation, or $1\frac{3}{10}$ inches.

Practice Problem On an electric bill, about $\frac{1}{8}$ of the energy is from solar energy and about $\frac{1}{10}$ is from wind power. How much of the total bill is from solar energy and wind power combined?

Example 3 In your body, $\frac{7}{10}$ of your muscle contractions are involuntary (cardiac and smooth muscle tissue). Smooth muscle makes $\frac{3}{15}$ of your muscle contractions. How many of your muscle contractions are made by cardiac muscle?

Step 1 Find the LCD of the fractions.

$\frac{7}{10} - \frac{3}{15}$ (LCD, 30)

Step 2 Rename each numerator and each denominator with the LCD.

$7 \times 3 = 21, \quad 10 \times 3 = 30$

$3 \times 2 = 6, \quad 15 \times 2 = 30$

Step 3 Subtract the numerators.

$\frac{21}{30} - \frac{6}{30} = \frac{(21 - 6)}{30} = \frac{15}{30}$

Step 4 Find the GCF.

$\frac{15}{30}$ (GCF, 15)

$\frac{1}{2}$

$\frac{1}{2}$ of all muscle contractions are cardiac muscle.

Example 4 Tony wants to make cookies that call for $\frac{3}{4}$ of a cup of flour, but he only has $\frac{1}{3}$ of a cup. How much more flour does he need?

Step 1 Find the LCD of the fractions.

$\frac{3}{4} - \frac{1}{3}$ (LCD, 12)

Step 2 Rename each numerator and each denominator with the LCD.

$3 \times 3 = 9, \quad 4 \times 3 = 12$

$1 \times 4 = 4, \quad 3 \times 4 = 12$

Step 3 Subtract the numerators.

$\frac{9}{12} - \frac{4}{12} = \frac{(9 - 4)}{12} = \frac{5}{12}$

$\frac{5}{12}$ of a cup of flour.

Practice Problem Using the information provided to you in Example 3 above, determine how many muscle contractions are voluntary (skeletal muscle).

Math Skill Handbook

206 ◆ F STUDENT RESOURCES

Multiply Fractions To multiply with fractions, multiply the numerators and multiply the denominators. Find the simplest form if necessary.

Example Multiply $\frac{3}{5}$ by $\frac{1}{3}$.

Step 1 Multiply the numerators and denominators.

$$\frac{3}{5} \times \frac{1}{3} = \frac{(3 \times 1)}{(5 \times 3)} = \frac{3}{15}$$

Step 2 Find the GCF.

$\frac{3}{15}$ (GCF, 3)

Step 3 Divide the numerator and denominator by the GCF.

$\frac{3}{3} = 1$, $\frac{15}{3} = 5$

$\frac{1}{5}$

$\frac{3}{5}$ multiplied by $\frac{1}{3}$ is $\frac{1}{5}$.

Practice Problem Multiply $\frac{3}{14}$ by $\frac{5}{16}$.

Find a Reciprocal Two numbers whose product is 1 are called multiplicative inverses, or reciprocals.

Example Find the reciprocal of $\frac{3}{8}$.

Step 1 Inverse the fraction by putting the denominator on top and the numerator on the bottom.

$\frac{8}{3}$

The reciprocal of $\frac{3}{8}$ is $\frac{8}{3}$.

Practice Problem Find the reciprocal of $\frac{4}{9}$.

Divide Fractions To divide one fraction by another fraction, multiply the dividend by the reciprocal of the divisor. Find the simplest form if necessary.

Example 1 Divide $\frac{1}{9}$ by $\frac{1}{3}$.

Step 1 Find the reciprocal of the divisor.

The reciprocal of $\frac{1}{3}$ is $\frac{3}{1}$.

Step 2 Multiply the dividend by the reciprocal of the divisor.

$$\frac{\frac{1}{9}}{\frac{1}{3}} = \frac{1}{9} \times \frac{3}{1} = \frac{(1 \times 3)}{(9 \times 1)} = \frac{3}{9}$$

Step 3 Find the GCF.

$\frac{3}{9}$ (GCF, 3)

Step 4 Divide the numerator and denominator by the GCF.

$\frac{3}{3} = 1$, $\frac{9}{3} = 3$

$\frac{1}{3}$

$\frac{1}{9}$ divided by $\frac{1}{3}$ is $\frac{1}{3}$.

Example 2 Divide $\frac{3}{5}$ by $\frac{1}{4}$.

Step 1 Find the reciprocal of the divisor.

The reciprocal of $\frac{1}{4}$ is $\frac{4}{1}$.

Step 2 Multiply the dividend by the reciprocal of the divisor.

$$\frac{\frac{3}{5}}{\frac{1}{4}} = \frac{3}{5} \times \frac{4}{1} = \frac{(3 \times 4)}{(5 \times 1)} = \frac{12}{5}$$

$\frac{3}{5}$ divided by $\frac{1}{4}$ is $\frac{12}{5}$ or $2\frac{2}{5}$.

Practice Problem Divide $\frac{3}{11}$ by $\frac{7}{10}$.

Math Skill Handbook

Multiply Fractions

$$\frac{3}{14} \times \frac{5}{16} = \frac{(3 \times 5)}{(14 \times 16)} = \frac{15}{224}$$

Find a Reciprocal

$\frac{9}{4}$

Divide Fractions

The reciprocal of $\frac{7}{10}$ is $\frac{10}{7}$.

$$\frac{3}{11} \times \frac{10}{7} = \frac{(3 \times 10)}{(11 \times 7)} = \frac{30}{77}$$

Math Skill Handbook

MATH SKILL HANDBOOK F ◆ 207

Use Ratios

$$\frac{100\text{ cm}}{144\text{ cm}} = \frac{100 \div 4}{144 \div 4} = \frac{25}{36}$$

25:36

Add or Subtract Decimals

$$\begin{array}{r} \scriptstyle 1 \\ 1.245 \\ +\ 3.842 \\ \hline 5.087 \end{array}$$

Use Ratios

When you compare two numbers by division, you are using a ratio. Ratios can be written 3 to 5, 3:5, or $\frac{3}{5}$. Ratios, like fractions, also can be written in simplest form.

Ratios can represent probabilities, also called odds. This is a ratio that compares the number of ways a certain outcome occurs to the number of outcomes. For example, if you flip a coin 100 times, what are the odds that it will come up heads? There are two possible outcomes, heads or tails, so the odds of coming up heads are 50:100. Another way to say this is that 50 out of 100 times the coin will come up heads. In its simplest form, the ratio is 1:2.

Example 1 A chemical solution contains 40 g of salt and 64 g of baking soda. What is the ratio of salt to baking soda as a fraction in simplest form?

Step 1 Write the ratio as a fraction.

$$\frac{\text{salt}}{\text{baking soda}} = \frac{40}{64}$$

Step 2 Express the fraction in simplest form.

The GCF of 40 and 64 is 8.

$$\frac{40}{64} = \frac{40 \div 8}{64 \div 8} = \frac{5}{8}$$

The ratio of salt to baking soda in the sample is 5:8.

Example 2 Sean rolls a 6-sided die 6 times. What are the odds that the side with a 3 will show?

Step 1 Write the ratio as a fraction.

$$\frac{\text{number of sides with a 3}}{\text{number of sides}} = \frac{1}{6}$$

Step 2 Multiply by the number of attempts.

$$\frac{1}{6} \times 6 \text{ attempts} = \frac{6}{6} \text{ attempts} = 1 \text{ attempt}$$

1 attempt out of 6 will show a 3.

Practice Problem Two metal rods measure 100 cm and 144 cm in length. What is the ratio of their lengths in simplest form?

Use Decimals

A fraction with a denominator that is a power of ten can be written as a decimal. For example, 0.27 means $\frac{27}{100}$. The decimal point separates the ones place from the tenths place.

Any fraction can be written as a decimal using division. For example, the fraction $\frac{5}{8}$ can be written as a decimal by dividing 5 by 8. Written as a decimal, it is 0.625.

Add or Subtract Decimals When adding and subtracting decimals, line up the decimal points before carrying out the operation.

Example 1 Find the sum of 47.68 and 7.80.

Step 1 Line up the decimal places when you write the numbers.

$$\begin{array}{r} 47.68 \\ +\ 7.80 \\ \hline \end{array}$$

Step 2 Add the decimals.

$$\begin{array}{r} 47.68 \\ +\ 7.80 \\ \hline 55.48 \end{array}$$

The sum of 47.68 and 7.80 is 55.48.

Example 2 Find the difference of 42.17 and 15.85.

Step 1 Line up the decimal places when you write the number.

$$\begin{array}{r} 42.17 \\ -15.85 \\ \hline \end{array}$$

Step 2 Subtract the decimals.

$$\begin{array}{r} 42.17 \\ -15.85 \\ \hline 26.32 \end{array}$$

The difference of 42.17 and 15.85 is 26.32.

Practice Problem Find the sum of 1.245 and 3.842.

Multiply Decimals To multiply decimals, multiply the numbers like any other number, ignoring the decimal point. Count the decimal places in each factor. The product will have the same number of decimal places as the sum of the decimal places in the factors.

Example Multiply 2.4 by 5.9.

Step 1 Multiply the factors like two whole numbers.
$24 \times 59 = 1416$

Step 2 Find the sum of the number of decimal places in the factors. Each factor has one decimal place, for a sum of two decimal places.

Step 3 The product will have two decimal places.
14.16

The product of 2.4 and 5.9 is 14.16.

Practice Problem Multiply 4.6 by 2.2.

Divide Decimals When dividing decimals, change the divisor to a whole number. To do this, multiply both the divisor and the dividend by the same power of ten. Then place the decimal point in the quotient directly above the decimal point in the dividend. Then divide as you do with whole numbers.

Example Divide 8.84 by 3.4.

Step 1 Multiply both factors by 10.
$3.4 \times 10 = 34$, $8.84 \times 10 = 88.4$

Step 2 Divide 88.4 by 34.

$$\begin{array}{r} 2.6 \\ 34\overline{)88.4} \\ -68 \\ \hline 20\,4 \\ -20\,4 \\ \hline 0 \end{array}$$

8.84 divided by 3.4 is 2.6.

Practice Problem Divide 75.6 by 3.6.

Use Proportions

An equation that shows that two ratios are equivalent is a proportion. The ratios $\frac{2}{4}$ and $\frac{5}{10}$ are equivalent, so they can be written as $\frac{2}{4} = \frac{5}{10}$. This equation is a proportion.

When two ratios form a proportion, the cross products are equal. To find the cross products in the proportion $\frac{2}{4} = \frac{5}{10}$, multiply the 2 and the 10, and the 4 and the 5. Therefore $2 \times 10 = 4 \times 5$, or $20 = 20$.

Because you know that both proportions are equal, you can use cross products to find a missing term in a proportion. This is known as solving the proportion.

Example The heights of a tree and a pole are proportional to the lengths of their shadows. The tree casts a shadow of 24 m when a 6-m pole casts a shadow of 4 m. What is the height of the tree?

Step 1 Write a proportion.

$$\frac{\text{height of tree}}{\text{height of pole}} = \frac{\text{length of tree's shadow}}{\text{length of pole's shadow}}$$

Step 2 Substitute the known values into the proportion. Let h represent the unknown value, the height of the tree.

$$\frac{h}{6} = \frac{24}{4}$$

Step 3 Find the cross products.
$h \times 4 = 6 \times 24$

Step 4 Simplify the equation.
$4h = 144$

Step 5 Divide each side by 4.

$$\frac{4h}{4} = \frac{144}{4}$$

$h = 36$

The height of the tree is 36 m.

Practice Problem The ratios of the weights of two objects on the Moon and on Earth are in proportion. A rock weighing 3 N on the Moon weighs 18 N on Earth. How much would a rock that weighs 5 N on the Moon weigh on Earth?

Multiply Decimals

Multiply 4.6 and 2.2 by 10.

$46 \times 22 = 1012$

Each factor had one decimal place.

10.12

Divide Decimals

Multiply both factors by 10.

Divide 756 by 36.

$$\begin{array}{r} 21 \\ 36\overline{)756} \\ 72 \\ \hline 36 \\ 36 \\ \hline 0 \end{array}$$

Use Proportions

$$\frac{3}{18} = \frac{5}{w}$$

$w \times 3 = 5 \times 18$

$$\frac{3w}{3} = \frac{90}{3}$$

$w = 30$

Math Skill Handbook

MATH SKILL HANDBOOK F ◆ 209

Use Percentages

$\frac{73}{365} = \frac{x}{100}$

$\frac{7300}{365} = \frac{365x}{365}$

$20\% = x$

Solve One-Step Equations

$h = gd$

$\frac{17.4}{12.3} = \frac{12.3d}{12.3}$

$1.41 = d$

Math Skill Handbook

Math Skill Handbook

Use Percentages

The word *percent* means "out of one hundred." It is a ratio that compares a number to 100. Suppose you read that 77 percent of the Earth's surface is covered by water. That is the same as reading that the fraction of the Earth's surface covered by water is $\frac{77}{100}$. To express a fraction as a percent, first find the equivalent decimal for the fraction. Then, multiply the decimal by 100 and add the percent symbol.

Example Express $\frac{13}{20}$ as a percent.

Step 1 Find the equivalent decimal for the fraction.

$$\begin{array}{r} 0.65 \\ 20\overline{)13.00} \\ \underline{12\,0} \\ 1\,00 \\ \underline{1\,00} \\ 0 \end{array}$$

Step 2 Rewrite the fraction $\frac{13}{20}$ as 0.65.

Step 3 Multiply 0.65 by 100 and add the % sign.

$0.65 \times 100 = 65 = 65\%$

So, $\frac{13}{20} = 65\%$.

This also can be solved as a proportion.

Example Express $\frac{13}{20}$ as a percent.

Step 1 Write a proportion.

$\frac{13}{20} = \frac{x}{100}$

Step 2 Find the cross products.

$1300 = 20x$

Step 3 Divide each side by 20.

$\frac{1300}{20} = \frac{20x}{20}$

$65\% = x$

Practice Problem In one year, 73 of 365 days were rainy in one city. What percent of the days in that city were rainy?

Solve One-Step Equations

A statement that two things are equal is an equation. For example, $A = B$ is an equation that states that A is equal to B.

An equation is solved when a variable is replaced with a value that makes both sides of the equation equal. To make both sides equal the inverse operation is used. Addition and subtraction are inverses, and multiplication and division are inverses.

Example 1 Solve the equation $x - 10 = 35$.

Step 1 Find the solution by adding 10 to each side of the equation.

$x - 10 = 35$

$x - 10 + 10 = 35 + 10$

$x = 45$

Step 2 Check the solution.

$x - 10 = 35$

$45 - 10 = 35$

$35 = 35$

Both sides of the equation are equal, so $x = 45$.

Example 2 In the formula $a = bc$, find the value of c if $a = 20$ and $b = 2$.

Step 1 Rearrange the formula so the unknown value is by itself on one side of the equation by dividing both sides by b.

$a = bc$

$\frac{a}{b} = \frac{bc}{b}$

$\frac{a}{b} = c$

Step 2 Replace the variables a and b with the values that are given.

$\frac{a}{b} = c$

$\frac{20}{2} = c$

$10 = c$

Step 3 Check the solution.

$a = bc$

$20 = 2 \times 10$

$20 = 20$

Both sides of the equation are equal, so $c = 10$ is the solution when $a = 20$ and $b = 2$.

Practice Problem In the formula $h = gd$, find the value of d if $g = 12.3$ and $h = 17.4$.

Math Skill Handbook

Use Statistics

The branch of mathematics that deals with collecting, analyzing, and presenting data is statistics. In statistics, there are three common ways to summarize data with a single number—the mean, the median, and the mode.

The **mean** of a set of data is the arithmetic average. It is found by adding the numbers in the data set and dividing by the number of items in the set.

The **median** is the middle number in a set of data when the data are arranged in numerical order. If there were an even number of data points, the median would be the mean of the two middle numbers.

The **mode** of a set of data is the number or item that appears most often.

Another number that often is used to describe a set of data is the range. The **range** is the difference between the largest number and the smallest number in a set of data.

A **frequency table** shows how many times each piece of data occurs, usually in a survey. **Table 2** below shows the results of a student survey on favorite color.

Table 2 Student Color Choice

Color	Tally	Frequency
red	𝍷𝍷𝍷𝍷	4
blue	𝍸	5
black	𝍷𝍷	2
green	𝍷𝍷𝍷	3
purple	𝍸 𝍷𝍷	7
yellow	𝍸 𝍷	6

Based on the frequency table data, which color is the favorite?

Example The speeds (in m/s) for a race car during five different time trials are 39, 37, 44, 36, and 44.

To find the mean:

Step 1 Find the sum of the numbers.
$39 + 37 + 44 + 36 + 44 = 200$

Step 2 Divide the sum by the number of items, which is 5.
$200 \div 5 = 40$

The mean is 40 m/s.

To find the median:

Step 1 Arrange the measures from least to greatest.
36, 37, 39, 44, 44

Step 2 Determine the middle measure.
36, 37, 39, 44, 44

The median is 39 m/s.

To find the mode:

Step 1 Group the numbers that are the same together.
44, 44, 36, 37, 39

Step 2 Determine the number that occurs most in the set.
44, 44, 36, 37, 39

The mode is 44 m/s.

To find the range:

Step 1 Arrange the measures from largest to smallest.
44, 44, 39, 37, 36

Step 2 Determine the largest and smallest measures in the set.
44, 44, 39, 37, 36

Step 3 Find the difference between the largest and smallest measures.
$44 - 36 = 8$

The range is 8 m/s.

Practice Problem Find the mean, median, mode, and range for the data set 8, 4, 12, 8, 11, 14, 16.

Use Statistics

mean

$8 + 4 + 12 + 8 + 11 + 14 + 16 = 73$

$73 \div 7 = 10.4$

median

4, 8, 8, 11, 12, 14, 16

mode

4, 8, 8, 11, 12, 14, 16

range

4, 8, 8, 11, 12, 14, 16

$16 - 4 = 12$

Math Skill Handbook

MATH SKILL HANDBOOK F ◆ 211

Use Geometry

The branch of mathematics that deals with the measurement, properties, and relationships of points, lines, angles, surfaces, and solids is called geometry.

Perimeter The **perimeter** (P) is the distance around a geometric figure. To find the perimeter of a rectangle, add the length and width and multiply that sum by two, or $2(l + w)$. To find perimeters of irregular figures, add the length of the sides.

Example 1 Find the perimeter of a rectangle that is 3 m long and 5 m wide.

Step 1 You know that the perimeter is 2 times the sum of the width and length.
$P = 2(3\text{ m} + 5\text{ m})$

Step 2 Find the sum of the width and length.
$P = 2(8\text{ m})$

Step 3 Multiply by 2.
$P = 16\text{ m}$

The perimeter is 16 m.

Example 2 Find the perimeter of a shape with sides measuring 2 cm, 5 cm, 6 cm, 3 cm.

Step 1 You know that the perimeter is the sum of all the sides.
$P = 2 + 5 + 6 + 3$

Step 2 Find the sum of the sides.
$P = 2 + 5 + 6 + 3$
$P = 16$

The perimeter is 16 cm.

Practice Problem Find the perimeter of a rectangle with a length of 18 m and a width of 7 m.

Practice Problem Find the perimeter of a triangle measuring 1.6 cm by 2.4 cm by 2.4 cm.

Area of a Rectangle The **area** (A) is the number of square units needed to cover a surface. To find the area of a rectangle, multiply the length times the width, or $l \times w$. When finding area, the units also are multiplied. Area is given in square units.

Example Find the area of a rectangle with a length of 1 cm and a width of 10 cm.

Step 1 You know that the area is the length multiplied by the width.
$A = (1\text{ cm} \times 10\text{ cm})$

Step 2 Multiply the length by the width. Also multiply the units.
$A = 10\text{ cm}^2$

The area is 10 cm^2.

Practice Problem Find the area of a square whose sides measure 4 m.

Area of a Triangle To find the area of a triangle, use the formula:

$$A = \frac{1}{2}(\text{base} \times \text{height})$$

The base of a triangle can be any of its sides. The height is the perpendicular distance from a base to the opposite endpoint, or vertex.

Example Find the area of a triangle with a base of 18 m and a height of 7 m.

Step 1 You know that the area is $\frac{1}{2}$ the base times the height.
$A = \frac{1}{2}(18\text{ m} \times 7\text{ m})$

Step 2 Multiply $\frac{1}{2}$ by the product of 18×7. Multiply the units.
$A = \frac{1}{2}(126\text{ m}^2)$
$A = 63\text{ m}^2$

The area is 63 m^2.

Practice Problem Find the area of a triangle with a base of 27 cm and a height of 17 cm.

Perimeter

Problem 1

$P = 2(18\text{ m} + 7\text{ m})$

$P = 2(25\text{ m})$

$P = 50\text{ m}$

Problem 2

$P = 1.6\text{ cm} + 2.4\text{ cm} + 2.4\text{ cm}$

$P = 6.4\text{ cm}$

Area of a Rectangle

$A = (4\text{ m} \times 4\text{ m})$

$A = 16\text{ m}^2$

Area of a Triangle

$A = \frac{1}{2}(27\text{ cm} \times 17\text{ cm})$

$A = \frac{1}{2}(459\text{ cm}^2)$

$A = 229.5\text{ cm}^2$

Circumference of a Circle The **diameter** (d) of a circle is the distance across the circle through its center, and the **radius** (r) is the distance from the center to any point on the circle. The radius is half of the diameter. The distance around the circle is called the **circumference** (C). The formula for finding the circumference is:

$C = 2\pi r$ *or* $C = \pi d$

The circumference divided by the diameter is always equal to 3.1415926... This nonterminating and nonrepeating number is represented by the Greek letter π (pi). An approximation often used for π is 3.14.

Example 1 Find the circumference of a circle with a radius of 3 m.

Step 1 You know the formula for the circumference is 2 times the radius times π.
$C = 2\pi(3)$

Step 2 Multiply 2 times the radius.
$C = 6\pi$

Step 3 Multiply by π.
$C = 19$ m

The circumference is 19 m.

Example 2 Find the circumference of a circle with a diameter of 24.0 cm.

Step 1 You know the formula for the circumference is the diameter times π.
$C = \pi(24.0)$

Step 2 Multiply the diameter by π.
$C = 75.4$ cm

The circumference is 75.4 cm.

Practice Problem Find the circumference of a circle with a radius of 19 cm.

Area of a Circle The formula for the area of a circle is:
$A = \pi r^2$

Example 1 Find the area of a circle with a radius of 4.0 cm.

Step 1 $A = \pi(4.0)^2$

Step 2 Find the square of the radius.
$A = 16\pi$

Step 3 Multiply the square of the radius by π.
$A = 50\ cm^2$

The area of the circle is 50 cm^2.

Example 2 Find the area of a circle with a radius of 225 m.

Step 1 $A = \pi(225)^2$

Step 2 Find the square of the radius.
$A = 50625\pi$

Step 3 Multiply the square of the radius by π.
$A = 158962.5$

The area of the circle is 158,962 m^2.

Example 3 Find the area of a circle whose diameter is 20.0 mm.

Step 1 You know the formula for the area of a circle is the square of the radius times π, and that the radius is half of the diameter.
$A = \pi\left(\frac{20.0}{2}\right)^2$

Step 2 Find the radius.
$A = \pi(10.0)^2$

Step 3 Find the square of the radius.
$A = 100\pi$

Step 4 Multiply the square of the radius by π.
$A = 314\ mm^2$

The area is 314 mm^2.

Practice Problem Find the area of a circle with a radius of 16 m.

Math Skill Handbook

Circumference of a Circle

$C = 2\pi r$
$C = 2\pi(19)$
$C = 38\pi$
$C = 119.3$

Area of a Circle

$A = \pi r^2$
$A = \pi(16\ m)^2$
$A = \pi\ 256\ m^2$
$A = 803.8\ m^2$

Math Skill Handbook

Volume

Problem 1

$V = 8\text{ m} \times 4\text{ m} \times 4\text{ m}$

$V = 128\text{ m}^3$

Problem 2

$V = \pi r^2 \times \text{height}$

$V = \left[\pi\left(\frac{1}{2} \times 7\right)^2\right] \times 16$

$V = [\pi(3.5)^2] \times 16$

$V = [\pi(12.25)] \times 16$

$V = 38.46 \times 16$

$V = 615.36$

Volume The measure of space occupied by a solid is the **volume** (V). To find the volume of a rectangular solid multiply the length times width times height, or $V = l \times w \times h$. It is measured in cubic units, such as cubic centimeters (cm^3).

Example Find the volume of a rectangular solid with a length of 2.0 m, a width of 4.0 m, and a height of 3.0 m.

Step 1 You know the formula for volume is the length times the width times the height.

$V = 2.0\text{ m} \times 4.0\text{ m} \times 3.0\text{ m}$

Step 2 Multiply the length times the width times the height.

$V = 24\text{ m}^3$

The volume is 24 m^3.

Practice Problem Find the volume of a rectangular solid that 8 m long, 4 m wide, and 4 m high.

To find the volume of other solids, multiply the area of the base times the height.

Example 1 Find the volume of a solid that has a triangular base with a length of 8.0 m and a height of 7.0 m. The height of the entire solid is 15.0 m.

Step 1 You know that the base is a triangle, and the area of a triangle is $\frac{1}{2}$ the base times the height, and the volume is the area of the base times the height.

$V = \left[\frac{1}{2}(b \times h)\right] \times 15$

Step 2 Find the area of the base.

$V = \left[\frac{1}{2}(8 \times 7)\right] \times 15$

$V = \left(\frac{1}{2} \times 56\right) \times 15$

Step 3 Multiply the area of the base by the height of the solid.

$V = 28 \times 15$

$V = 420\text{ m}^3$

The volume is 420 m^3.

Example 2 Find the volume of a cylinder that has a base with a radius of 12.0 cm, and a height of 21.0 cm.

Step 1 You know that the base is a circle, and the area of a circle is the square of the radius times π, and the volume is the area of the base times the height.

$V = (\pi r^2) \times 21$

$V = (\pi 12^2) \times 21$

Step 2 Find the area of the base.

$V = 144\pi \times 21$

$V = 452 \times 21$

Step 3 Multiply the area of the base by the height of the solid.

$V = 9490\text{ cm}^3$

The volume is 9490 cm^3.

Example 3 Find the volume of a cylinder that has a diameter of 15 mm and a height of 4.8 mm.

Step 1 You know that the base is a circle with an area equal to the square of the radius times π. The radius is one-half the diameter. The volume is the area of the base times the height.

$V = (\pi r^2) \times 4.8$

$V = \left[\pi\left(\frac{1}{2} \times 15\right)^2\right] \times 4.8$

$V = (\pi 7.5^2) \times 4.8$

Step 2 Find the area of the base.

$V = 56.25\pi \times 4.8$

$V = 176.63 \times 4.8$

Step 3 Multiply the area of the base by the height of the solid.

$V = 847.8$

The volume is 847.8 mm^3.

Practice Problem Find the volume of a cylinder with a diameter of 7 cm in the base and a height of 16 cm.

Math Skill Handbook

214 ◆ F STUDENT RESOURCES

Science Applications

Measure in SI

The metric system of measurement was developed in 1795. A modern form of the metric system, called the International System (SI), was adopted in 1960 and provides the standard measurements that all scientists around the world can understand.

The SI system is convenient because unit sizes vary by powers of 10. Prefixes are used to name units. Look at **Table 3** for some common SI prefixes and their meanings.

Table 3 Some SI Prefixes

Prefix	Symbol	Meaning	
kilo-	k	1,000	thousand
hecto-	h	100	hundred
deka-	da	10	ten
deci-	d	0.1	tenth
centi-	c	0.01	hundredth
milli-	m	0.001	thousandth

Example How many grams equal one kilogram?

Step 1 Find the prefix *kilo* in **Table 3.**

Step 2 Using **Table 3,** determine the meaning of *kilo.* According to the table, it means 1,000. When the prefix *kilo* is added to a unit, it means that there are 1,000 of the units in a "*kilo*unit."

Step 3 Apply the prefix to the units in the question. The units in the question are grams. There are 1,000 grams in a kilogram.

Practice Problem Is a milligram larger or smaller than a gram? How many of the smaller units equal one larger unit? What fraction of the larger unit does one smaller unit represent?

Dimensional Analysis

Convert SI Units In science, quantities such as length, mass, and time sometimes are measured using different units. A process called dimensional analysis can be used to change one unit of measure to another. This process involves multiplying your starting quantity and units by one or more conversion factors. A conversion factor is a ratio equal to one and can be made from any two equal quantities with different units. If 1,000 mL equal 1 L then two ratios can be made.

$$\frac{1{,}000 \text{ mL}}{1 \text{ L}} = \frac{1 \text{ L}}{1{,}000 \text{ mL}} = 1$$

One can covert between units in the SI system by using the equivalents in **Table 3** to make conversion factors.

Example 1 How many cm are in 4 m?

Step 1 Write conversion factors for the units given. From **Table 3,** you know that 100 cm = 1 m. The conversion factors are

$$\frac{100 \text{ cm}}{1 \text{ m}} \text{ and } \frac{1 \text{ m}}{100 \text{ cm}}$$

Step 2 Decide which conversion factor to use. Select the factor that has the units you are converting from (m) in the denominator and the units you are converting to (cm) in the numerator.

$$\frac{100 \text{ cm}}{1 \text{ m}}$$

Step 3 Multiply the starting quantity and units by the conversion factor. Cancel the starting units with the units in the denominator. There are 400 cm in 4 m.

$$4 \cancel{\text{m}} \times \frac{100 \text{ cm}}{1 \cancel{\text{m}}} = 400 \text{ cm}$$

Practice Problem How many milligrams are in one kilogram? (Hint: You will need to use two conversion factors from **Table 3.**)

Measure in SI

smaller; 1000; one thousandth

Dimensional Analysis

$x \text{ mg} = 1 \cancel{\text{kg}} \times \frac{1000 \cancel{\text{g}}}{1 \cancel{\text{kg}}} \times \frac{1000 \text{ mg}}{1 \cancel{\text{g}}} =$

1,000,000 mg

1,000,000 mg = 1 kg

Convert Between Unit Systems

$$\frac{(1\ \text{in})^3}{(2.54\ \text{cm})^3}$$

$$= \frac{1\ \text{in} \times 1\ \text{in} \times 1\ \text{in}}{2.54\ \text{cm} \times 2.54\ \text{cm} \times 2.54\ \text{cm}}$$

$$= \frac{1\ \text{in}^3}{16.39\ \text{cm}^3}$$

Table 4 Unit System Equivalents

Type of Measurement	Equivalent
Length	1 in = 2.54 cm 1 yd = 0.91 m 1 mi = 1.61 km
Mass and Weight*	1 oz = 28.35 g 1 lb = 0.45 kg 1 ton (short) = 0.91 tonnes (metric tons) 1 lb = 4.45 N
Volume	1 in^3 = 16.39 cm^3 1 qt = 0.95 L 1 gal = 3.78 L
Area	1 in^2 = 6.45 cm^2 1 yd^2 = 0.83 m^2 1 mi^2 = 2.59 km^2 1 acre = 0.40 hectares
Temperature	$°\text{C} = \frac{(°\text{F} - 32)}{1.8}$ K = °C + 273

*Weight is measured in standard Earth gravity.

Convert Between Unit Systems **Table 4** gives a list of equivalents that can be used to convert between English and SI units.

Example If a meterstick has a length of 100 cm, how long is the meterstick in inches?

Step 1 Write the conversion factors for the units given. From **Table 4,** 1 in = 2.54 cm.

$$\frac{1\ \text{in}}{2.54\ \text{cm}} \quad \textit{and} \quad \frac{2.54\ \text{cm}}{1\ \text{in}}$$

Step 2 Determine which conversion factor to use. You are converting from cm to in. Use the conversion factor with cm on the bottom.

$$\frac{1\ \text{in}}{2.54\ \text{cm}}$$

Step 3 Multiply the starting quantity and units by the conversion factor. Cancel the starting units with the units in the denominator. Round your answer based on the number of significant figures in the conversion factor.

$$100\ \cancel{\text{cm}} \times \frac{1\ \text{in}}{2.54\ \cancel{\text{cm}}} = 39.37\ \text{in}$$

The meterstick is 39.4 in long.

Practice Problem A book has a mass of 5 lbs. What is the mass of the book in kg?

Practice Problem Use the equivalent for in and cm (1 in = 2.54 cm) to show how 1 in^3 = 16.39 cm^3.

Math Skill Handbook

Math Skill Handbook

216 ◆ F STUDENT RESOURCES

Precision and Significant Digits

When you make a measurement, the value you record depends on the precision of the measuring instrument. This precision is represented by the number of significant digits recorded in the measurement. When counting the number of significant digits, all digits are counted except zeros at the end of a number with no decimal point such as 2,050, and zeros at the beginning of a decimal such as 0.03020. When adding or subtracting numbers with different precision, round the answer to the smallest number of decimal places of any number in the sum or difference. When multiplying or dividing, the answer is rounded to the smallest number of significant digits of any number being multiplied or divided.

Example The lengths 5.28 and 5.2 are measured in meters. Find the sum of these lengths and record your answer using the correct number of significant digits.

Step 1 Find the sum.

5.28 m	2 digits after the decimal
+ 5.2 m	1 digit after the decimal
10.48 m	

Step 2 Round to one digit after the decimal because the least number of digits after the decimal of the numbers being added is 1.

The sum is 10.5 m.

Practice Problem How many significant digits are in the measurement 7,071,301 m? How many significant digits are in the measurement 0.003010 g?

Practice Problem Multiply 5.28 and 5.2 using the rule for multiplying and dividing. Record the answer using the correct number of significant digits.

Scientific Notation

Many times numbers used in science are very small or very large. Because these numbers are difficult to work with scientists use scientific notation. To write numbers in scientific notation, move the decimal point until only one non-zero digit remains on the left. Then count the number of places you moved the decimal point and use that number as a power of ten. For example, the average distance from the Sun to Mars is 227,800,000,000 m. In scientific notation, this distance is 2.278×10^{11} m. Because you moved the decimal point to the left, the number is a positive power of ten.

The mass of an electron is about 0.000 000 000 000 000 000 000 000 000 000 911 kg. Expressed in scientific notation, this mass is 9.11×10^{-31} kg. Because the decimal point was moved to the right, the number is a negative power of ten.

Example Earth is 149,600,000 km from the Sun. Express this in scientific notation.

Step 1 Move the decimal point until one non-zero digit remains on the left.
1.496 000 00

Step 2 Count the number of decimal places you have moved. In this case, eight.

Step 3 Show that number as a power of ten, 10^8.

The Earth is 1.496×10^8 km from the Sun.

Practice Problem How many significant digits are in 149,600,000 km? How many significant digits are in 1.496×10^8 km?

Practice Problem Parts used in a high performance car must be measured to 7×10^{-6} m. Express this number as a decimal.

Practice Problem A CD is spinning at 539 revolutions per minute. Express this number in scientific notation.

Precision and Significant Digits

Problem 1
7; 4

Problem 2
$5.28 \times 5.2 = 27.456$

5.28 has 3 significant digits.

5.2 has 2 significant digits.

When multiplying and dividing, the answer is rounded to the smallest number of significant digits of the numbers being multiplied or divided—in this case, 2.

27.456 is rounded to 27.

Scientific Notation

Problem 1
4; 4

Problem 2
0.000007

Problem 3
5.39×10^2

Math Skill Handbook

MATH SKILL HANDBOOK F ◆ 217

Line Graph

x	y
3	52
6	72
9	83
12	86

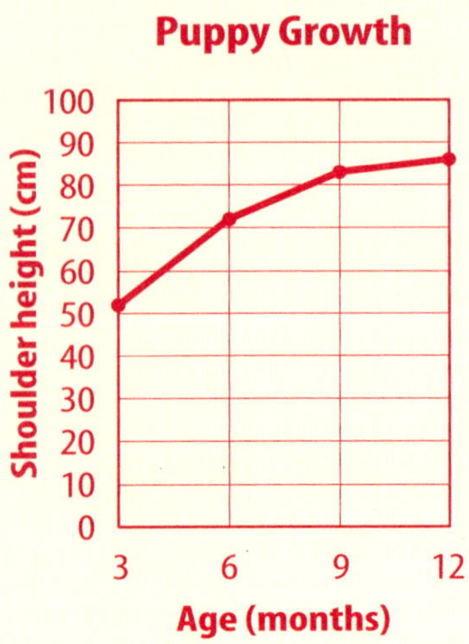

Math Skill Handbook

Make and Use Graphs

Data in tables can be displayed in a graph—a visual representation of data. Common graph types include line graphs, bar graphs, and circle graphs.

Line Graph A line graph shows a relationship between two variables that change continuously. The independent variable is changed and is plotted on the x-axis. The dependent variable is observed, and is plotted on the y-axis.

Example Draw a line graph of the data below from a cyclist in a long-distance race.

Table 5 Bicycle Race Data

Time (h)	Distance (km)
0	0
1	8
2	16
3	24
4	32
5	40

Step 1 Determine the x-axis and y-axis variables. Time varies independently of distance and is plotted on the x-axis. Distance is dependent on time and is plotted on the y-axis.

Step 2 Determine the scale of each axis. The x-axis data ranges from 0 to 5. The y-axis data ranges from 0 to 40.

Step 3 Using graph paper, draw and label the axes. Include units in the labels.

Step 4 Draw a point at the intersection of the time value on the x-axis and corresponding distance value on the y-axis. Connect the points and label the graph with a title, as shown in **Figure 20.**

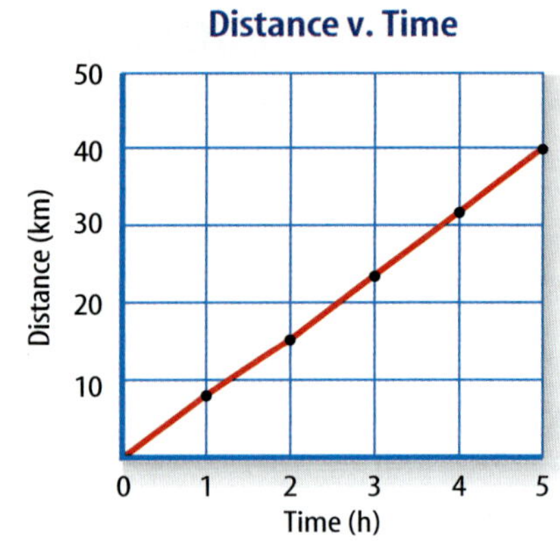

Figure 20 This line graph shows the relationship between distance and time during a bicycle ride.

Practice Problem A puppy's shoulder height is measured during the first year of her life. The following measurements were collected: (3 mo, 52 cm), (6 mo, 72 cm), (9 mo, 83 cm), (12 mo, 86). Graph this data.

Find a Slope The slope of a straight line is the ratio of the vertical change, rise, to the horizontal change, run.

$$\text{Slope} = \frac{\text{vertical change (rise)}}{\text{horizontal change (run)}} = \frac{\text{change in } y}{\text{change in } x}$$

Example Find the slope of the graph in **Figure 20.**

Step 1 You know that the slope is the change in y divided by the change in x.

$$\text{Slope} = \frac{\text{change in } y}{\text{change in } x}$$

Step 2 Determine the data points you will be using. For a straight line, choose the two sets of points that are the farthest apart.

$$\text{Slope} = \frac{(40-0)\text{ km}}{(5-0)\text{ hr}}$$

Step 3 Find the change in y and x.

$$\text{Slope} = \frac{40\text{ km}}{5\text{h}}$$

Step 4 Divide the change in y by the change in x.

$$\text{Slope} = \frac{8\text{ km}}{\text{h}}$$

The slope of the graph is 8 km/h.

Math Skill Handbook

Bar Graph To compare data that does not change continuously you might choose a bar graph. A bar graph uses bars to show the relationships between variables. The x-axis variable is divided into parts. The parts can be numbers such as years, or a category such as a type of animal. The y-axis is a number and increases continuously along the axis.

Example A recycling center collects 4.0 kg of aluminum on Monday, 1.0 kg on Wednesday, and 2.0 kg on Friday. Create a bar graph of this data.

Step 1 Select the x-axis and y-axis variables. The measured numbers (the masses of aluminum) should be placed on the y-axis. The variable divided into parts (collection days) is placed on the x-axis.

Step 2 Create a graph grid like you would for a line graph. Include labels and units.

Step 3 For each measured number, draw a vertical bar above the x-axis value up to the y-axis value. For the first data point, draw a vertical bar above Monday up to 4.0 kg.

Aluminum Collected During Week

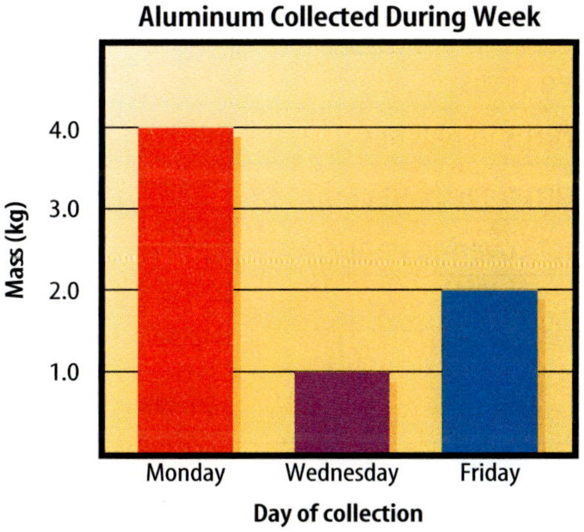

Practice Problem Draw a bar graph of the gases in air: 78% nitrogen, 21% oxygen, 1% other gases.

Circle Graph To display data as parts of a whole, you might use a circle graph. A circle graph is a circle divided into sections that represent the relative size of each piece of data. The entire circle represents 100%, half represents 50%, and so on.

Example Air is made up of 78% nitrogen, 21% oxygen, and 1% other gases. Display the composition of air in a circle graph.

Step 1 Multiply each percent by 360° and divide by 100 to find the angle of each section in the circle.

$$78\% \times \frac{360°}{100} = 280.8°$$

$$21\% \times \frac{360°}{100} = 75.6°$$

$$1\% \times \frac{360°}{100} = 3.6°$$

Step 2 Use a compass to draw a circle and to mark the center of the circle. Draw a straight line from the center to the edge of the circle.

Step 3 Use a protractor and the angles you calculated to divide the circle into parts. Place the center of the protractor over the center of the circle and line the base of the protractor over the straight line.

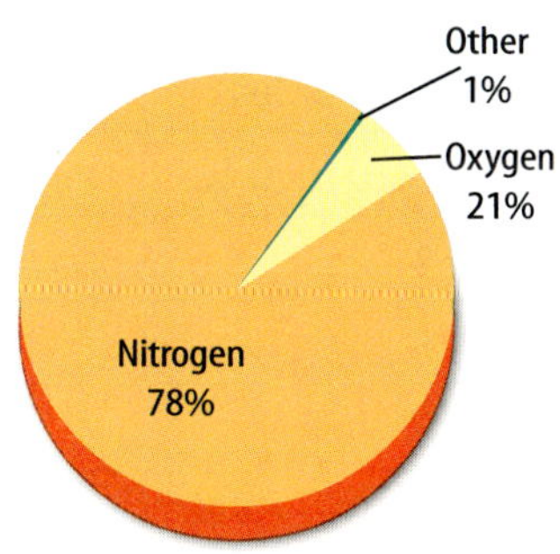

Practice Problem Draw a circle graph to represent the amount of aluminum collected during the week shown in the bar graph to the left.

Bar Graph

Composition of Air

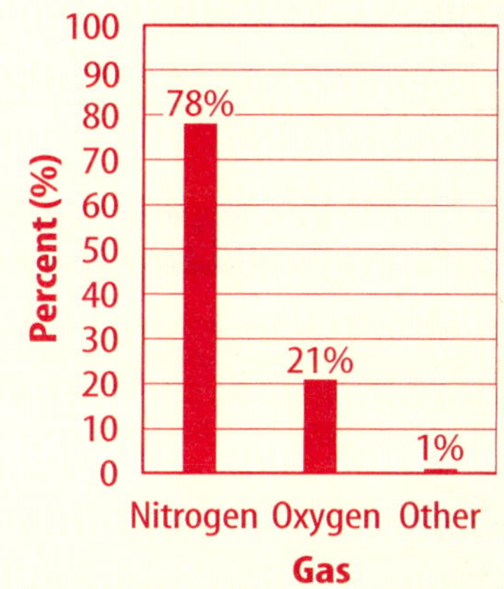

Circle Graph

The total amount of aluminum collected is:

$$4.0\text{ kg} + 1.0\text{ kg} + 2.0\text{ kg} = 7.0\text{ kg}$$

$$\frac{4.0\text{ kg}}{7.0\text{ kg}} = \frac{x}{360°};\ x = 206°$$

$$\frac{1.0\text{ kg}}{7.0\text{ kg}} = \frac{x}{360°};\ x = 51°$$

$$\frac{2.0\text{ kg}}{7.0\text{ kg}} = \frac{x}{360°};\ x = 103°$$

Weather Map Symbols

Sample Station Model

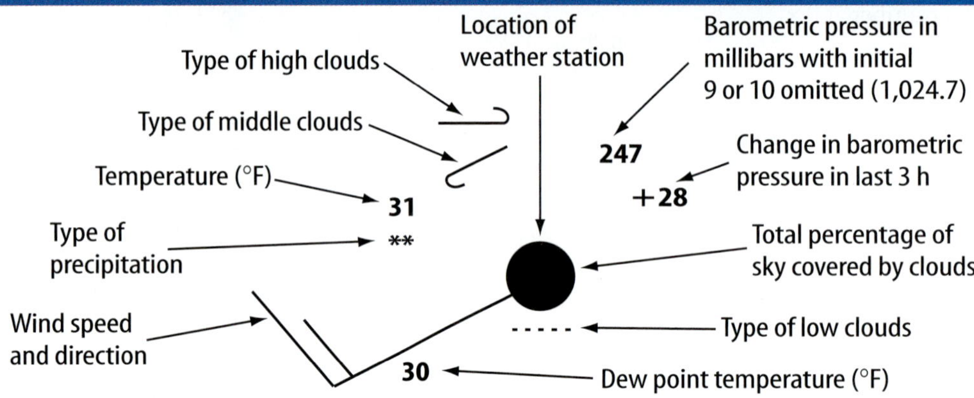

Sample Plotted Report at Each Station

Precipitation	Wind Speed and Direction	Sky Coverage	Some Types of High Clouds
≡ Fog	○ 0 calm	○ No cover	Scattered cirrus
★ Snow	1–2 knots	1/10 or less	Dense cirrus in patches
● Rain	3–7 knots	2/10 to 3/10	Veil of cirrus covering entire sky
Thunderstorm	8–12 knots	4/10	Cirrus not covering entire sky
, Drizzle	13–17 knots	–	
▽ Showers	18–22 knots	6/10	
	23–27 knots	7/10	
	48–52 knots	Overcast with openings	
	1 knot = 1.852 km/h	● Completely overcast	

Some Types of Middle Clouds	Some Types of Low Clouds	Fronts and Pressure Systems	
Thin altostratus layer	Cumulus of fair weather	(H) or High (L) or Low	Center of high- or low-pressure system
Thick altostratus layer	Stratocumulus	▲▲▲▲	Cold front
Thin altostratus in patches	----- Fractocumulus of bad weather		Warm front
Thin altostratus in bands	— Stratus of fair weather		Occluded front
			Stationary front

Rocks

Rock Type	Rock Name	Characteristics
Igneous (intrusive)	Granite	Large mineral grains of quartz, feldspar, hornblende, and mica. Usually light in color.
	Diorite	Large mineral grains of feldspar, hornblende, and mica. Less quartz than granite. Intermediate in color.
	Gabbro	Large mineral grains of feldspar, augite, and olivine. No quartz. Dark in color.
Igneous (extrusive)	Rhyolite	Small mineral grains of quartz, feldspar, hornblende, and mica, or no visible grains. Light in color.
	Andesite	Small mineral grains of feldspar, hornblende, and mica or no visible grains. Intermediate in color.
	Basalt	Small mineral grains of feldspar, augite, and possibly olivine or no visible grains. No quartz. Dark in color.
	Obsidian	Glassy texture. No visible grains. Volcanic glass. Fracture looks like broken glass.
	Pumice	Frothy texture. Floats in water. Usually light in color.
Sedimentary (detrital)	Conglomerate	Coarse grained. Gravel or pebble-size grains.
	Sandstone	Sand-sized grains 1/16 to 2 mm.
	Siltstone	Grains are smaller than sand but larger than clay.
	Shale	Smallest grains. Often dark in color. Usually platy.
Sedimentary (chemical or organic)	Limestone	Major mineral is calcite. Usually forms in oceans and lakes. Often contains fossils.
	Coal	Forms in swampy areas. Compacted layers of organic material, mainly plant remains.
Sedimentary (chemical)	Rock Salt	Commonly forms by the evaporation of seawater.
Metamorphic (foliated)	Gneiss	Banding due to alternate layers of different minerals, of different colors. Parent rock often is granite.
	Schist	Parallel arrangement of sheetlike minerals, mainly micas. Forms from different parent rocks.
	Phyllite	Shiny or silky appearance. May look wrinkled. Common parent rocks are shale and slate.
	Slate	Harder, denser, and shinier than shale. Common parent rock is shale.
Metamorphic (nonfoliated)	Marble	Calcite or dolomite. Common parent rock is limestone.
	Soapstone	Mainly of talc. Soft with greasy feel.
	Quartzite	Hard with interlocking quartz crystals. Common parent rock is sandstone.

Minerals

Minerals					
Mineral (formula)	**Color**	**Streak**	**Hardness**	**Breakage Pattern**	**Uses and Other Properties**
Graphite (C)	black to gray	black to gray	1–1.5	basal cleavage (scales)	pencil lead, lubricants for locks, rods to control some small nuclear reactions, battery poles
Galena (PbS)	gray	gray to black	2.5	cubic cleavage perfect	source of lead, used for pipes, shields for X rays, fishing equipment sinkers
Hematite (Fe_2O_3)	black or reddish-brown	reddish-brown	5.5–6.5	irregular fracture	source of iron; converted to pig iron, made into steel
Magnetite (Fe_3O_4)	black	black	6	conchoidal fracture	source of iron, attracts a magnet
Pyrite (FeS_2)	light, brassy, yellow	greenish-black	6–6.5	uneven fracture	fool's gold
Talc ($Mg_3 Si_4O_{10} (OH)_2$)	white, greenish	white	1	cleavage in one direction	used for talcum powder, sculptures, paper, and tabletops
Gypsum ($CaSO_4{\cdot}2H_2O$)	colorless, gray, white, brown	white	2	basal cleavage	used in plaster of paris and dry wall for building construction
Sphalerite (ZnS)	brown, reddish-brown, greenish	light to dark brown	3.5–4	cleavage in six directions	main ore of zinc; used in paints, dyes, and medicine
Muscovite ($KAl_3Si_3 O_{10}(OH)_2$)	white, light gray, yellow, rose, green	colorless	2–2.5	basal cleavage	occurs in large, flexible plates; used as an insulator in electrical equipment, lubricant
Biotite ($K(Mg,Fe)_3 (AlSi_3O_{10}) (OH)_2$)	black to dark brown	colorless	2.5–3	basal cleavage	occurs in large, flexible plates
Halite (NaCl)	colorless, red, white, blue	colorless	2.5	cubic cleavage	salt; soluble in water; a preservative

Minerals

Minerals					
Mineral (formula)	**Color**	**Streak**	**Hardness**	**Breakage Pattern**	**Uses and Other Properties**
Calcite ($CaCO_3$)	colorless, white, pale blue	colorless, white	3	cleavage in three directions	fizzes when HCl is added; used in cements and other building materials
Dolomite ($CaMg(CO_3)_2$)	colorless, white, pink, green, gray, black	white	3.5–4	cleavage in three directions	concrete and cement; used as an ornamental building stone
Fluorite (CaF_2)	colorless, white, blue, green, red, yellow, purple	colorless	4	cleavage in four directions	used in the manufacture of optical equipment; glows under ultraviolet light
Hornblende ($(CaNa)_{2\text{-}3}(Mg,Al,Fe)_5\text{-}(Al,Si)_2Si_6O_{22}(OH)_2$)	green to black	gray to white	5–6	cleavage in two directions	will transmit light on thin edges; 6-sided cross section
Feldspar ($KAlSi_3O_8$) ($NaAlSi_3O_8$), ($CaAl_2Si_2O_8$)	colorless, white to gray, green	colorless	6	two cleavage planes meet at 90° angle	used in the manufacture of ceramics
Augite ($(Ca,Na)(Mg,Fe,Al)(Al,Si)_2O_6$)	black	colorless	6	cleavage in two directions	square or 8-sided cross section
Olivine ($(Mg,Fe)_2SiO_4$)	olive, green	none	6.5–7	conchoidal fracture	gemstones, refractory sand
Quartz (SiO_2)	colorless, various colors	none	7	conchoidal fracture	used in glass manufacture, electronic equipment, radios, computers, watches, gemstones

PERIODIC TABLE OF THE ELEMENTS

Columns of elements are called groups. Elements in the same group have similar chemical properties.

Element — Hydrogen
Atomic number — 1
Symbol — H
Atomic mass — 1.008
State of matter

Gas
Liquid
Solid
Synthetic

The first three symbols tell you the state of matter of the element at room temperature. The fourth symbol identifies elements that are not present in significant amounts on Earth. Useful amounts are made synthetically.

	1	2	3	4	5	6	7	8	9
1	Hydrogen 1 H 1.008								
2	Lithium 3 Li 6.941	Beryllium 4 Be 9.012							
3	Sodium 11 Na 22.990	Magnesium 12 Mg 24.305							
4	Potassium 19 K 39.098	Calcium 20 Ca 40.078	Scandium 21 Sc 44.956	Titanium 22 Ti 47.867	Vanadium 23 V 50.942	Chromium 24 Cr 51.996	Manganese 25 Mn 54.938	Iron 26 Fe 55.845	Cobalt 27 Co 58.933
5	Rubidium 37 Rb 85.468	Strontium 38 Sr 87.62	Yttrium 39 Y 88.906	Zirconium 40 Zr 91.224	Niobium 41 Nb 92.906	Molybdenum 42 Mo 95.94	Technetium 43 Tc (98)	Ruthenium 44 Ru 101.07	Rhodium 45 Rh 102.906
6	Cesium 55 Cs 132.905	Barium 56 Ba 137.327	Lanthanum 57 La 138.906	Hafnium 72 Hf 178.49	Tantalum 73 Ta 180.948	Tungsten 74 W 183.84	Rhenium 75 Re 186.207	Osmium 76 Os 190.23	Iridium 77 Ir 192.217
7	Francium 87 Fr (223)	Radium 88 Ra (226)	Actinium 89 Ac (227)	Rutherfordium 104 Rf (261)	Dubnium 105 Db (262)	Seaborgium 106 Sg (266)	Bohrium 107 Bh (264)	Hassium 108 Hs (277)	Meitnerium 109 Mt (268)

The number in parentheses is the mass number of the longest-lived isotope for that element.

Rows of elements are called periods. Atomic number increases across a period.

Lanthanide series	Cerium 58 Ce 140.116	Praseodymium 59 Pr 140.908	Neodymium 60 Nd 144.24	Promethium 61 Pm (145)	Samarium 62 Sm 150.36
Actinide series	Thorium 90 Th 232.038	Protactinium 91 Pa 231.036	Uranium 92 U 238.029	Neptunium 93 Np (237)	Plutonium 94 Pu (244)

The arrow shows where these elements would fit into the periodic table. They are moved to the bottom of the table to save space.

Metal

Metalloid

Nonmetal

The color of an element's block tells you if the element is a metal, nonmetal, or metalloid.

Science Online
Visit bookf.msscience.com for updates to the periodic table.

10	11	12	13	14	15	16	17	18
								Helium 2 **He** 4.003
			Boron 5 **B** 10.811	Carbon 6 **C** 12.011	Nitrogen 7 **N** 14.007	Oxygen 8 **O** 15.999	Fluorine 9 **F** 18.998	Neon 10 **Ne** 20.180
			Aluminum 13 **Al** 26.982	Silicon 14 **Si** 28.086	Phosphorus 15 **P** 30.974	Sulfur 16 **S** 32.065	Chlorine 17 **Cl** 35.453	Argon 18 **Ar** 39.948
Nickel 28 **Ni** 58.693	Copper 29 **Cu** 63.546	Zinc 30 **Zn** 65.409	Gallium 31 **Ga** 69.723	Germanium 32 **Ge** 72.64	Arsenic 33 **As** 74.922	Selenium 34 **Se** 78.96	Bromine 35 **Br** 79.904	Krypton 36 **Kr** 83.798
Palladium 46 **Pd** 106.42	Silver 47 **Ag** 107.868	Cadmium 48 **Cd** 112.411	Indium 49 **In** 114.818	Tin 50 **Sn** 118.710	Antimony 51 **Sb** 121.760	Tellurium 52 **Te** 127.60	Iodine 53 **I** 126.904	Xenon 54 **Xe** 131.293
Platinum 78 **Pt** 195.078	Gold 79 **Au** 196.967	Mercury 80 **Hg** 200.59	Thallium 81 **Tl** 204.383	Lead 82 **Pb** 207.2	Bismuth 83 **Bi** 208.980	Polonium 84 **Po** (209)	Astatine 85 **At** (210)	Radon 86 **Rn** (222)
Darmstadtium 110 **Ds** (281)	Unununium * 111 **Uuu** (272)	Ununbium * 112 **Uub** (285)		Ununquadium * 114 **Uuq** (289)		** 116		** 118

* The names and symbols for elements 111–114 are temporary. Final names will be selected when the elements' discoveries are verified.

** Elements 116 and 118 were thought to have been created. The claim was retracted because the experimental results could not be repeated.

Europium 63 **Eu** 151.964	Gadolinium 64 **Gd** 157.25	Terbium 65 **Tb** 158.925	Dysprosium 66 **Dy** 162.500	Holmium 67 **Ho** 164.930	Erbium 68 **Er** 167.259	Thulium 69 **Tm** 168.934	Ytterbium 70 **Yb** 173.04	Lutetium 71 **Lu** 174.967
Americium 95 **Am** (243)	Curium 96 **Cm** (247)	Berkelium 97 **Bk** (247)	Californium 98 **Cf** (251)	Einsteinium 99 **Es** (252)	Fermium 100 **Fm** (257)	Mendelevium 101 **Md** (258)	Nobelium 102 **No** (259)	Lawrencium 103 **Lr** (262)

Topographic Map Symbols

Topographic Map Symbols	
Primary highway, hard surface	Index contour
Secondary highway, hard surface	Supplementary contour
Light-duty road, hard or improved surface	Intermediate contour
Unimproved road	Depression contours
Railroad: single track	
Railroad: multiple track	Boundaries: national
Railroads in juxtaposition	State
	County, parish, municipal
Buildings	Civil township, precinct, town, barrio
cem Schools, church, and cemetery	Incorporated city, village, town, hamlet
Buildings (barn, warehouse, etc.)	Reservation, national or state
Wells other than water (labeled as to type)	Small park, cemetery, airport, etc.
Tanks: oil, water, etc. (labeled only if water)	Land grant
Located or landmark object; windmill	Township or range line, U.S. land survey
Open pit, mine, or quarry; prospect	
	Township or range line, approximate location
Marsh (swamp)	
Wooded marsh	Perennial streams
Woods or brushwood	Elevated aqueduct
Vineyard	Water well and spring
Land subject to controlled inundation	Small rapids
Submerged marsh	Large rapids
Mangrove	Intermittent lake
Orchard	Intermittent stream
Scrub	Aqueduct tunnel
Urban area	Glacier
	Small falls
x7369 Spot elevation	Large falls
670 Water elevation	Dry lake bed

extrusive: describes fine-grained igneous rock that forms when magma cools quickly at or near Earth's surface. (p. 41)

extrusivo: describe rocas ígneas de grano fino que se forman cuando el magma se enfría rápidamente en o cerca de la superficie terrestre. (p. 41)

F

fault: surface along which rocks move when they pass their elastic limit and break. (p. 126)

focus: in an earthquake, the point below Earth's surface where energy is released in the form of seismic waves. (p. 130)

foliated: describes metamorphic rock, such as slate and gneiss, whose mineral grains line up in parallel layers. (p. 47)

fossil fuel: nonrenewable energy resource, such as oil and coal, formed over millions of years from the remains of dead plants and other organisms. (p. 66)

fracture: physical property of some minerals that causes them to break with uneven, rough, or jagged surfaces. (p. 17)

falla: área a lo largo de la cual las rocas se mueven cuando sobrepasan su límite elástico y se rompen. (p. 126)

foco: en un terremoto, el punto bajo la superficie terrestre donde se libera la energía en forma de ondas sísmicas. (p. 130)

foliado: describe rocas metamórficas, como pizarra y gneis, cuyas vetas minerales se alinean en capas paralelas. (p. 47)

combustible fósil: recurso energético no renovable, como el petróleo y el carbón mineral, formado durante millones de años a partir de restos de plantas y otros organismos muertos. (p. 66)

fractura: propiedad física de algunos minerales que causa que se rompan formando superficies irregulares, ásperas o dentadas. (p. 17)

G

gem: beautiful, rare, highly prized mineral that can be worn in jewelry. (p. 19)

geothermal energy: inexhaustible energy resource that uses hot magma or hot, dry rocks from below Earth's surface to generate electricity. (p. 78)

granitic: describes generally light-colored, silica-rich igneous rock that is less dense than basaltic rock. (p. 43)

gema: mineral hermoso, raro y altamente valorado que puede usarse como joya. (p. 19)

energía geotérmica: recurso energético inagotable que utiliza el magma caliente o las piedras secas calientes encontradas debajo de la superficie terrestre para producir electricidad. (p. 78)

granítica: roca ígnea rica en sílice, generalmente de color claro y menos densa que la rocas basáltica. (p. 43)

H

hardness: measure of how easily a mineral can be scratched. (p. 15)

hot spot: the result of an unusually hot area at the boundary between Earth's mantle and core that forms volcanoes when melted rock is forced upward and breaks through the crust. (p. 160)

hydroelectric energy: electricity produced by waterpower using large dams in a river. (p. 78)

dureza: medida de la facilidad con que un mineral puede ser rayado. (p. 15)

punto caliente: el resultado de un área extraordinariamente caliente en los límites entre el manto y el núcleo de la Tierra; forma volcanes cuando la roca fundida es empujada hacia arriba y se abre paso hacia la corteza. (p. 160)

energía hidroeléctrica: electricidad producida por la energía hidráulica generada mediante represas grandes construidas en los ríos. (p. 78)

I

igneous rock: rock formed when magma or lava cools and hardens. (p. 40)

intrusive: describes a type of igneous rock that generally contains large crystals and forms when magma cools slowly beneath Earth's surface. (p. 41)

roca ígnea: roca formada cuando se enfría y endurece el magma o la lava. (p. 40)

intrusivo: describe un tipo de roca ígnea que generalmente contiene cristales grandes y se forma cuando el magma se enfría lentamente por debajo de la superficie terrestre. (p. 41)

L

lava: molten rock that flows from volcanoes onto Earth's surface. (p. 40)

liquefaction: occurs when wet soil acts more like a liquid during an earthquake. (p. 141)

lithosphere (LIH thuh sfihr): rigid layer of Earth about 100 km thick, made of the crust and a part of the upper mantle. (p. 106)

luster: describes the way a mineral reflects light from its surface; can be metallic or nonmetallic. (p. 16)

lava: roca derretida que fluye de los volcanes hacia la superficie terrestre. (p. 40)

licuefacción: ocurre cuando el suelo húmedo se comporta como un líquido durante un terremoto. (p. 141)

litosfera: capa rígida de la Tierra de unos 100 kilómetros de profundidad, comprende la corteza y una parte del manto superior. (p. 106)

brillo: describe la forma en que un mineral refleja la luz desde su superficie; puede ser metálicos o no metálicos. (p. 16)

M

magma: hot, melted rock material beneath Earth's surface. (p. 11)

magnitude: measure of the energy released during an earthquake. (p. 140)

metamorphic rock: forms when heat, pressure, or fluids act on igneous, sedimentary, or other metamorphic rock to change its form or composition, or both. (p. 45)

mineral: naturally occurring inorganic solid that has a definite chemical composition and an orderly internal atomic structure. (p. 8)

mineral resources: resources from which metals are obtained. (p. 83)

magma: material rocoso fundido y caliente que se encuentra por debajo de la superficie terrestre. (p. 11)

magnitud: medida de la energía liberada durante un terremoto. (p. 140)

roca metamórfica: se forma cuando el calor, la presión o los fluidos actúan sobre una roca ígnea, sedimentaria u otra roca metamórfica para cambiar su forma, composición o ambas. (p. 45)

mineral: sólido inorgánico que se encuentra en la naturaleza, tiene una composición química definida y una estructura atómica ordenada. (p. 8)

recursos minerales: recursos a partir de los cuales pueden obtenerse metales. (p. 83)

N

natural gas: fossil fuel formed from marine organisms that is often found in tilted or folded rock layers and is used for heating and cooking. (p. 69)

nonfoliated: describes metamorphic rock, such as quartzite or marble, whose mineral grains grow and rearrange but generally do not form layers. (p. 48)

gas natural: combustible fósil formado a partir de organismos marinos y que a menudo se encuentra en capas rocosas inclinadas o plegadas; se usa para calefacción y para cocinar. (p. 69)

no foliado: describe rocas metamórficas, como la cuarcita o el mármol, cuyas vetas minerales se acumulan y reestructuran pero rara vez forman capas. (p. 48)

normal fault: break in rock caused by tension forces, where rock above the fault surface moves down relative to the rock below the fault surface. (p. 128)

nuclear energy: alternative energy source that is based on atomic fission. (p. 73)

falla normal: ruptura en la roca causada por fuerzas de tensión, donde la roca sobre la superficie de la falla se mueve hacia abajo con respecto a la roca debajo de la superficie de la falla. (p. 128)

energía nuclear: fuente de energía alternativa que se basa en la fisión atómica. (p. 73)

O

oil: liquid fossil fuel formed from marine organisms that is burned to obtain energy and used in the manufacture of plastics. (p. 69)

ore: deposit in which a mineral exists in large enough amounts to be mined at a profit. (pp. 23, 83)

petróleo: combustible fósil líquido formado a partir de organismos marinos; es quemado para obtener energía y se usa en la manufactura de plásticos. (p. 69)

mena: depósito en el que existen cantidades suficientes de un mineral para que la actividad minera sea rentable. (pp. 23, 83)

P

Pangaea (pan JEE uh): large, ancient landmass that was composed of all the continents joined together. (p. 98)

plate tectonics: theory that Earth's crust and upper mantle are broken into plates that float and move around on a plasticlike layer of the mantle. (p. 106)

plate: a large section of Earth's oceanic or continental crust and rigid upper mantle that moves around on the asthenosphere. (p. 106)

primary wave: seismic wave that moves rock particles back-and-forth in the same direction that the wave travels. (p. 131)

Pangea: masa terrestre extensa y antigua que estaba compuesta por todos los continentes unidos. (p. 98)

placa: gran sección de la corteza terrestre u oceánica y del manto rígido superior que se mueve sobre la astenosfera. (p. 106)

tectónica de placas: teoría respecto a que la corteza terrestre y el manto superior están fraccionados en placas que flotan y se mueven sobre una capa plástica del manto. (p. 106)

onda primaria: onda sísmica que mueve partículas rocosas en la misma dirección en que viaja la onda. (p. 131)

R

recycling: conservation method in which old materials are processed to make new ones. (p. 87)

reserve: amount of a fossil fuel that can be extracted from Earth at a profit using current technology. (p. 71)

reverse fault: break in rock caused by compressive forces, where rock above the fault surface moves upward relative to the rock below the fault surface. (p. 128)

rock: mixture of one or more minerals, rock fragments, volcanic glass, organic matter, or other natural materials; can be igneous, metamorphic, or sedimentary. (p. 36)

reciclaje: método de conservación en el que los materiales usados son procesados para producir otros nuevos. (p. 87)

reserva: depósito de un combustible fósil que puede extraerse de la Tierra y del cual, utilizando la tecnología actual, se obtienen utilidades. (p. 71)

falla inversa: ruptura en la roca causada por fuerzas de compresión, donde la roca sobre la superficie de la falla se mueve hacia arriba con respecto a la roca debajo de la superficie de la falla. (p. 128)

roca: mezcla de uno o más minerales, fragmentos de roca, obsidiana, materia orgánica u otros materiales naturales; puede ser ígnea, metamórfica o sedimentaria. (p. 36)

rock cycle: model that describes how rocks slowly change from one form to another through time. (p. 37)

ciclo de la roca: modelo que describe cómo cambian lentamente las rocas de una forma a otra a través del tiempo. (p. 37)

S

seafloor spreading: Hess's theory that new seafloor is formed when magma is forced upward toward the surface at a mid-ocean ridge. (p. 103)

expansión del suelo oceánico: teoría de Hess respecto a que se forma un nuevo suelo oceánico cuando el magma es empujado hacia la superficie a través de un surco en la mitad del océano. (p. 103)

secondary wave: seismic wave that moves rock particles at right angles to the direction of the wave. (p. 131)

onda secundaria: onda sísmica que mueve partículas rocosas en ángulos rectos respecto a la dirección de la onda. (p. 131)

sedimentary rock: forms when sediments are compacted and cemented together or when minerals form from solutions. (p. 49)

roca sedimentaria: se forma cuando los sedimentos son compactados y cementados o cuando se forman minerales a partir de soluciones. (p. 49)

sediments: loose materials, such as rock fragments, mineral grains, and the remains of once-living plants and animals, that have been moved by wind, water, ice, or gravity. (p. 49)

sedimentos: materiales sueltos, como fragmentos de roca, granos minerales y restos de animales y plantas, que han sido arrastrados por el viento, el agua, el hielo o la gravedad. (p. 49)

seismic (SIZE mihk) wave: wave generated by an earthquake. (p. 130)

onda sísmica: onda generada por un terremoto. (p. 130)

seismograph: instrument used to register earthquake waves and record the time that each arrived. (p. 133)

sismógrafo: instrumento utilizado para registrar las ondas sísmicas y la hora a la que llega cada una. (p. 133)

shield volcano: broad, gently sloping volcano formed by quiet eruptions of basaltic lava. (p. 166)

volcán de escudo: volcán levemente inclinado y de gran extensión, formado por erupciones de baja intensidad de lava basáltica. (p. 166)

silicate: mineral that contains silicon and oxygen and usually one or more other elements. (p. 12)

silicato: mineral que contiene sílice y oxígeno y generalmente uno o varios elementos distintos. (p. 12)

sill: igneous rock feature formed when magma is squeezed into a horizontal crack between layers of rock and hardens underground. (p. 173)

alféizar: roca ígnea característica formada cuando el magma es comprimido en una grieta horizontal entre capas de roca y se endurece en el subsuelo. (p. 173)

solar energy: energy from the Sun that is clean, inexhaustible, and can be transformed into electricity by solar cells. (p. 76)

energía solar: energía del sol, la cual es limpia e inagotable y puede transformarse en electricidad a través de celdas solares. (p. 76)

specific gravity: ratio of a mineral's weight compared with the weight of an equal volume of water. (p. 16)

gravedad específica: cociente del peso de un mineral comparado con el peso de un volumen igual de agua. (p. 16)

streak: color of a mineral when it is in powdered form. (p. 17)

veta: color de un mineral en forma de polvo. (p. 17)

strike-slip fault: break in rock caused by shear forces, where rocks move past each other without much vertical movement. (p. 129)

falla deslizante: ruptura en la roca causada por fuerzas opuestas, donde las rocas se mueven una tras otra sin mucho movimiento vertical. (p. 129)

surface wave: seismic wave that moves rock particles up-and-down in a backward rolling motion and side-to-side in a swaying motion. (p. 131)

onda de superficie: onda sísmica que mueve partículas rocosas en forma ascendente y descendente en un movimiento circular en retroceso y de un lado a otro en un movimiento oscilante. (p. 131)

T

tephra (TEFF ruh): bits of rock or solidified lava dropped from the air during an explosive volcanic eruption; ranges in size from volcanic ash to volcanic bombs and blocks. (p. 166)

tsunami (soo NAH mee): seismic sea wave that begins over an earthquake focus and can be highly destructive when it crashes on shore. (p. 142)

tefra: trozos de roca o lava solidificada que caen del aire durante una erupción volcánica explosiva; su tamaño oscila desde la ceniza volcánica hasta las bombas o bloques volcánicos. (p. 166)

maremoto: onda sísmica marina que comienza sobre el foco del terremoto y que puede ser altamente destructiva cuando se estrella en la costa. (p. 142)

V

vent: opening where magma is forced up and flows out onto Earth's surface as lava, forming a volcano. (p. 158)

volcanic neck: solid igneous core of a volcano left behind after the softer cone has been eroded. (p. 173)

volcano: opening in Earth's surface that erupts sulfurous gases, ash, and lava; can form at Earth's plate boundaries, where plates move apart or together, and at hot spots. (p. 156)

chimenea: apertura donde el magma es empujado hacia arriba y fluye sobre la superficie terrestre como lava, formando un volcán. (p. 158)

cuello volcánico: núcleo ígneo sólido de un volcán que queda después de que el cono más blando ha sido erosionado. (p. 173)

volcán: apertura en la superficie terrestre que arroja gases sulfurosos, ceniza y lava; puede formarse en los límites de las placas continentales, donde las placas se separan o encuentran y en los puntos calientes. (p. 156)

W

wind farm: area where many windmills use wind to generate electricity. (p. 77)

granja de energía eólica: área en donde muchos molinos usan el viento para generar electricidad. (p. 77)

Index

Italic numbers = *illustration/photo* **Bold numbers = vocabulary term**
lab = a page on which the entry is used in a lab
act = a page on which the entry is used in an activity

Index

Index

Index

Credits

Magnification Key: Magnifications listed are the magnifications at which images were originally photographed.
LM–Light Microscope
SEM–Scanning Electron Microscope
TEM–Transmission Electron Microscope

Acknowledgments: Glencoe would like to acknowledge the artists and agencies who participated in illustrating this program: Absolute Science Illustration; Andrew Evansen; Argosy; Articulate Graphics; Craig Attebery, represented by Frank & Jeff Lavaty; CHK America; John Edwards and Associates; Gagliano Graphics; Pedro Julio Gonzalez, represented by Melissa Turk & The Artist Network; Robert Hynes, represented by Mendola Ltd.; Morgan Cain & Associates; JTH Illustration; Laurie O'Keefe; Matthew Pippin, represented by Beranbaum Artist's Representative; Precision Graphics; Publisher's Art; Rolin Graphics, Inc.; Wendy Smith, represented by Melissa Turk & The Artist Network; Kevin Torline, represented by Berendsen and Associates, Inc.; WILDlife ART; Phil Wilson, represented by Cliff Knecht Artist Representative; Zoo Botanica.

Photo Credits

Cover PhotoDisc; **i ii** PhotoDisc; **iv** (bkgd)John Evans, (inset)PhotoDisc; **v** (t)PhotoDisc, (b)John Evans; **vi** (l)John Evans, (r)Geoff Butler; **vii** (l)John Evans, (r)PhotoDisc; **viii** PhotoDisc; **ix** Aaron Haupt Photography; **x** Inga Spence/ Visuals Unlimited; **xi** Soames Summerhays/Photo Researchers; **xii** (tl)Albert J. Copley/Visuals Unlimited, (tc bcr)Mark A. Schneider/Visuals Unlimited, (tr bcl)Doug Martin, (bl)Visuals Unlimited, (br)Jose Manuel Sanchis Calvete/ CORBIS; **1** David J. Cross/Peter Arnold, Inc.; **2** (t)AP/Wide World Photos/Jack Smith, (b)AP/Wide World Photos/Gary Stewart; **3** (t)Francois Gohier/Photo Researchers, (b)David Muench/CORBIS; **5** (t)Chlaus Lotscher/Stock Boston, (b)USGS; **6–7** SuperStock; **8** Matt Meadows; **9** (inset)John R. Foster/Photo Researchers, (l)Mark A. Schneider/Visuals Unlimited; **10** (tr)Mark A. Schneider/Visuals Unlimited, (cl)A.J. Copley/Visuals Unlimited, (cr bl)Harry Taylor/DK Images, (bc)Mark A. Schneider/Photo Researchers, (br)Mark A. Schneider/Visuals Unlimited; **11** (inset)Patricia K. Armstrong/Visuals Unlimited, (r)Dennis Flaherty Photography/Photo Researchers; **13** KS Studios; **14** (l)Mark Burnett/Photo Researchers, (c)Dan Suzio/Photo Researchers, (r)Breck P. Kent/Earth Scenes; **15** (inset)Icon Images, (t)Bud Roberts/Visuals Unlimited, (b)Charles D. Winters/Photo Researchers; **16** (l)Andrew McClenaghan/Science Photo Library/Photo Researchers, (r)Charles D. Winters/Photo Researchers; **17** (t)Goeff Butler, (bl)Doug Martin, (br)Photo Researchers; **18** Matt Meadows; **19** Reuters NewMedia, Inc./ CORBIS; **20** (Beryl, Spinel)Biophoto Associates/Photo Researchers, (Emerald, Topaz)H. Stern/Photo Researchers, (Ruby Spinel, Tanzanite)A.J. Copley/Visuals Unlimited, (Zoisite)Visuals Unlimited, (uncut Topaz)Mark A. Schneider/ Visuals Unlimited; **21** (Olivine)University of Houston, (Peridot)Charles D. Winters/Photo Researchers, (Garnet) Arthur R. Hill/Visuals Unlimited, (Almandine)David Lees/ CORBIS, (Quartz, Corundum)Doug Martin, (Amethyst)A.J. Copley/Visuals Unlimited, (Blue Sapphire)Vaughan Fleming/ Science Photo Library/Photo Researchers; **22** (l)Francis G. Mayer/CORBIS, (r)Smithsonian Institution; **23** (inset)Doug Martin, (l)Fred Whitehead/Earth Scenes; **24** (t)Matt Meadows, (bl)Paul Silverman/Fundamental Photographs, (br)Biophoto Associates/Photo Researchers; **25** Jim Cummins/Getty Images; **26** Matt Meadows; **27** (inset)José Manuel Sanchis Calvete/CORBIS, (t)Doug Martin, (bl)Andrew J. Martinez/Photo Researchers, (br)Charles D. Winter/Photo Researchers; **28** (bkgd)Science Photo Library/ Custom Medical Stock Photo, (bl)Bettmann/CORBIS; **29** José Manuel Sanchis Calvete/CORBIS; **30** R. Weller/Cochise College; **32** José Manuel Sanchis Calvete/CORBIS; **33** Breck P. Kent/Earth Scenes; **34–35** Michael T. Sedam/CORBIS; **36** (l)CORBIS, (r)Doug Martin; **37** (tl)Steve Hoffman, (cl)Brent Turner/BLT Productions, (r)Breck P. Kent/Earth Scenes; **38** (bkgd)CORBIS/PictureQuest, (t)CORBIS, (bl)Martin Miller, (bc)Jeff Gnass, (br)Doug Sokell/Tom Stack & Assoc.; **39** Russ Clark; **40** USGS/HVO; **41** (t)Breck P. Kent/ Earth Scenes, (b)Doug Martin; **42** (basalt)Mark Steinmetz, (scoria, obsidian)Doug Martin, (pumice)Tim Courlas, (others)Breck P. Kent/Earth Scenes; **44** (l)Breck P. Kent/Earth Scenes, (r)Doug Martin/Photo Researchers; **45** (t)Breck P. Kent/Earth Scenes, (l)Breck P. Kent/Earth Scenes, (bl)Courtesy Kent Ratajeski & Dr. Allen Glazner, University of North Carolina, (br)Alfred Pasieka/Photo Researchers; **47** (l)Aaron Haupt, (r)Robert Estall/CORBIS; **48** Paul Rocheleau/Index Stock; **49** (l)Timothy Fuller, (r)Steve McCutcheon/Visuals Unlimited; **51** (l)Icon Images, (cl)Doug Martin, (cr)Andrew Martinez/Photo Researchers, (r)John R. Foster/Photo Researchers; **52** (l)Breck P. Kent/Earth Scenes, (r)Aaron Haupt; **53** (bkgd)Georg Gerster/Photo Researchers, Icon Images; **55** Beth Davidow/Visuals Unlimited; **56** (l)Icon Images, (r)Breck P. Kent/Earth Scenes; **57** (l)Jack Sekowski, (r)Tim Courlas; **58** (bkgd)Y. Kawasaki/Photonica, (inset)Matt Turner/Liaison Agency; **60** Breck P. Kent/Earth Scenes; **61** Jeremy Woodhouse/DRK Photo; **64–65** Bill Ross/CORBIS; **67** Visuals Unlimited; **70** (l)George Lepp/CORBIS, (r)Carson Baldwin Jr./Earth Scenes; **71** Paul A. Souders/CORBIS; **72** (bkgd)Ian R. MacDonald/Texas A&M University, (l)Emory Kristof, (r)National Energy Technology Laboratory; **73** Hal Beral/Visuals Unlimited; **75** Roger Ressmeyer/ CORBIS; **76** Spencer Grant/PhotoEdit, Inc.; **77** Inga Spence/ Visuals Unlimited; **78** Robert Cameron/Stone/Getty Images; **79** Vince Streano/CORBIS; **80** (t)David Young-Wolff/ PhotoEdit, Inc., (b)Earl Young/Archive Photos; **81** Peter Holden/Visuals Unlimited; **83** Aaron Haupt; **84** Joseph Nettis/Photo Researchers; **85** (t)Mark Joseph/Stone/Getty Images, (bl)Aaron Haupt, (br)Wyoming Mining Association; **88** (t)Aaron Haupt, (b)Joel W. Rogers/CORBIS; **89** Aaron Haupt; **90** (t)Ed Clark, (bl)Brown Brothers, (br)Shell Oil Co.; **91** (l)Andrew J. Martinez/Photo Researchers, (r)Coco McCoy/Rainbow; **95** Mark Joseph/Stone/Getty Images; **96–97** Bourseiller/Durieux/Photo Researchers; **100** Martin Land/Science Source/Photo Researchers; **103** Ralph White/ CORBIS; **109** Davis Meltzer; **110** Craig Aurness/CORBIS; **112** Craig Brown/Index Stock; **113** Ric Ergenbright/CORBIS; **114** Roger Ressmeyer/CORBIS; **116** AP/Wide World Photos; **118** L. Lauber/Earth Scenes; **124–125** Chuck Nacke/TimeLife Pictures/Getty Images; **126** Tom & Therisa Stack; **128** (t)Tom Bean/DRK Photo, (b)Lysbeth Corsi/Visuals Unlimited; **129** David Parker/Photo Researchers; **130** Tom & Therisa Stack; **132** Robert W. Tope/Natural Science Illustrations; **139** (l)Steven D. Starr/Stock Boston, (r)Berkeley

Seismological Laboratory; **140** AP/Wide World Photos; **141** David J. Cross/Peter Arnold, Inc.; **144** James L. Stanfield/ National Geographic Image Collection; **145** David Young-Wolff/PhotoEdit, Inc.; **147** Reuters NewMedia Inc./CORBIS; **148** (tr)Richard Cummins/CORBIS, (l)Bettmann/CORBIS, (br)RO-MA Stock/Index Stock; **149** (l)Science VU/Visuals Unlimited, (r)Peter Menzel/Stock Boston; **152** Vince Streano/ CORBIS; **154–155** Art Wolfe/Getty Images; **155** KS Studios; **156** Sigurjon Sindrason; **157** (t)John Cancalosi/DRK Photo, (b)Deborah Brosnan, Sustainable Ecosystems Institute; **161** Image courtesy NASA/GSFC/JPL, MISR Team; **162 163** Gary Rosenquist; **164** (bkgd)API/Explorer/Photo Researchers, (t)Krafft/HOA-QUI/Photo Researchers, (bl)Robert Hessler/Planet Earth Pictures, (br)Paul Chesley; **165** (l)Steve Kaufman/DRK Photo, (r)Dee Breger/Photo Researchers; **167** (t)Krafft/Explorer/Science Source/Photo Researchers, (b)Darrell Gulin/DRK Photo; **171** (tl)Joyce Photo/Photo Researchers, (tr)Doug Martin, (b)Brent Turner; **172** (t)Dick Canby/DRK Photo, (b)Tom Bean/DRK Photo; **174** Larry Ulrich/DRK Photo; **175** Amanita Pictures; **176** (t)Spencer Grant/PhotoEdit, Inc., (b)Darrell Gulin/DRK Photo; **177** Mimmo Jodice/CORBIS; **178** (l)Soames Summerhays/Photo Researchers, (r)Photri/The Stock Market/CORBIS; **179** Krafft/Explorer/Science Source/Photo Researchers; **181** Kerrick James/Getty Images; **184** PhotoDisc; **186** Tom Pantages; **190** Michell D. Bridwell/PhotoEdit, Inc.; **181** (t)Mark Burnett, (b)Dominic Oldershaw; **192** StudiOhio; **193** Timothy Fuller; **194** Aaron Haupt; **196** KS Studios; **197** Matt Meadows; **198** (t)Matt Meadows, (b)Doug Martin; **199** Doug Martin; **201** Amanita Pictures; **202** Bob Daemmrich; **204** Davis Barber/PhotoEdit, Inc.

PERIODIC TABLE OF THE ELEMENTS

Columns of elements are called groups. Elements in the same group have similar chemical properties.

Element — Hydrogen
Atomic number — 1
Symbol — H
Atomic mass — 1.008
State of matter

Gas
Liquid
Solid
Synthetic

The first three symbols tell you the state of matter of the element at room temperature. The fourth symbol identifies elements that are not present in significant amounts on Earth. Useful amounts are made synthetically.

	1	2	3	4	5	6	7	8	9
1	Hydrogen 1 **H** 1.008								
2	Lithium 3 **Li** 6.941	Beryllium 4 **Be** 9.012							
3	Sodium 11 **Na** 22.990	Magnesium 12 **Mg** 24.305							
4	Potassium 19 **K** 39.098	Calcium 20 **Ca** 40.078	Scandium 21 **Sc** 44.956	Titanium 22 **Ti** 47.867	Vanadium 23 **V** 50.942	Chromium 24 **Cr** 51.996	Manganese 25 **Mn** 54.938	Iron 26 **Fe** 55.845	Cobalt 27 **Co** 58.933
5	Rubidium 37 **Rb** 85.468	Strontium 38 **Sr** 87.62	Yttrium 39 **Y** 88.906	Zirconium 40 **Zr** 91.224	Niobium 41 **Nb** 92.906	Molybdenum 42 **Mo** 95.94	Technetium 43 **Tc** (98)	Ruthenium 44 **Ru** 101.07	Rhodium 45 **Rh** 102.906
6	Cesium 55 **Cs** 132.905	Barium 56 **Ba** 137.327	Lanthanum 57 **La** 138.906	Hafnium 72 **Hf** 178.49	Tantalum 73 **Ta** 180.948	Tungsten 74 **W** 183.84	Rhenium 75 **Re** 186.207	Osmium 76 **Os** 190.23	Iridium 77 **Ir** 192.217
7	Francium 87 **Fr** (223)	Radium 88 **Ra** (226)	Actinium 89 **Ac** (227)	Rutherfordium 104 **Rf** (261)	Dubnium 105 **Db** (262)	Seaborgium 106 **Sg** (266)	Bohrium 107 **Bh** (264)	Hassium 108 **Hs** (277)	Meitnerium 109 **Mt** (268)

The number in parentheses is the mass number of the longest-lived isotope for that element.

Rows of elements are called periods. Atomic number increases across a period.

Lanthanide series	Cerium 58 **Ce** 140.116	Praseodymium 59 **Pr** 140.908	Neodymium 60 **Nd** 144.24	Promethium 61 **Pm** (145)	Samarium 62 **Sm** 150.36
Actinide series	Thorium 90 **Th** 232.038	Protactinium 91 **Pa** 231.036	Uranium 92 **U** 238.029	Neptunium 93 **Np** (237)	Plutonium 94 **Pu** (244)

The arrow shows where these elements would fit into the periodic table. They are moved to the bottom of the table to save space.